Tundra-Taiga Biology

To Edith — Richard
with best wishes
Bob

Tundra-Taiga Biology

Human, Plant, and Animal Survival in the Arctic

R. M. M. Crawford
*Professor Emeritus,
University of St Andrews, Scotland*

Tundra-Taiga Biology. First Edition. R.M.M. Crawford © R.M.M. Crawford 2013.
© Published 2013 by Oxford University Press.

OXFORD
UNIVERSITY PRESS

Great Clarendon Street, Oxford, OX2 6DP,
United Kingdom

Oxford University Press is a department of the University of Oxford.
It furthers the University's objective of excellence in research, scholarship,
and education by publishing worldwide. Oxford is a registered trade mark of
Oxford University Press in the UK and in certain other countries

© R. M. M. Crawford 2014

The moral rights of the author have been asserted

First Edition published in 2014

Impression: 1

All rights reserved. No part of this publication may be reproduced, stored in
a retrieval system, or transmitted, in any form or by any means, without the
prior permission in writing of Oxford University Press, or as expressly permitted
by law, by licence or under terms agreed with the appropriate reprographics
rights organization. Enquiries concerning reproduction outside the scope of the
above should be sent to the Rights Department, Oxford University Press, at the
address above

You must not circulate this work in any other form
and you must impose this same condition on any acquirer

Published in the United States of America by Oxford University Press
198 Madison Avenue, New York, NY 10016, United States of America

British Library Cataloguing in Publication Data
Data available

Library of Congress Control Number: 2013938571

ISBN 978–0–19–955940–4 (hbk.)
   978–0–19–955941–1 (pbk.)

Printed in the UK by
Bell & Bain Ltd, Glasgow

Links to third party websites are provided by Oxford in good faith and
for information only. Oxford disclaims any responsibility for the materials
contained in any third party website referenced in this work.

Preface

The Arctic is changing. How it will change is uncertain. In contemplating the future in the far north a distinction has to be made between the maritime and terrestrial Arctic. The Arctic Ocean is a relatively monotonous region whether it be frozen or unfrozen. By contrast the terrestrial Arctic, the Tundra and Taiga, are ecologically very diverse with habitats ranging from polar mountains and deserts to frozen expanses of Tundra, giving way in the south to the bogs and forests of the Taiga. Now that the Earth is warming there is a need to examine how these varied regions will respond to a changing world. All life in polar regions has to contend with similar problems irrespective of whether it is plant, animal, or human. This book is therefore an attempt to provide a circumpolar perspective of evolution and adaptation in a region with short growing seasons and variable temperature conditions.

The terrestrial Arctic has undergone many changes in the past. There is nevertheless one common feature that is inescapable, irrespective of climate change and that is the capacity to survive the long polar winter. Fundamental to all residents of polar regions is the need for the provision of reserves for long adverse seasons. This is common to both plants and animals, just as it demands foresight and planning for human settlers. This facet of arctic life has left an imprint on the evolution of the present-day flora and fauna, with a biota that has specialized in adapting to a range of very different habitats. Just how these different Tundra and Taiga habitats will respond ecologically to change will depend in large measure on the adaptive ability of the inhabitants, plant, animal, and human. The facility for adaptation in arctic life is therefore discussed in terms of physiology, reproduction, migration, and dormancy, together with relevant genetic case histories from the northern flora and fauna and its human settlers.

Recent advances in genetics, and molecular biology are providing a new assessment of the evolutionary and ecological history of the plants, animals, as well as the human inhabitants in this region. Every effort has therefore been made to discuss these exciting developments in plain English.

Technical terms when needed are defined within the text when this can be done briefly. Terms that require more extensive definition are explained in the Glossary.

Acknowledgements

This book would never have materialized had it not been for the patience and help of those who have given me the benefit of their specialist knowledge of northern lands. I am especially grateful to colleagues who have either read chapters or given detailed comments. In Scotland this has included Professor R.J. Abbott, Dr Jean Balfour, Dr Torben Ballin, Mr J.R. Stapleton, and Dr Fiona Vincent. From further afield I am indebted to Professor Inger Greve Alsos (Tromsø), Professor Ch. Brochmann (Oslo), Professor K. Fagerstedt (Helsinki), Dr Eva Fuglei (Tromsø), Dr B.E. Gudleifsson (Akureyri), Dr Anne Gunn (British Columbia), Professor K.-E. Holtmeier (Münster), Professor Ch. Körner (Bale), Dr F.H. Schwarzenbach (Bern), and Dr K.P. Timoney (Alberta). Their combined knowledge has added greatly to my own understanding of northern phenomena and may also have saved me from error; if not the fault is entirely mine.

I am particularly appreciative of having received detailed documentation as well as access to extensive collections of images from distant places from Professor Bruce Forbes (Rovaniemi), Dr F. Fyodoroov (Petrozavodsk), Dr A. Gerlach (Oldenburg), Dr T. Hajék (Trebon), Professor S.N. Kirpotin (Tomsk), Dr H. Körner (Freiburg), Dr A.H. Jahren (Hawai'i), Dr Már Jonsson (Reykjavik), Clare Kines (Baffin Island), Michael Pealow (Yukon), Dr G. Rees (Cambridge), Charles Woodley (Ottawa), and Dr N. Warnock (Anchorage). The privilege of using these northern images is acknowledged in the legends. Unattributed images are the copyright of the author. My own opportunities for studying plants in the North have been greatly aided by research assistance from Dr Hazel M. Chapman and Dr Lisa C. Smith, together with generous financial assistance from the Natural Environment Research Council and The Carnegie Trust for the Universities of Scotland.

A work of this nature would not have been undertaken had it not been for the stimulation and encouragement provided by Oxford University Press and I am particularly indebted to Ian Sherman for the initial imaginative prompting that made me attempt this task, and for bringing it to completion. In this final task I am immeasurably indebted to the production staff for their patience and painstaking care in the final stages of preparation for publication.

Contents

1. **Arctic climate history** — 1
    1.1 The concept of the Arctic — 1
    1.2 Characteristics of polar climates — 2
    1.3 Polar climatic history — 9
    1.4 Cenozoic temperature changes — 13
    1.5 Pleistocene climate changes — 16
    1.6 Conclusions — 20

2. **The Holocene at high latitudes** — 21
    2.1 After the ice age — 21
    2.2 Glacial refugia — 23
    2.3 Reconstructing past plant distributions — 28
    2.4 Pleistocene megafauna decline — 34
    2.5 Pleistocene megafauna survival — 41
    2.6 The Hypsithermal, or Xerothermic, or Climatic Optimum — 42
    2.7 Late Holocene climate fluctuations — 44

3. **Human arrival in the Arctic** — 51
    3.1 Prehistoric arctic peoples — 51
    3.2 Indigenous peoples of Arctic Eurasia — 58
    3.3 Indigenous peoples of Arctic America — 61
    3.4 Adapting to unpredictable environments — 63
    3.5 Eurasian Reindeer herding — 64
    3.6 Norse settlements across the North Atlantic — 64
    3.7 Northern peoples in recent times — 68

4. **Tundra diversity** — 70
    4.1 Defining the Tundra — 70
    4.2 Tundra classification — 71
    4.3 Arctic bryophytes — 79
    4.4 Sequestration and refixation of soil carbon — 82
    4.5 Disturbance and diversity — 82
    4.6 Tundra animal diversity — 85
    4.7 Conclusions — 92

5. Taiga and bog — 93

- 5.1 Boreal forest diversity — 93
- 5.2 The Tundra-Taiga Interface — 94
- 5.3 Fire in the boreal forest — 97
- 5.4 Taiga regeneration and the seed bank — 98
- 5.5 Northern mires — 100
- 5.6 Forest grazing — 106
- 5.7 Boreal insects — 113
- 5.8 Boreal forest carnivores — 118
- 5.9 Conclusions — 119

6. Arctic survival in mammals, birds and insects — 121

- 6.1 Advantages and disadvantages of arctic habitats — 121
- 6.2 Low temperature survival — 121
- 6.3 Low temperature protection — 129
- 6.4 Adaptation case studies — 130
- 6.5 Arctic terrestrial carnivores — 136
- 6.6 Arctic terrestrial birds — 138
- 6.7 Conclusions — 145

7. Plant survival in cold habitats — 146

- 7.1 Plant tolerance of cold — 146
- 7.2 Cold climate adaptations — 147
- 7.3 Metabolism and low temperatures — 149
- 7.4 Cold tolerance adaptations — 151
- 7.5 Ice-encasement injury — 154
- 7.6 Plant form and climate — 156
- 7.7 Conclusions — 163

8. Demography and reproduction — 164

- 8.1 Reproduction at high latitudes — 164
- 8.2 Human demography in the Arctic — 164
- 8.3 Longevity at high latitudes — 167
- 8.4 Predator–prey interactions — 170
- 8.5 Phenology and reproduction — 173
- 8.6 Plant reproduction — 180
- 8.7 Conclusions — 187

9. Evolution in the Arctic — 190

- 9.1 Pre-adaptation for arctic survival — 190
- 9.2 Arctic evolutionary influences — 190
- 9.3 Evolution of seasonal bird migrations — 191
- 9.4 Plant diversity and evolution at high latitudes — 198
- 9.5 Arctic and boreal mammal lineages — 206
- 9.6 Human evolution at high latitudes — 212
- 9.7 Conclusions — 216

10. Disturbance, pollution, conservation, and the future 217

    10.1 Disturbance at high latitudes 217
    10.2 Conservation case histories 219
    10.3 Pollution in the Arctic 233
    10.4 Radioactivity in the Arctic 241
    10.5 Conclusions 243
    10.6 Coda—the future 244

Glossary 246
References 250
Index 264

CHAPTER 1

Arctic climate history

1.1 The concept of the Arctic

The Arctic is a human concept with its origins in antiquity. Over 2,000 years ago Aristotle (384–322 BC) postulated the existence of a northerly Frigid Zone where human life would be impossible. Unlike the Antarctic, the concept of the frozen Arctic began and remains an idea. It originated as a supposition denoting a region that was imagined but unseen by the civilized world, despite the fact that human beings had been living there for 27,000 years (see Chapter 3).

For the Ancient Greeks this imaginary Frigid Zone lay under the constellation of the Great Bear (Latin *Ursa major*, Greek *Arctuous*—bear) a circumpolar constellation circling the pole star (Figure 1.1). Ptolemy, the Greek Astronomer (AD 87–*ca.* 170), named and recorded, amongst many other celestial objects, the elevations of the paws of the Great Bear.

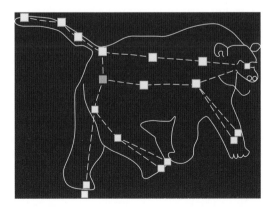

Figure 1.1 The classical conception of the constellation of the Great Bear. This constellation is known in the British Isles as the Plough, in North America as the Big Dipper (due to the resemblance to a ladle), and in Germanic and Slavic languages by a variety of names meaning a large wagon or cart. (Image courtesy of Dr F.R. Zabriskie).

The Arctic was ever a place of mystery and in Greek mythology was the supposed home of the *Hyperboreans*, a fortunate race who were thought to avoid the rigours of the Frigid Zone by inhabiting a sunny land beyond the source of the north winds (*Boreas*—god of the north wind).

1.1.1 Defining the Arctic

As with all concepts, the Arctic can be subject to variations in interpretation. Even today, defining the Arctic presents problems. For those who dislike geographical uncertainty, the Arctic may be conveniently delineated as that region of the Earth that lies within the Arctic Circle, the latitude at which the sun fails to rise for at least one day in the year (66.56°N). However, the land that lies north of the Arctic Circle contains a wide variety of habitats over a large area and in many places can support commercial forestry and even agriculture. Equally, there are regions south of the Arctic Circle that resemble regions to the north of this somewhat arbitrary boundary.

A consequence of the ability of some northern tree species to tolerate partially frozen soil profiles is seen in the nature of the arctic treeline which shows considerable geographic variation (Figure 1.2). In many regions there is also a marked and extensive transition zone, the *Tundra-Taiga Interface*. In places this zone of ecological uncertainty can be up to 600 km wide (see Chapter 5). In Eurasia this is commonly referred to as *Forest-Tundra* (Russian—*lesotundra*), as in many regions, and particularly in Russia, the vegetation resembles areas to the north of this somewhat arbitrary treeline boundary. The actual position of the Arctic Circle is therefore not of great interest ecologically as it denotes little but its

Tundra-Taiga Biology. First Edition. R.M.M. Crawford © R.M.M. Crawford 2013.
Published 2013 by Oxford University Press.

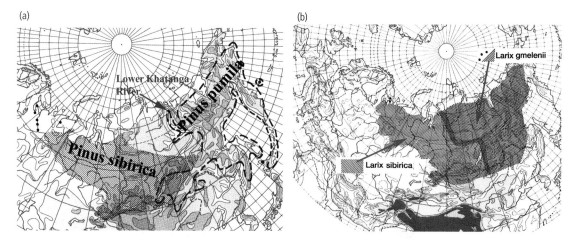

Figure 1.2 Northern tree distribution limits: (a) Siberian Dwarf Pine (*Pinus pumila*) and East Siberian Pine (*P. sibirica*). *P. pumila* has the record for most northerly tree survival, reaching 72°32′N in the lower Khatanga River; (b) Siberian Larch (*Larix sibirica*) and Dahurian Larch (*L. gmelenii*). (Reproduced with permission from Hulten and Fries, 1986).

own exactness, and, moreover, its position changes slightly over the millennia with the variation in the tilt of the Earth's axis (see section 1.2.3). Nevertheless, the Arctic Circle does mark a day-length phenomenon that was known to the ancient Greek astronomers.

Climatically, the Arctic is sometimes considered as that region of the northern hemisphere that is underlain by permafrost (permanently frozen ground). However, this is also not an entirely satisfactory boundary, as permafrost underlies approximately 24% of the terrestrial surface of the northern hemisphere (Figure 1.3) and occurs not only at high latitudes but also in some clearly non-arctic locations at high elevations (Zhang *et al.*, 1999). A more ecologically comprehensive definition is obtained by dividing the Arctic into its forested and non-forested components, namely the *Tundra* and *Taiga*, the non-forested region being the *Tundra*, while *Taiga* is a general term used to describe the northern regions of the circumpolar *boreal forest*.

If the Arctic and its tundra habitats are to be defined phytogeographically then a classification based on plant communities derived from the absence of upright trees provides a system that offers the least biological ambiguity in determining its boundaries. Nevertheless, the treeline, despite its apparent convenience and suggestion of unambiguous clarity, can also cause confusion. There are notable tree species (Figure 1.2) that can survive within the permafrost region, e.g. Siberian Dwarf Pine (*Pinus pumila*), Dahurian Larch (*Larix gmelenii*), and Siberian Larch (*L. sibirica*).

1.2 Characteristics of polar climates

The polar lands lie in a ring around the Arctic Ocean where the northern convergence of three continents brings their respective biota into a thermally challenging environment (Figure 1.4). Each of these land masses contributes its own particular flora, fauna, and evolutionary history, thus increasing the biodiversity of the Arctic. In addition, the continents differ climatically, particularly with regard to the modifying effects of proximity to the oceans. It is therefore not surprising that throughout the northern regions of the world there are wide variations in the latitudinal distributions of both plant and animal communities. Some of these differences can appear counterintuitive, particularly in relation to the distribution of trees. In some very cold regions with harsh continental climates the polar limits of the boreal forest lie surprisingly far to the north of the Arctic Circle, while in more oceanic areas they can be several hundreds of kilometres to the south. Thus, in North America the *arctic treeline* can be found at 69°N in Alaska and the Yukon, but falls to 55°N in Québec (see Chapter 5).

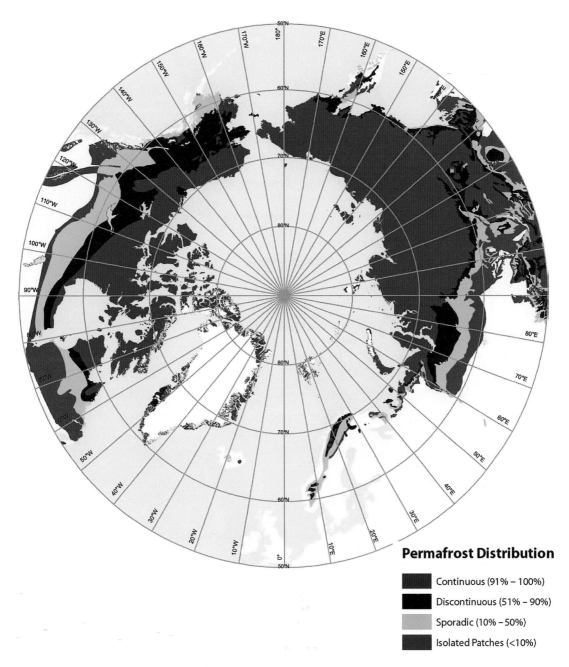

Figure 1.3 Distribution of the permafrost zones in the northern circumpolar region. Permafrost zones occupy up to 24% of the exposed land area of the northern hemisphere. The continuous permafrost zone occupies nearly 50% of the area affected by a permanently frozen sub-soil. Permafrost also exists in mountainous areas, due to high elevation cold climate. (Reproduced with permission from *Permafrost in the Northern Hemisphere*—Hugo Ahlenius, UNEP/GRID, Arendal).

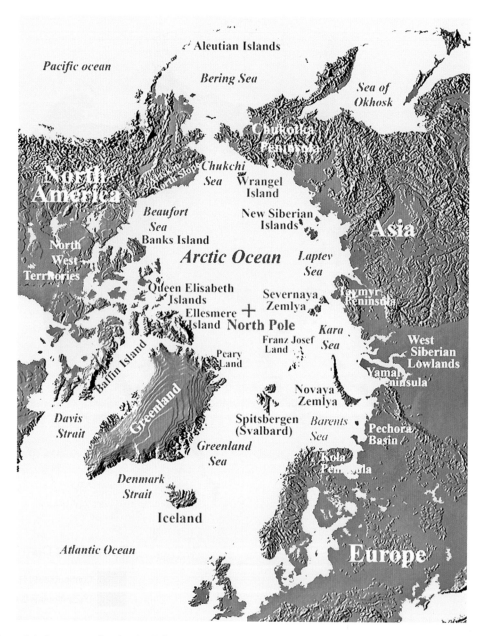

Figure 1.4 Convergence of northern hemisphere continents around the Arctic Ocean and names of locations referred to in the text.

In north-east Siberia the most northerly tree is the Siberian Dwarf Pine (*Pinus pumila*) which reaches 72°32′N in one of the most continental and coldest climates of eastern Siberia, the lower Khatanga River (Figure 1.2). By contrast, in the West Siberian Lowlands, forest cover does not advance beyond 57°N due to the cool moist air that comes in summer from the Arctic Ocean. Equally, many sub-arctic regions have plant and animal communities that are essentially arctic in their species composition (see Chapters 6 and 7).

There are therefore many ecological zones within the Arctic, and in examining the environments of the circumpolar biota a degree of inclusiveness is necessary when discussing both the Tundra and the adjacent boreal forest in terms of related arctic biomes: hence the title of this book, 'Tundra-Taiga Biology'.

1.2.1 Temperature regime variations

Over the annual seasonal cycle the Tundra and Taiga have a small positive radiation balance. In winter even the coldest regions are still warmer than space and lose heat by radiation into the polar night. When the sun does rise above the horizon the Tundra receives only about 40% of the solar radiation that reaches equatorial regions and of this, 80–90% can be reflected back into space from snow and ice, a phenomenon that is termed *albedo*. The forested regions fare better. The foliage of the evergreen conifers sheds much of the fallen snow and the darker colour of the emerging foliage absorbs more radiation.

The thermal deficiencies of these northern areas are, however, partially relieved by heat that is transferred from lower latitudes by ocean currents and winds. Consequently, in any examination of the climatic characteristics of these northern lands, geographical location, and especially the degree of exposure to maritime influences, are highly significant ecological factors in promoting biological diversity at high latitudes (Figure 1.5). The relatively mild conditions of Arctic Norway and the archipelagoes in the Barents Sea (Spitsbergen and Franz Josef Land) as compared with the Canadian High Arctic Islands are a consequence of differing maritime conditions.

At the highest latitudes, the sparse plant-cover of the Tundra gives way to the even sparser plant-cover of the Polar Desert. The latter would have been more extensive in the late Pleistocene when the sea level was as much as 120 m lower than it is today, and snow and ice-cover would be less than further south due to the arid conditions imposed by a high-pressure zone over the North Pole.

1.2.2 Sea levels and oceanicity

A major climatic change took place when the sea level rose 6,000 years ago and increased the impact of oceanic conditions at high latitudes in both North America and Eurasia. This was particularly marked in the low-lying lands of northern Siberia where the sea advanced over a large area of low-lying land and imposed an oceanic current circulation on the climate of the Arctic Ocean with a very marked effect on the West Siberian Lowlands that persists to this day (Figure 1.5).

Oceanic winters with high levels of precipitation, together with cool, moist summers, have over the past 6,000 years encouraged bog growth to spread south from the Tundra and force the northern limit of the forest to retreat southwards, leaving in the West Siberian Lowlands the largest expanse of bog in the world (see Chapter 5).

In the West Siberian Lowlands the northern limit of the boreal forest is therefore not caused by low temperatures but is instead an edaphic limitation due to bog expansion (*paludification*). These very noticeable differences at northern latitudes between continental and oceanic regions have a profound effect on the distribution of the plant and animal communities and make the Arctic and the forested sub-arctic regions biologically diverse. Such diversity in marginal areas should not be surprising. In any situation where conditions are severe, minor differences can have significant ecological impacts. Thus, despite the seemingly monotonous, cold, and prolonged winters of the North, geographical location influences ecological diversity. This, together with the ever-changing states of oceanic currents, makes predicting the future climate and ecology of the Arctic a topic fraught with uncertainty.

1.2.3 The polar night

Despite marked local variations, polar regions have one feature in common, namely the marked seasonality that is imposed by the tilt of the axis of rotation of the Earth relative to its orbit round the sun. As long as the Earth continues to spin on a rotational axis that is not perpendicular to its orbital plane (Figure 1.6), but makes an angle known as the *obliquity of the ecliptic*, then the polar regions will continue to have a climate that is distinct from that of lower latitudes in both periodicity and thermal input.

The *obliquity of the ecliptic* (ε epsilon) varies between ε 22.1° and ε 24.5° with a 41,000-year period. At present, the tilt is 23.44° and decreasing.

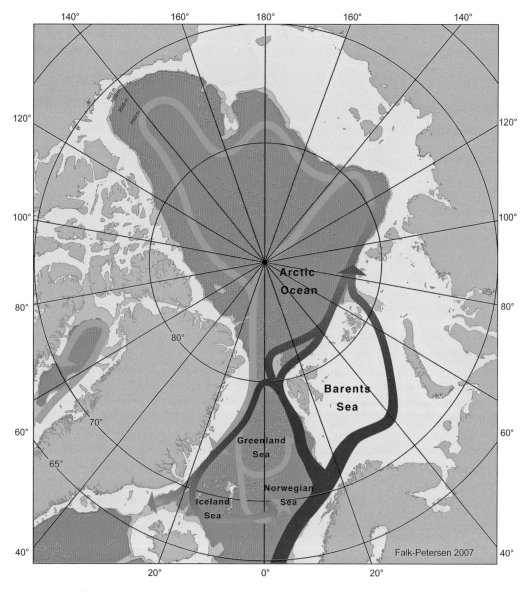

Figure 1.5 The paths of ocean currents in the Arctic. Warm Atlantic water moves northwards along the coast of Norway. It divides into two main branches and continues northwards, with one branch on either side of Spitsbergen. In the Arctic Ocean, the Atlantic water is cooled, becomes heavier, and sinks. After circulating in the North Polar Basin, the now cold Arctic water leaves the Arctic Ocean, mainly through the Fram Strait between Svalbard and Greenland. (Reproduced with permission from *The Arctic System*—a project of the Norwegian Polar Institute. Ed. G.S. Jaklin).

As a result of these minimal changes in inclination of the Earth's axis, the *polar night* is an astronomical fixture subject only to minor variations, in the past, and in the future. The Arctic Circle, currently at 66.56°N (ε = 23.44°) marks the southern limit of the so-called polar night where the sun at the winter solstice disappears below the horizon for only 24 hours, or put more positively where it does not set on midsummer's day.

Moving north, the length of the polar night increases, and in Spitsbergen (77°N) the sun disappears below the horizon on 22nd October and does not reappear until 20th February, resulting in a polar night of 122 days. At 80°N, which is approximately

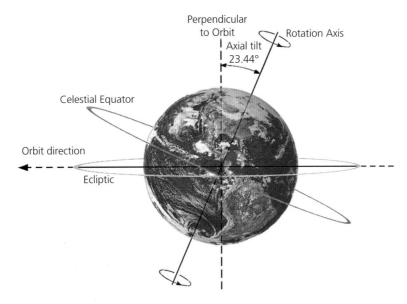

Figure 1.6 The obliquity of the ecliptic—the most permanent feature of the arctic environment.

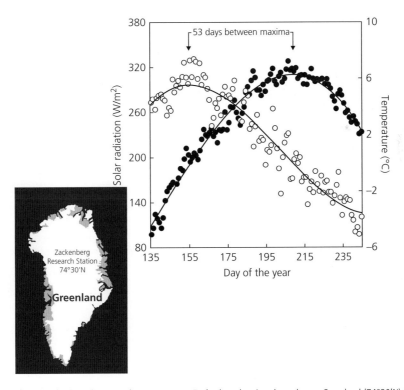

Figure 1.7 Seasonal dynamics of solar radiation and temperature at Zackenberg (see inset), north-east Greenland (74°30′N). The level of average daily incoming solar radiation (open dots) and temperature (filled dots) in relation to day of the year averaged for the years 1996–2005. (Reproduced with permission from Høye and Forchhammer, 2008).

the latitudinal limit of the world's most northerly lands and includes Ellesmere Island, Peary Land, Franz Josef Land, and Severnaya Zemlya, the length of the polar night increases to 135 days—just over one-third of the year!

The duration of the polar night, coupled with the low angle of incidence of solar radiation when the sun is above the horizon, results in long winters and brief summers, as the winter snow and ice has to melt before the brief growing season can begin (Figure 1.7). In these more northerly regions this growing season frequently lies between 45 and 55 days, depending on local conditions.

1.2.4 Climatic consequences of winter ice

The extent of sea ice at the end of the summer at northern latitudes is currently declining (Figure 1.7). Nevertheless, the high *albedo* of the remaining ice still contributes to keeping the Arctic cold. As a result, much of the summer warmth is still used for melting the winter snow and ice, and it is not until late June or even early July that plants can begin to grow and flower, and migrating geese are able to feed and nest. Thus, by the time the high-latitude regions of the Arctic have escaped from the jaws of winter, the elevation of the sun is already declining (see Figure 1.7). As a result of climatic warming, this loss of productive thermal time (day degrees) due to late snowmelt may now be changing.

One very noticeable change that is already apparent is the earlier disappearance of snow and ice-cover (Figure 1.8) and an advancement of plant growth in spring (Figure 10.24). Despite the short duration of the growing season, at high latitudes the 24-hour light regime in summer provides a remarkable source of energy. At ground level the microclimate can be relatively warm, promoting photosynthesis and plant growth.

Climatic warming in suitable areas may therefore be expected to increase the productivity of these regions, especially if nutrient cycling is facilitated by grazing (see Chapter 6). Nevertheless, despite

Figure 1.8 Satellite data revealing a new record low Arctic sea ice extent as seen on 12th September 2012. The average minimum extent over the past 30 years is indicated by the yellow line. The frozen cap of the Arctic Ocean appears to have reached its annual summertime minimum extent and broken a new record low. This new record minimum measures almost 300,000 sq miles less than the previous lowest extent in the satellite record, set in mid-September 2007, of 1.61 million sq miles (4.17 million km^2). (Reproduced with permission of NASA/Goddard Scientific Visualization Studio).

these possible climatic improvements, the unavoidable polar night will continue to enforce a degree of seasonality that is not experienced at lower latitudes and the Arctic will retain its own distinct climatic identity, even in a warmer world.

1.3 Polar climatic history

1.3.1 Before the ice

Despite the inescapable polar night, the Arctic and adjacent sub-arctic regions have not always been in the grip of snow and ice. In the Cretaceous, 145–65 million years ago, dinosaurs roamed the North Slope of Alaska (Figure 1.9). The latitude of the North Slope of Alaska which at that time was at least 70°N is also its present latitude (Brouwers et al., 1987). Admittedly, these animals were not of a size that compared with the giants found elsewhere. Nevertheless, where dinosaurs roamed, the climate must have been one that could support a growth of vegetation sufficient for their grazing appetites.

When the first dinosaur remains were found on the North Slope of Alaska in the 1980s it was conjectured that they must have migrated out of the region during the polar winter. Alaska was already well north of the Arctic Circle by the late Cretaceous and the winter would have been prolonged. However, more discoveries along the Colville River (Figure 1.9) have thrown doubt on the migration theory.

Several of the more recently found dinosaur types were small flesh-eaters and it has been concluded (Rich et al., 2002) that neither the adults nor juveniles could have physically migrated a round-trip distance estimated at more than 5,000 miles (8,000 km).

It is possible that the plant-eating dinosaurs could have been the over-wintering food source of the meat-eaters. All reptiles can hibernate, and evidence from cold climate habitats in southern Australia suggests some dinosaurs lived in burrows where presumably they retreated in winter (Varricchio et al., 2007). Recent fossil discoveries have noted very large eye sockets which might suggest that these animals were sufficiently sight-adapted to be able to graze during the polar night if it was not too prolonged (Constantine et al., 1998).

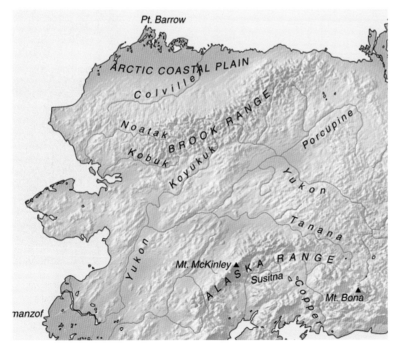

Figure 1.9 Map of main Alaskan mountain ranges. The North Slope region lies between Brook Range and the Colville River, where numerous dinosaur remains dating from the late Cretaceous have been found. Point Barrow is 71°N and Mt McKinley 63°N.

1.3.2 Ancient arctic forests

An indication of past plant productivity at high latitudes is seen in the remains of middle Eocene forests (*ca.* 45 million years ago) found in arctic Canada and Greenland (Figures 1.10–1.13). The most northerly sites for the occurrence of former ancient arctic forests are in Peary Land, north-east Greenland (Figure 1.10). The geologist Lauge Koch first discovered these ancient tree trunks in the 1920s at an altitude of up to 165 m above sea level in the hills of the Kap København area in eastern Peary Land. Some were so well preserved that they looked like postglacial driftwood that had been stranded on subsequently raised beaches. However, the Kap København Formation is now dated to the Plio-Pleistocene (Bennike, 1998). Although no trees have been found in a growth position in the Kap København area, it was concluded that the trees grew locally, as leaves, needles, seeds, and cones were also found, together with a rich fossil insect fauna comprising numerous species that are dependent on trees.

Another indication of the past productivity at high latitudes is seen in the remains of middle Eocene wood (*ca.* 45 million years ago) found on Axel Heiberg Island (Figures 1.11–1.13). This High Arctic Canadian site at 79°55′N has yielded a particularly rich assemblage of plant macro- and microfossils, indicating diverse conifer forests with a rich angiosperm understorey that successfully endured 3 months of continuous light and 3 months of continuous darkness.

Palaeo-environmental research for this region suggests a warm, ice-free environment, with high growing-season relative humidity and high rates of soil methanogenesis (Jahren, 2007). These ancient timbers reveal the existence of very substantial trees. Their spectacular state of preservation has made it possible to apply stable isotope analyses of carbon, oxygen, and hydrogen and thus reconstruct the palaeo-seasonal environment.

The analysis of the carbon isotopes in their organic and cellulose remains has also revealed striking annual growth patterns which have been interpreted as evidence of the deciduous habit in these unusual conifers. *Metasequoia* sp. was the most abundant tree and the forest had an understorey of fringe vegetation containing maple, alder, birch, hickory, chestnut, beech, ash, holly, walnut, sweetgum, sycamore, oak, willow, and elm. Smaller flowering plants included Honeysuckle and Sumac.

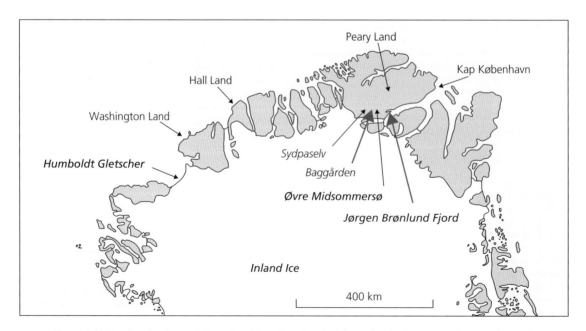

Figure 1.10 Location of ancient arctic forests found in northern Greenland. (Reproduced with permission from Bennike, 1998).

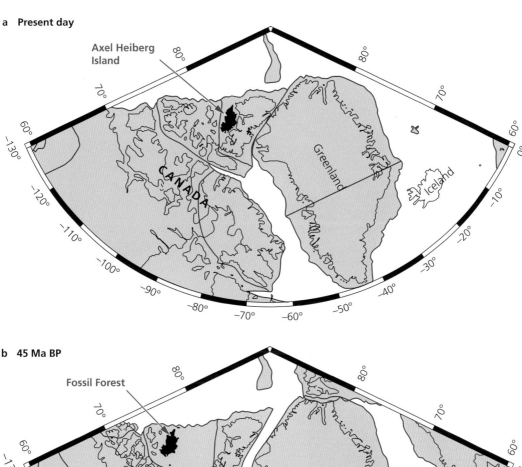

Figure 1.11 Position of tectonic plates (a) at present and (b) during the middle Eocene. The forest location is well above the Arctic Circle (66.56°N), where remnants of deciduous conifer forests including leaves, cones, and other fossils were found and dated to *ca.*45 million BP. Plate boundaries are shown in blue, black lines indicate modern shorelines. (Reproduced with permission from Jahren, 2007).

1.3.3 Recently extinct arctic forests

Concentrations of pre-Holocene wood have been found at two sites in Peary Land, namely at Jørgen Brønlund Fjord and at Baggården, where trees were still growing up to 2.0–2.5 million years ago, at the Pliocene–Pleistocene transition when northern Greenland was already situated at a high latitude and before

Figure 1.12 Wood excavated from the fossil forest on Axel Heiberg Island, Canada, showing the exceptional preservation. (a) The largest trees found measured close to 3 m in diameter. (b) Juvenile trees were also found, sometimes still attached to the root bole. (c) Continuous lengths were also found sometimes several metres in length. (Reproduced with permission from Jahren, 2007).

the subsequent land rise in the early Holocene. This forest was mixed with heathlands and well-vegetated lakes. The woody genera found in these high- latitude sites included *Larix*, *Picea*, *Thuja*, *Taxus*, *Betula*, and *Salix* (Bennike and Böcher, 1990). The lack of pines in the Kap København Formation distinguishes this flora from the *Late Cenozoic* floras in northern Canada.

The floristic composition and the size of the wood fragments from Washington Land (for location see Figure 1.10) showed a strong resemblance to that of the Beaufort Formation on Meighen Island (just west of Axel Heiberg Island, Figure 1.11) and the high-level alluvium site on Ellesmere Island, for which a mid or late Pliocene age is thought to be about 3 million BP.

The persistence of forest in southwestern Greenland up to 1 million years ago has been demonstrated from biomolecular studies of 2-mile-deep ice cores at a high altitude in southern Greenland. Examination of DNA and amino acids from buried organisms recovered from the basal sections of the ice cores has identified the species present which have included a diverse array of conifer trees as well as birch and insects, all of which lived at this site within the past 1 million years (Willerslev *et al.*, 2007).

A question often arises as to how large trees, such as the *Metasequoia* sp., survived long warm winters as they existed at the end of the Tertiary in polar North America, with a probable polar night of 3 months. It

Figure 1.13 (a) The thick litter layer that blanketed the floor of a middle fossil Miocene forest on Axel Heiberg Island, Canada, within which leaf, stem, cones, spores, and pollen were abundantly preserved. (b) Fossilized *Metasequoia* leaves. (c) Fossilized *Betula* leaves. (Reproduced with permission from Jahren, 2007).

has been suggested that the energy demands of high respiration rates of the over-wintering tissues might have posed a risk of spring starvation due to winter carbohydrate depletion. This was a much-debated point in the first half of the 20th century (Beerling, 2007). However, recent research into the question of carbon balance of trees at the treeline has shown that where summer productivity is adequate, woody plants are more than capable of storing enough carbohydrate to survive prolonged winters.

At the treeline in the Austrian Alps, a characteristic species, the Swiss Stone Pine (*Pinus cembra*), maintains a positive annual carbon balance even though the daily carbon balance can be negative for 5 months in the year. The estimated total winter-time respiratory losses amount to only 9% of the amount of carbon fixed during the growing season (Wieser *et al.*, 2005). It is probable therefore that a similar state of summer–winter carbon balance existed also in the past and the long polar night would not have been an obstacle for woody plants, even when winter temperatures were warm.

1.4 Cenozoic temperature changes

The past 45 million years have witnessed dramatic changes in the temperature regime of the Earth. The early Eocene (~56–49 million years ago) had the highest prolonged global temperatures of the past 65 million years, and also levels of atmospheric carbon dioxide (CO_2) that were probably in excess of 2,000 ppm in the late Palaeocene and approximately five times greater than at present. The Palaeocene–Eocene thermal maximum, 55 million years ago, was a brief period of widespread, extreme climatic warming that was associated with a massive atmospheric greenhouse gas input. Sea surface temperatures near the North Pole increased from 18°C to over 23°C during this event (Pearson and Palmer, 2000).

The cooling that caused the gradual onset of a colder Arctic began approximately 60 million years ago and is termed the *Cenozoic Temperature Decline* (Figure 1.19). The sudden, widespread glaciation of Antarctica and the associated shift towards colder temperatures at the Eocene–Oligocene boundary 34 million years ago was one of the most outstanding changes in global climate ever recorded.

The long-term features of Cenozoic temperature decline have been attributed to a variety of causes, including the movement of tectonic plates, alterations in ocean currents, and a decrease in atmospheric CO_2 concentrations.

The freezing of Antarctica has been thought to have resulted from the tectonic opening of Southern Ocean gateways, which enabled the formation of the *Antarctic Circumpolar Current* and the subsequent thermal isolation of the Antarctic continent. (DeConto and Pollard, 2003).

1.4.1 Causes of climatic oscillations

In searching for possible causes of repeated periods of climatic cooling and warming it is necessary to consider possible interactions between the CO_2 content of the oceans and that of the atmosphere. It is possible that the oceans may act as a 'biological pump', causing the sequestration of carbon in the ocean interior by the burial of calcium carbonate in marine sediments. Such mechanisms have been invoked for the fall of atmospheric CO_2 levels during glacial periods. In particular, attention has been focused on the open ocean surrounding Antarctica, involving both the biology and physics of that region (Sigman and Boyle, 2000).

The incidence of climatic changes has to be considered also in relation to geological processes. Apart from their effects on plate tectonics and continental migration phenomena, they can also bring about increases and decreases in atmospheric CO_2. During the late Palaeocene and early Eocene, there was large-scale volcanic activity and rifting that accompanied the expansion of the North Atlantic Ocean.

The enhanced volcanic outgassing of CO_2 that this would have produced may have been accompanied by magmatism (the formation of igneous rock from magma) and regional metamorphism in parts of the Himalayan belt and North America. Similar subsequent events occurred in the early Eocene (see Pearson and Palmer, 2000). The termination of North Atlantic volcanism at about 54–53 million years ago corresponds approximately to an initial drop in ocean bottom water temperatures (Figure 1.14).

Another source of CO_2 may have been the oxidation of methane released from storage either in wetlands or from clathrate hydrate (sometimes referred to as *methane ice*) reservoirs lying under the oceans. CO_2 from these sources may have contributed to the short-lived late Palaeocene thermal maximum event at 55 million years ago (Figure 1.14). Similar subsequent events occurred in the early Eocene (see Pearson and Palmer, 2000).

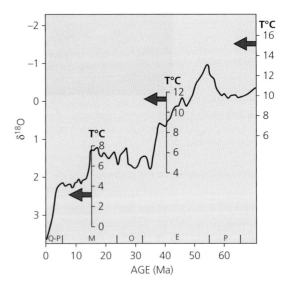

Figure 1.14 The decline in ocean bottom water temperatures in the Atlantic indicated by oxygen isotope ratios from deep-sea benthic Foraminifera. The interpreted decline in bottom water temperatures is assumed to reflect decline of polar surface temperatures during the Cenozoic. Three temperature scales are shown: *left* is for the younger period Late Miocene to Recent and assumes that the $\delta^{18}O$ of seawater is −0.28‰. The scale on the *right* applies to the older part of the diagram late Cretaceous to Eocene and assumes an ice-free earth, with a $\delta^{18}O$ of seawater of −1.2‰. The scale in the *middle* applies to the mid-Tertiary–Oligocene up to the middle Miocene and assumes an intermediate $\delta^{18}O$ of seawater. (Adapted from Hay *et al.*, 2002).

The breaking away of the Antarctic continent from India to its present position over the South Pole, where it is now covered in a very deep layer of ice, has increased terrestrial albedo, thus reducing the capacity of the sun to heat a very substantial landmass.

Geological explanations have also been put forward to account for the subsequent cooling effect. CO_2 could be removed from the atmosphere by limestone formation or organic carbon burial. It is equally possible that the tectonic activity that brought about the rising of the Tibetan Plateau, the Himalaya, the European Alps, the Rockies, and the Andes would have brought fresh rock to the surface of the Earth. The newly exposed rock would then react with atmospheric CO_2 to form carbonates which eventually eroded into the sea, depleting the atmosphere of CO_2 and thus gradually imposing a global climatic cooling.

There is still an active scientific debate concerning the exact relationship between the various physical and biological factors that may have brought about this sustained period of global cooling that caused the *Cenozoic climatic decline*. One of the most intriguing of past climatic events is the possible role of a small aquatic fern, *Azolla*, in reducing atmospheric CO_2 concentrations.

1.4.2 The Azolla event

A biological factor that could have played a role in reducing atmospheric levels of CO_2 has been attributed to the freshwater aquatic fern *Azolla* (Figure 1.15). *Azolla* contains within its leaf cavities a symbiotic, heterocystous, cyanobacterium, *Anabaena azollae*, with the ability to fix atmospheric nitrogen. Due to the nitrogen-fixing properties of the cyanobacteria, this aquatic fern is capable of high growth rates even in arctic waters that would otherwise have limited growth due to a lack of nitrogen.

An analysis of Palaeogene sediments obtained during the 2004 Arctic Coring Expedition (Brinkhuis *et al.*, 2006) found that large quantities of the free-floating fern *Azolla* were growing and reproducing in the Arctic Ocean by the onset of the middle Eocene epoch (~50 million years ago) and that it caused the deposition of fossilized matter into the anaerobic conditions of the ocean floor throughout the Arctic Basin.

Figure 1.15 *Azolla* sp., a freshwater aquatic fern which contains a symbiotic, heterocystous, blue-green alga, *Anabaena azollae*, within cavities in its leaves.

The *Azolla event* coincided precisely with an unparalleled decline in atmospheric CO_2 levels, which fell from 3,500 ppm in the early Eocene 50 million years ago to 650 ppm over a period of 800,000 years.

The polar *Azolla* populations appear to have been able to exploit the extraordinary high CO_2 concentrations and temperatures in the Arctic Ocean at this time due to having access through symbiosis to unlimited supplies of nitrogen. It is astonishing to calculate that this one phenomenon alone, of *Azolla* bloom episodes over a period of 800,000 years across a 4,000,000 km² ocean basin, may have been sufficient for the sequestration of carbon by plant burial to account for 80% of the observed drop in atmospheric CO_2.

Other factors would probably also have had a role in the drop of temperature which continued for millions of years as the Arctic cooled from an average sea-surface temperature of 13°C in the early Eocene to today's −9°C (Brinkhuis *et al.*, 2006). It is noticeable that a rapid decrease in temperature between 49 and 47 million years ago coincided with the *Azolla* event, as is evident from dropstones (isolated fragments of rock—see Table 1.1) which range in size from small pebbles to boulders, dropped vertically through the water column and indicating the onset of colder weather by the presence of melting of icebergs.

Figure 1.16 shows the early Eocene arctic basin and marks with dots the sites used by the Arctic

Table 1.1 Terms used in describing geological epochs that are relevant to periods of climatic change in arctic and boreal regions.

Cenozoic temperature decline	The Cenozoic era is divided into two periods, the *Palaeogene* and *Neogene*, and each is further divided into epochs. The *Palaeogene* consists of the *Palaeocene, Eocene,* and *Oligocene* epochs, and the *Neogene* consists of the Miocene, Pliocene, Pleistocene, and Holocene epochs, the last of which is the present epoch.
Younger *Dryas* stadial	The Younger Dryas stadial, named after the alpine/tundra wildflower *Dryas octopetala*, was a brief cold period (approximately 1,300 ± 70 years) following the Bölling/Allerød interstadial at the end of the Pleistocene between approximately 12,800 and 11,500 BP. In the UK this is known as the Loch Lomond Stadial and most recently the Greenland Stadial.
Dropstones	Isolated fragments of rock found within finer-grained water-deposited sedimentary rocks. They range in size from small pebbles to boulders. The critical distinguishing feature is that there is evidence that they were not transported by normal water currents, but were dropped in vertically through the water column from melting ice.
Clathrate hydrate	Crystalline water-based solids physically resembling ice, in which small non-polar molecules (typically gases) are trapped inside 'cages' of hydrogen-bonded water molecules.
Magmatism	The formation of igneous rock from magma.
North Atlantic Oscillation (NAO)	An indicator of winter surface air pressure difference between Iceland and the Azores (latterly Gibraltar).

Coring Expedition. As *Azoll*a is intolerant of salt concentration greater than 1–1.6% there would have to have been a layer of fresh water on the surface of the Arctic Ocean. It has been postulated (Brinkhuis *et al.*, 2006) that at the time of the *Azolla* event the Arctic may have been almost completely cut off

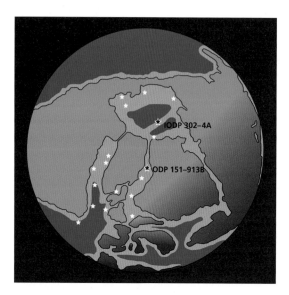

Figure 1.16 The early Eocene Arctic basin. The white dots mark the coring locations of the Arctic Coring Expedition. (Reproduced with permission from Brinkhuis *et al.*, 2006).

from the inflow of warm and salty currents originating from lower latitudes.

Due to the lack of salty currents flowing into the Arctic Basin, the local excess of precipitation over evaporation would have created a surface freshwater environment over the deeper stratified salt-rich layer. During this unique interval, pulses of fresh water even overflowed from the Arctic and carried remains of the ferns into the surrounding ocean basins. The reign of the freshwater ferns in the Arctic ended abruptly 48.3 million years ago when waters became saline.

1.5 Pleistocene climate changes

1.5.1 The Croll–Milankovitch theory

Ever since the discovery of the *Great Ice Age* and its history of glacial and interglacial periods there has been a plethora of theories attempting to explain the astonishing waxing and waning of the ice sheets. An early theory, which still claims attention, was first put forward by a Scottish self-taught scientist James Croll (Figure 1.17). During Croll's lifetime the notion of multiple glacial and interglacial epochs was much debated. Croll suggested that the periodic changes in the Earth's orbit around the sun would alter the amount of sunshine reaching the planet at various times of the year. Less sunshine in winter would lead to snow accumulating, and with the ice sheets expanding, albedo would increase the effect of the orbital changes, thus leading to ice ages. He also added that there would be other positive feedbacks such as changes in ocean currents. Croll's work was widely discussed, but his dates for the end of the ice age at 80,000 years ago did not match the emerging knowledge of the timing of the glacial periods, and by the end of the 19th century his theory was generally disbelieved.

The concept that variations in the Earth's position and orientation in relation to the sun having a bearing on climate change was revived by the Serbian mathematician Milutin Milankovitch (1879–1958) (Figure 1.17) who dedicated the greater part of his scientific life to developing a mathematical theory, now known as the Milankovitch or sometimes the *Croll–Milankovitch theory*, which states that as the Earth travels through space around the sun, cyclical variations in four elements of Earth–sun geometry combine to produce variations in the amount of solar energy that reaches Earth.

The variations comprise:

(1) *Variations in the Earth's orbital eccentricity*, i.e. the shape of the elliptical orbit of the Earth around the sun which varies between a minimal and a maximal eccentricity with a periodicity of approximately 100,000 years.
(2) *Changes in obliquity*—changes in the angle that the Earth's axis makes with the plane of the Earth's orbit with a periodicity of 41,000 years.
(3) *Precession*—the Earth's axis also has a slow rate of rotation (cf the wobble of a gyroscope), completing a rotation in approximately 26,000 years. This change of rotational axis alters the dates of perihelion (closest distance from the sun) and

Figure 1.17 (*Left*) James Croll (1821–90), the first person to suggest that periodic changes in the Earth's orbit could influence climate. In 1875 he published his seminal book *Climate and Time in Their Geological Relations: A Theory of Secular Changes of the Earth's Climate*. In 1876 he was elected a Fellow of the Royal Society and awarded an honorary degree by the University of St Andrews. (*Right*) Portrait by P. Jovanovic (1943) of Milutin Milankovitch (1879–1958), the Serbian civil engineer and mathematician who developed the mathematical predictions for estimating the eccentricity, axial tilt, and precession of the Earth's orbit and the hypothesis that this could underlie the Pleistocene glacial fluctuations. He dedicated his entire career to developing a mathematical theory, now known as the Milankovitch or sometimes the Croll–Milankovitch theory.

aphelion (farthest distance from the sun), which increases the seasonal contrast in one hemisphere while decreasing it in the other. Currently, the Earth is closest to the sun in the northern hemisphere winter, which makes the winters there less severe (Figure 1.18).

Despite the apparent coincidences between the Milankovitch peaks and glacial activity there is still the problem of whether the changes in solar energy reaching the Earth with these fluctuating periodicities are sufficient to account for the changes in thermal environment that are needed to create and end an ice age.

Within the last million years, glaciation patterns have changed. The last 100,000 years has seen a series of more severe glacial periods. Although the Earth's orbit alters slightly over periods of 95,000 and 125,000 years, this has a weaker effect on the seasons than the other orbital cycles, and it is not clear from the Milankovitch cycles as to why the deepest ice ages should have been driven by the smallest changes in summer sunshine. Recent discussions of this dilemma have tended to look upon the Croll–Milankovitch cycles as acting like pacemakers.

1.5.2 Oceans and climate change

For some time it has been evident that the oceans of both hemispheres are connected with the abrupt changes in temperature that take place both at the time of the initiation of glacial periods and with their termination (Rickaby and Elderfield, 2005). There is still a great degree of uncertainty as to how these changes operate, in particular in relation to the changes in atmospheric levels of CO_2.

Data derived from two Antarctic ice cores (Schmitt et al., 2012) have succeeded in producing the best record to date of the glacial to interglacial variations in stable carbon isotopes of CO_2. Their findings show cyclic changes for the past 24,000 years. They have detected a rapid 0.3 per millennium depletion in ^{13}C between about 17,500 and 14,000 years, a time when the CO_2 concentration rose by about 60 ppm. It is suggested that this is consistent with the release of CO_2 from a previously isolated deep-ocean reservoir that accumulated carbon due to oxidation of organic detritus sinking from the ocean surface. A summary depiction of these findings (Brook, 2012) is shown in Figure 1.19.

A possible mechanism has been postulated to try to explain the interactions between ocean tempera-

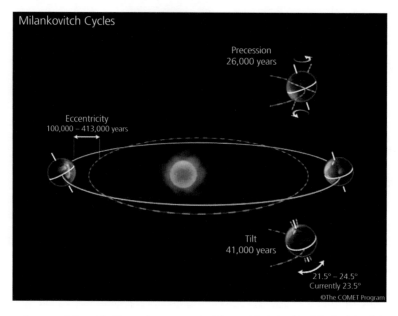

Figure 1.18 Diagrammatic representation of the three major components of the spatial relationship of the Earth in relation to the sun as defined by the Milankovitch cycles. (Reproduced with permission from UCAR—University Corporation for Atmospheric Research, Boulder, Colorado).

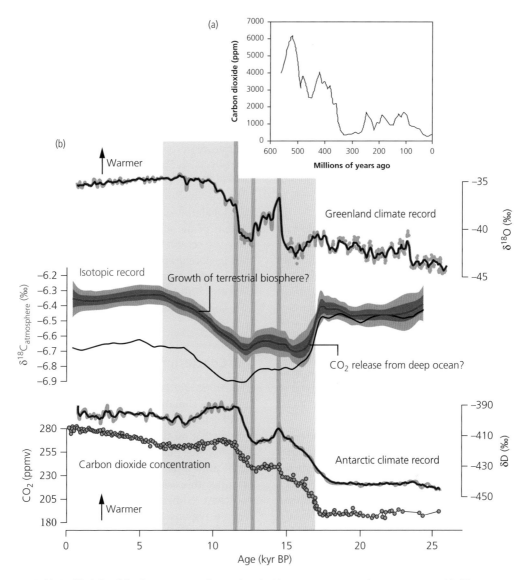

Figure 1.19 (a) Simplified plot of the changes in atmospheric carbon dioxide concentrations over Phanerozoic time, modified from Berner (2006). (b) Glacial–interglacial climatic changes over the the past 24,000 years. CO2 concentrations have risen (black curve) as the Earth emerged from glaciation, as shown by climate records from Greenland and Antarctica (blue curves) and change in the $_{13}C/_{12}C$ ratio of CO2 during this time (red curve). The isotopic ratio is expressed in delta notation, where $\delta_{13}C$ is the deviation of a sample ratio from that of an internationally expected standard. (Reproduced from Brook 2012).

ture and atmospheric CO_2 concentrations, and why, after a long interval with a cooling climate, each late Quaternary ice age ended with a relatively short warming leg called a *termination* (Denton *et al.*, 2010). Terminations have invariably begun when ice sheets were at or close to their greatest area and volume. When at their largest, these sheets expanded across continental shelves, and with their increasing weight induced maximum isostatic depression beneath them. The bigger ice sheets grow, the lower the continental crust beneath them sinks. As the crust sinks, much of the ice sheet will end up below sea level. Ice sheets resting on the seabed are more vulnerable to warming. Being somewhat buoyed up, the ice can flow more easily and thus disintegrate faster. When the Laurentide ice sheet covering North America started to melt, vast amounts of water and ice would have poured into the North Atlantic. This freshwater would reduce the density of the surface layer, stopping it sinking and thus shutting down the *Atlantic meridional overturning circulation* (AMOC) which carries warm upper waters into far-northern latitudes and returns cold deep waters southward across the Equator. If less heat is carried north, the southern oceans warm and as CO_2 is less soluble in warm water, this leads to the release of CO_2 into the atmosphere.

It is suggested therefore that ice sheets build up until they near the brink of stability, at which point a modest rise in summer sunshine is enough to tip them over the edge. As the ice sheets melt, fresh water is released into the Atlantic, shutting down ocean circulation and pumping CO_2 into the atmosphere. As long as the combined effect of extra summer sunshine and rising CO_2 outweighs the regional cooling produced by the shutdown of ocean circulation, the ice keeps melting, pouring more fresh water into the Atlantic and eventually pumping so much CO_2 into the atmosphere that the ice sheets melt away in just a few thousand years.

Initially, this suggested sequence of events for the termination of a glacial epoch received little support because the rise in CO_2 was thought to precede the rise in sea level. Furthermore, the source of the CO_2 was not clear or how it would fit together in a plausible sequence, starting with a small rise in insolation. However, the hypothesis, as it now stands, has been clarified by further research on the history of atmospheric CO_2 levels as recorded in ancient cave stalagmites and coral reefs. These repositories of fossilized carbon have provided evidence of the necessary higher atmospheric CO_2, and suggest that the periodic temperature oscillations were driven by the reorganization of ocean and atmospheric circulation that accompanied the northern stadials, and thus acted as a key factor to complete the deglaciation.

Another possible source of CO_2 may have been the oxidation of methane released from storage either in wetlands or from large sea or clathrate hydrate reservoirs, as may have occurred in the short-lived late Palaeocene thermal maximum event at 55 million years ago.

Further research into the relationships between the global upwelling of ocean currents and the CO_2 content of the atmosphere will be needed before there is a complete understanding of the history of the Pleistocene epoch. It would appear that this will only be achieved when there is an integrated account of both northern and southern hemisphere influences on the oceans and their role as CO_2 sinks and sources.

1.6 Conclusions

The Arctic and its adjacent lands have endured a turbulent climatic history, especially throughout the Pleistocene. Celestial geometry will ensure that the Earth's orbital variations will continue. It is also probable that there will be higher levels of CO_2 in the atmosphere, and it may well be that at high latitudes this amelioration of the climate will be sufficient to liberate the Arctic and adjacent cold-climate regions from the threat of returning ice ages. This will be likely to bring about many ecological changes. Whether or not future generations of northern peoples will regard these changes as beneficial or damaging for wildlife and human prosperity cannot be judged by those who are alive today.

CHAPTER 2

The Holocene at high latitudes

2.1 After the ice age

The transition from a glacial to a post-glacial world at northern latitudes began after ice volumes peaked about 21,000 years ago. Depending on location, there then followed a gradual ice retreat. About 16,000 BP the post-glacial recession of the ice sheets had accelerated in response to increasing summer warmth. During the final stages of the glacial period there were a number of climatic oscillations referred to as the Bølling-Allerød interstadial (14,700–12,900 BP) which began with the Oldest Dryas and ended abruptly with the onset of the Younger Dryas. This last cold period of the Pleistocene from *ca.* 12,900–11,600 BP reduced temperatures back to near-glacial levels within a decade. Altogether there were three Dryas periods, the Oldest, the Old, and the Younger, so named as they were associated with the spread of the Mountain Avens (*Dryas octopetala*) (Figure 2.1).

The fluctuating climate of the early post-glacial period has been linked with the periodic outpouring of cold, glacial, melt-water into the North Atlantic. The last cold period, the *Younger Dryas* or *Big Freeze* (12,900–11,600 BP) appears to have been triggered by a catastrophic release of fresh water stored in Lake Agassiz, an immense glacial lake located in the middle of the northern part of North America (Figure 2.2). This release was initiated when the retreating margins of the Laurentide ice sheet opened a lower outlet, allowing much of the lake's stored water to flood across the region now occupied by the northern Great Lakes into the St Lawrence valley. From there it entered into the North Atlantic where it affected the thermo-haline circulation of the ocean and brought about the *Big Freeze* of the *Younger Dryas period* as well as accelerating the post-glacial rise in sea level with a number of melt-water pulses (Figure 2.3).

On the basis of reconstructions of the pre- and post-diversion shorelines of Lake Agassiz, it has been estimated that *ca.* 9,500 km^3 of water was released. If this outpouring took place over the course of a single year, this flood would have matched today's net annual input of fresh water to the North Atlantic (Broecker, 2003). In most predictive oceanic models, an input of this magnitude would shut down the model's ocean conveyor circulation. It would also have created a major rise in sea levels. The amount and timing of change in sea level due to glacial melting is still being debated and varies with location. Data from the Netherlands, however, indicate that a later sea-level rise that commenced 8,458 ± 44 BP caused an increase in sea level of 3.0 ± 1.2 m per 200 years (Hijma and Cohen, 2010).

The beginning of the Holocene is generally considered to be the termination of the Younger Dryas period which in northern-western Europe was around 11,200–11,600 BP. One of the most remarkable events in the early Holocene was the rapidity of the recolonization of the ice-liberated regions by diverse plant and animal communities. As might be expected, a climate that was warm enough to melt the Pleistocene ice would also have provided sufficient warmth to encourage the renewal of the arctic vegetation. It can often be surprising how warmth-demanding species can grow in close proximity to retreating ice sheets (Figure 2.4). The immediate post-glacial plant communities are generally assumed to have had a level of productivity that would have been adequate for the support of the still extant members of the Pleistocene megafauna.

Tundra-Taiga Biology. First Edition. R.M.M. Crawford © R.M.M. Crawford 2013.
Published 2013 by Oxford University Press.

Figure 2.1 The Mountain Avens (*Dryas octopetala*) forming an arctic heath on top of *thufúr mounds* (see Chapter 4) at Petuniabukta 78°40'N in Billeforden Spitsbergen. This arctic-alpine species became widespread at the interface between the Pleistocene and Holocene and gave its name to the last cold epoch at the end of the Pleistocene glaciations.

2.1.1 The 8,200 BP event

In northern Europe and on the eastern North Atlantic seaboard the relatively stable climate of the early to mid Holocene was punctuated by an abrupt, transient cold event at *ca.* 8,200 BP which lasted for about 400 years between 8,400 and 8,000 calendar years BP when a remnant of the Laurentide ice sheet collapsed in the Hudson Bay, allowing Lakes Agassiz and Ojibway, which had previously discharged over spillways south-eastwards to the St Lawrence estuary (discussed in section 2.1), to drain swiftly northwards. This time the ice water passed through the Hudson Strait to the Labrador Sea (Barber *et al.*, 1999).

This event caused a major perturbation in the Holocene climate history, the effects of which were felt widely across the Arctic from Canada to Finland (Sarmaja-Korjonen and Seppa, 2007), as well as affecting Europe as far south as the Mediterranean. Such events are attracting considerable interest as their causative mechanisms might provide an

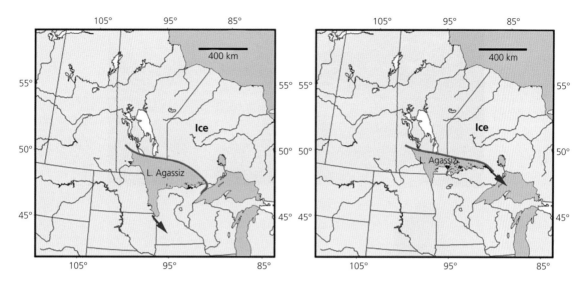

Figure 2.2 (*Left*) The outline of Lake Agassiz just before the catastrophic flood. At that time its outlet was to the south into the Mississippi drainage basin. (*Right*) The outline after the opening of the eastward outlet. A volume of 9,500 km^3 of water was suddenly released to the northern Atlantic through the St Lawrence Valley. (Redrawn with permission from Broecker, 2003).

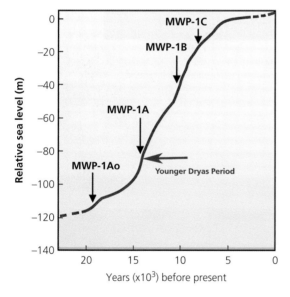

Figure 2.3 Generalized curve of sea level rise since the last ice age. MWP, Melt water pulse; MWP-1A0, *ca.* 19,000 years ago; MWP-1A, 14,600–13,500 years ago; MWP-1B, 11,500–11,000 years ago; MWP-1C, ~8,200–7,600 years ago. (Reproduced with permission from Gornitz V. Science Briefs, NASA, 2007).

insight into how sudden changes in the North Atlantic Ocean circulation may take place with climatic warming as a result of the de-icing of the Arctic Ocean.

2.2 Glacial refugia

The rapid spread of vegetation in the early post-glacial period could have taken place only if there had been a suitable provision of propagules. This has led to a long-standing debate in biogeography on the question of whether or not arctic and high alpine organisms survived the *Last Glacial Maximum* in favourable localities within the general region of ice cover. Possible glacial refugia could be the so-called *nunataks*—mountain peaks and outcrops that project above an ice-covered landscape (Figure 2.5), Alternatively, glacial refugia can be found in favoured regions where land is free of ice without having to be a projecting nunatak. Such sites are usually described as *thermal oases*. A particularly outstanding polar oasis is in Greenland at Lake Hazen (81°50′N, 70°00′W) which in spite of its high latitude has a flora of about 117 vascular species (Figure 2.6).

In oceanic areas a more frequent type of refugium is the semi-nunatak, where a mountain is free of ice on its seaward side while being ice-covered to landward (Andersson *et al.*, 2000). One notable example in Spitsbergen is on the west coast of Prince Charles Foreland (Figure 2.7).

A study of debris-covered glaciers in different regions of the world has prompted the notion that

Figure 2.4 Birch tree growing beside the retreating Svartisen glacier, Moløy, northern Norway (66°40′N) illustrating the proximity of a warmth-requiring species to glacial fronts during a period of retreating ice. (Reproduced with permission from Elsevier).

Figure 2.5 Nunatak on the southern shore of Horn Sound (77°N), Spitsbergen. The word *nunatak* is derived via Danish from the Eastern Canada Inuit language *Inuktitut*.

Figure 2.6 Aerial view looking to Lake Hazen, Canada's most northern lake (82°N). This high-latitude thermal oasis supports a flowering plant flora of 117 species. (Photo Professor J. Svoboda).

Figure 2.7 View and location of the west-facing cliffs on Prince Charles Foreland, Spitsbergen (see inset marked in red). The island was named in 1616 after the Prince of Wales, later Charles I of Great Britain and Ireland, 1600–49. This region is considered to have been free of ice at the Last Glacial Maximum and these cliffs would probably have been semi-nunataks, being free of ice on their westward aspect and therefore providing a possible potential biological refugium (Andersson et al., 2000).

debris lying on the top of glaciers may have served as possible Pleistocene biological refugia. Debris cover on glaciers creates special microclimatic conditions that make this type of terrain habitable for flowering plants. As glaciers descend to a 'milder climate' (by Pleistocene standards), the glacier surface, together with a suitable depth of debris, can be thermally more favourable for plant survival than high-elevation nunataks. Such thermal microsites exist or are even surpassed today in Ny Ålesund (Spitsbergen), Thule (Greenland), and Barrow (Alaska), where present-day climatic conditions allow a diverse assemblage of hardy plants growing on top of the ice to be gradually transported to a more favourable environment. This hypothesis could be complementary to the nunatak notion of plant survival in glacial landscapes (Fickert et al., 2007).

Examination of the flora of polar hotspots reveals that the plants are predominantly aggregates of neo-endemic species, having a limited range due to their recent evolution as compared with palaeo-endemics, or else ancient relict species, which suggests that their existence is a consequence of local moderation of environmental extremes, allowing unique species populations to survive through changing climatic periods.

Up to the 1960s, there was nearly a complete consensus that disjunctions and endemism in the North Atlantic could not be explained without *in situ* survival during the glaciations (the 'nunatak hypothesis'). However, the students of the Scandinavian botanists who originally proposed the nunatak theory became generally hostile to their instructors' theory and described it as unnecessary and making no significant contribution to the study of biogeography. The more extreme position, which claimed that the Last Glacial Maximum caused a *tabula rasa*, i.e. created a clean slate by the total removal of the arctic biota, was particularly championed by those who were concerned more with animal survival than plants.

Such a sentiment can be found in a paper entitled *'If this is a refugium why are my feet so bloody cold?'* (Buckland and Dugmore, 1991). Gradually, however, due to more accurate dating of the extent of past ice cover at high latitudes it has become evident that there were places sufficiently free of ice at least for plants to survive during the Last Glacial Maximum (Figure 2.8). Some of the doubt about the extent of the Pleistocene ice sheets was due to the earlier practice of assessing the extent of ice cover from the deposition of coastal sediments. More accurate methods have now been used for dating the emergence of rocks from above the ice fields, beginning with amino-acid analysis and subsequently thermo-luminescence. Recent advances in dating rock exposure based on cosmic-ray-produced isotopes in rock

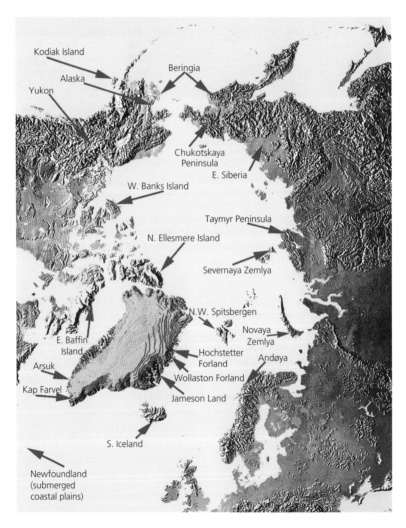

Figure 2.8 Location of some sites within the Arctic that were ice-free and may have served as refugia for flowering plants during at least part of the Last Glacial Maximum. All examples are based on geological or palaeological evidence with the exception of S. Greenland (Arsuk and Kap Farvel) and S. Iceland, where the evidence is phytogeographical. (For sources of information see Crawford, 2004).

Figure 2.9 Two rare arctic alpine species, (a) *Arenaria humifusa* and (b) *Sagina caespitosa*, which are able to survive in very cold habitats with short growing seasons and appear to have survived in the Arctic since the Last Glacial Maximum. (Photo (a) Dr Inger Greve Alsos; photo (b) Kjell Ivar Flatberg).

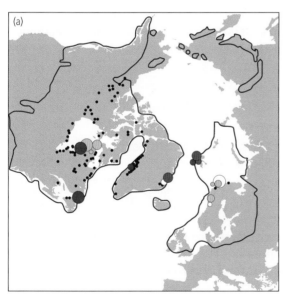
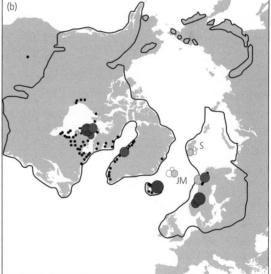

Figure 2.10 Total geographic distribution (black dots) and sampling localities of (a) *Arenaria humifusa* and (b) *Sagina caespitosa* together with the distribution of their chloroplast DNA haplotypes. The size of the circles is proportional to genetic diversity within populations. The extent of ice cover at the Last Glacial Maximum is outlined in blue. The geographic locations of Spitsbergen (S) and Jan Mayen Island (JM) are indicated in b. (Reproduced with permission from Westergaard *et al.*, 2011).

surfaces have now shown that in many polar regions there are rock surfaces that were free of ice at the time of the Last Glacial Maximum (Nesje *et al.*, 2007). Nevertheless, the fact that no fossil evidence for glacial survival by plants had been found at this time served to preserve for a while a sceptical attitude in relation to plant glacial survival.

2.2.1 Genetic evidence for glacial refugia

Developments in molecular genetics are now providing new evidence that can unravel the past migrations of arctic species as well as indicating that glacial refugia have had a role in contributing to the genetic inheritance of the modern arctic flora. The case for

Figure 2.11 View looking east in Kongsfjorden, one of the localities from which *Arenaria humifusa* was sampled in Spitsbergen. (Photo Dr Kristine Bakke Westergaard).

genetic evidence for glacial survival was greatly strengthened in 2011 with evidence for the persistence of two rare arctic species, *Arenaria humifusa* and *Sagina caespitosa*, within the area of glacial cover at the Last Glacial Maximum (Westergaard *et al.*, 2011). Both species are known for their ability to survive in cold habitats. These two species with similar distributions (Figures 2.9–2.11) are diminutive members of the carnation family (Caryophyllaceae) found on both sides of the Atlantic from north-eastern America via Greenland to Scandinavia and Spitsbergen. There are no occurrences in southern or central European mountain ranges.

This biomolecular study was a landmark, as for the first time it presented evidence strongly favouring the *in situ* glacial Pleistocene persistence of two species within a glaciated region. Both of these rare arctic-alpine pioneer species have their few and only European occurrences well within the limits of the last glaciation. Sequencing of non-coding regions of chloroplast DNA revealed only limited variation. However, two very distinct and partly diverse genetic groups, one East and one West Atlantic, detected amplified fragment length polymorphism (AFLP) dispersal from North America as an explanation for their European occurrences.

Patterns of diversity and distinctiveness indicate that glacial populations of these species existed in East Greenland and/or Svalbard (*A. humifusa*) and in southern Scandinavia (*S. caespitosa*). Despite their presumed lack of long-distance dispersal adaptations, intermixed populations in several regions indicate post-glacial contact zones. Both species are declining in Nordic countries, probably due to climate change-induced habitat losses.

2.3 Reconstructing past plant distributions

Genetic evidence alone is at best only a circumstantial, yet nevertheless provides a powerful argument for the persistence of species within any specific area. As in all the best murder stories the presence of the corpse is essential for intellectual satisfaction. Here again, new techniques are providing just such evidence for the actual presence and dating of remains of past floras at high latitudes.

In northern Norway the island of Andøya has long been the subject of both geological and botanical investigations as its situation close to the continental shelf (Figure 2.12) has prompted the suggestion that it supported a possible late Weichselian unglaciated enclave. Early deglaciation of the island has been suggested (Alm and Birks, 1991).

Geomorphological mapping has shown that although the upper surface of the late Weichselian ice sheet at Narvik lay at *ca.* 1,200 m, nevertheless on the northern Lofoten island of Andøya there were ice-free areas with possible nunataks on the neighbouring islands of Senja and Grytøya (Vorren et al., 1988). Radiocarbon datings of a lacustrine sequence from a lake on Andøya have revealed a continuous sediment accumulation during most of the late Weichselian, starting before 21,800 ± 410 BP. The pollen record for the period 21,800–12,800 BP was uniform and the vegetation was strongly dominated by the Poaceae (Alm, 1993). Recent research has now discovered the presence of a rare mitochondrial DNA haplotype of Spruce (*Picea* sp.) that appears to be unique to Scandinavia and with its highest frequency in the western area is believed to have been an ice-free refugium during most of the last ice age. The survival of DNA from this haplotype in lake sediments in Andøya from 22,000 years BP matches pollen from the Trøndelag in central Norway dating back 10,300 years. This is further confirmed from chloroplast DNA of pine and spruce

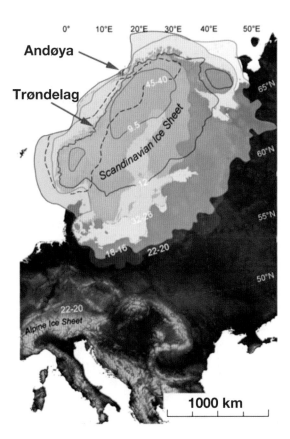

Figure 2.12 Reconstruction of stages in the extent of the Scandinavian Ice Sheet 45,000–9,500 calendar BP, compiled from a range of sources. (Reproduced with permission from Parducci et al., 2012).

in lake sediments adjacent to the ice-free Andøya refugium in north-western Norway as early as 22,000 and 17,700 years ago. These findings suggest that coniferous trees survived in ice-free refugia of Scandinavia during the last glaciation, challenging current views on survival and spread of trees as a response to climate change (Parducci et al., 2012).

The validity of these findings has been critically questioned as failing to present convincing evidence for glacial survival of *Pinus* and *Picea* in northern Scandinavia, as the methodology did not exclude contamination and consideration should have been given to alternative hypotheses (Birks et al., 2012). However, although the authors made a robust reply the absence of these tree species at present in this region tends to leave this matter unresolved.

It is, however, now evident that the simplistic notion that plant and animal life in the Arctic was extinguished by the Pleistocene glaciations (the *tabula rasa* hypothesis) is a gross over-simplification. The Pleistocene glaciations without doubt presented polar plant and animal communities with an environmental challenge that drastically reduced their presence at high latitudes. As the ice sheets spread south, the high-pressure zone to the north resulted in a reduction in precipitation to levels that were not sufficient to maintain permanent snow and ice cover, which together with lower sea levels would have created potential habitats for species able to survive in polar deserts.

It has also been possible from macrofossil assemblages sampled at several sites in this buried landscape to make a detailed description of the vegetation that had been growing there at the time of the *Last Glacial Maximum* (see section 2.2.1).

Even during the coldest periods of the Pleistocene, glaciation did not extend to all areas of Arctic Russia (Möller et al., 1999). When viewed from a circumpolar perspective, the Arctic was never entirely covered by ice at any one time (Figures 2.13 and 2.14). The occurrence of rare species in Beringia and parts of the north-east Siberian coast is a further indication of the existence of arctic refugia (see Chapter 10).

2.3.1 Dating the ice retreat

With the advent of *cosmogenic radionuclide dating* it now appears that many more places in the Arctic were free of ice at the time of the Last Glacial maximum than had hitherto seemed possible (see Fig. 2.8). Bearing in mind that in the later stages of the last glaciation the sea level would have been as much as 120 m lower than at present, and there could therefore have been substantial ice-free localities which would have provided suitable habitats for cold-tolerant species of both plants and some animal species. This last point is borne out from recent archaeological excavations in a cave in north-west Scotland which found the ulna of a Brown Bear dated to 45,000 ± 1,000 BP (Birch and Young, 2009). Such finds necessitate a major reconsideration of the extent of the Devensian ice sheet and suggest that the generalized depiction of blanket ice cover over Western Europe at the time of the Last Glacial Maximum 18,000 years ago is yet another gross simplification of Pleistocene history.

An example of the application of this rock-surface dating technique (cosmogenic radio nucleotide dating) is shown in Figure 2.14 where the measurement of cosmic-generated isotopes has provided geological evidence that nunataks existed on the islands of Danskøya and Amsterdamøya of north-west Spitsbergen. The exposure ages for the glacial erratics laid down in block-field-covered plateaus more than 300 m above sea level showed that the last ice sheet that completely covered the islands deglaciated over 80,000 years ago. Dating of marine sediments close to sea level also showed that full ice-free conditions were achieved by *ca*. 50,000 years ago.

2.3.2 Early Holocene presence of plants at high latitudes

The detection of ice-free terrain does not in itself mean that plants survived in these nunataks and rocky outcrops. Ice scouring removes plant remains and in the Arctic evidence from buried pollen and macrofossils is found only in a few localities. Nevertheless, evidence is accumulating from both diligent exploration and improved biomolecular technology, making it evident that the re-establishment of arctic plant communities after the Last Glacial Maximum was more than just a gradual re-advance of individual species from more southerly locations.

Actual evidence of the existence of species-rich plant communities in the more northern parts of Alaska in the latter stages of the last ice age has now come to light. In 1968, there was discovered an ancient buried vegetated land surface covering

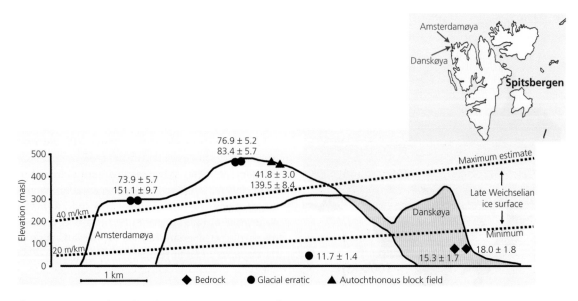

Figure 2.13 Topographic profiles of Amsterdamøya and Danskøya off the north coast of Spitsbergen (79°50′N), with dates for the exposure ages of sampled rocks in thousands of years (ka). The dotted lines are estimates of the minimum and maximum elevations of the Weichselian ice sheet. (Reproduced with permission from Landvik et al., 2003).

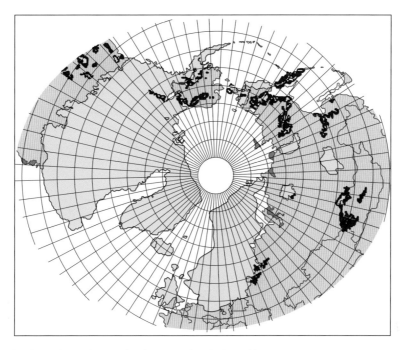

Figure 2.14 The northern hemisphere at the Last Glacial Maximum (Weichselian, Valdaian, Würmian, Devensian, Wisconsinan, MIS 2). (Compiled by Jürgen Ehlers for 2005 from data and maps assembled and published in *Quaternary Glaciations—Extent and Chronology* by Jürgen Ehlers and Philip Gibard (published by Elsevier, 2004). Reproduced with permission.).

approximately 2,500 km² on the Seward Peninsula (Figure 2.15). The area had been blanketed by more than 1 m of tephra *ca.* 21,500 calendar years BP, and the former land surface was preserved in the permafrost. This discovery of the land surface provided a unique opportunity to study a fossil ecosystem preserved *in situ* (Goetcheus and Birks, 2001). It was also possible from macrofossil assemblages sampled at several sites on this buried landscape to make a detailed description of the vegetation that would have been growing there at the time of the Last Glacial Maximum.

The plant communities included graminoids as well as other herbaceous flowering plant species. There were also occasional occurrences of the Arctic Willow (*Salix arctica*—Figure 2.16). The vegetation was dominated by the Bellardi Bog Sedge (*Kobresia myosuroides*), a widespread circumpolar species, as well as other Sedges (*Carex* spp.). Overall, the vegetation was a closed, dry, herb-rich tundra-grassland with a continuous moss layer, growing on calcareous soil that was continuously supplied with loess. Nutrient renewal by loess deposition was probably responsible for the relatively fertile vegetation, and the occurrence of a continuous mat of acrocarpous mosses (mosses that bear their capsules terminally on stalks).

A re-examination of the question of plant survival in Northern Iceland during the last glacial period has revealed that many of the species present in the area before the onset of the cold Younger Dryas stadial, 11,300–9,000 BP, continued to produce pollen during that cold event. The more or less immediate reappearance of other pollen taxa at the Younger Dryas–Pre-Boreal boundary suggests that these plants also survived, even if they did not produce sufficient pollen to be recorded during the Younger Dryas stadial.

It can therefore be justifiably concluded that the relatively high plant diversity found in some high arctic areas and present-day nunataks is due to the survival of species that have a high tolerance of cold, fluctuating climates. This conclusion is supported by recent palaeoclimatic data from ice cores and deep-sea sediments, indicating that the Icelandic climate during the last glacial period was only

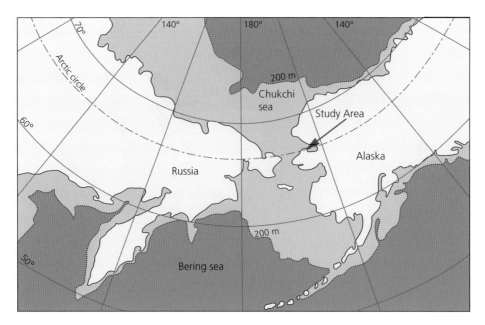

Figure 2.15 Map of Beringia showing the site of the discovery of the tephra-covered landscape dating from the time of the Last Glacial Maximum *ca.* 21,500 BP. The full-glacial Bering Land Bridge is bounded by Russia on the west, Alaska on the east, and the 200-m isobath to the north and south. (Reproduced with permission from Goetcheus and Birks, 2001).

Figure 2.16 *Salix arctica* growing at 73°N in north-east Greenland. This species found in the 21,500 BP tephra-buried landscape grew also in Beringia at the time of the Last Glacial Maximum.

occasionally slightly colder than during the Younger Dryas stadial (Rundgren and Ingolfsson, 1999).

2.3.3 Vegetation history from pollen analysis and biome reconstruction

Pollen analysis has proved a very useful tool in following the history of precisely named taxa over thousands of years and relating changes in species composition to climatic alterations. The Beringian region of the Arctic with its minimal exposure to glaciation has allowed peat deposits to accumulate which have preserved the pollen record back to the Last Glacial Maximum.

By comparing the modern pollen spectra of Beringia with the actual vegetation cover by a process described as *objective biomization* (Prentice and Webb, 1998), it is possible to predict the nature of past biomes from their pollen spectra. This method has been used to compare the vegetation of Beringia as it is at present with that which existed 6,000–18,000 years BP (Figures 2.17 and 2.18).

The reconstructed distribution of biomes across Beringia at 6,000 ^{14}C years BP (Figure 2.18) is similar to the modern distribution. Except for a possible northward advance in the Mackenzie Delta region, there is almost no change in the northern position of the Tundra-Taiga Interface location and also little

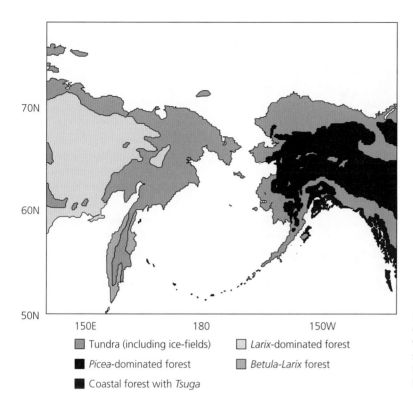

Figure 2.17 Modern biome distribution of vegetation of Beringia as reconstructed with data from various authors. (For sources see Edwards et al., 2000. Reproduced with permission from Edwards et al., 2000).

movement in the northern forest limit. At 18,000 ^{14}C years BP the western forest limit in Alaska was probably east of its modern position.

2.3.4 DNA and vegetation history

Another significant advance that is already shedding light on the vegetation history of the Arctic is the ability to identify species from ancient DNA (Sonstebo et al., 2010). Hitherto, pollen analysis has been the main source for tracing past species. However, pollen can be difficult to find in the Arctic and when amounts are present only in low quantities the pollen content of the soil may not necessarily reflect the vegetation of the area in which it is found. Furthermore, pollen analysis is time-consuming and suffers from low taxonomic resolution. Macrofossils are more reliable but rare in the Arctic for both plants and animals.

Molecular biology has recently produced a significant advance in reconstructing biogeographical history. It is now possible to identify both plant and animal species from ancient DNA. Here again, with plants it is possible to use conservative maternally inherited chloroplast DNA. One particular section of the P6 loop in the chloroplast *trnL* (UAA) intron is a spacer which has been found to be particularly suitable as it has no known functional genetic properties. Thus, although short, it is a highly variable region with highly conserved flanking sequences.

A study carried out on two permafrost soil samples from southern Chukotka, Russia (Figure 2.19; 64°17′N; 171°15′E), was able to radiocarbon date the samples in addition to identifying the species from the chloroplast DNA extract. The permafrost samples were dated to 22,960 ± 120 and 15,810 ± 75 uncalibrated radiocarbon years BP.

For taxonomic reference, a whole relative sequence database was constructed from recently collected material of 842 species, representing all widespread and ecologically important taxa of this region's species-poor arctic flora (Sonstebo et al., 2010). The P6 loop alone allowed identification of all families, most of the genera (more than 75%), and one-third of the species, thus providing a much higher taxonomic

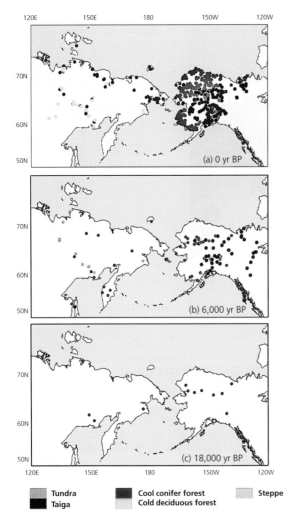

Figure 2.18 Biomes reconstructed from pollen data at (a) 0 BP, (b) 6,000 ^{14}C BP, and (c) 18,000 ^{14}C BP. (Adapted from Edwards et al., 2000).

resolution than would have been obtained from pollen records (Table 2.1).

2.4 Pleistocene megafauna decline

At the end of the Pleistocene, permanent ice cover gave way to open ground with a steeper temperature gradient across the ice–Tundra interface than that which now exists. This enabled the development of herb-rich pasture with numerous grass species capable of supporting the appetites of the large herbivores of the Pleistocene. This vegetation was a mixture of the species that are found today in the Tundra as well as others that are more typical of the *Steppe* (a Russian word of unknown origin, cf *prairie*).

This mixed community is therefore referred to as the *Tundra-Steppe*. This type of tundra plant community can still be seen today in northern Yakutia in north-east of Siberia and also most notably on Wrangel Island (Figure 2.20) The Tundra-Steppe plants are adapted to cold winters and hot summers and create a more productive biome than the normal Tundra as it now exists at high latitudes (Chernov, 1988). As the Taiga and its trees advanced, these productive pastures disappeared and with them the herbivores that could not adapt to forest-dominated habitats or the less productive Tundra vegetation. Thus the Pleistocene–Holocene boundary witnessed the decline and the extinction of some of the dominant members of Pleistocene megafauna (Figures 2.21–2.23).

The renowned large herbivores, mammoth (*Mammuthus primigenius*) and Woolly Rhinoceros (*Coelodonta antiquitatis*), became extinct, while others such as Muskoxen (*Ovibus moschatus*) (Figure 2.21), horse (*Equus caballus*), Saiga (*Saiga tatarica*), Caribou (*Rangifer tarandus*), and Bison (*Bison bison*) survived, although with much reduced distributions and populations that were less numerous.

The basic causes of the extinctions and reductions of these large herbivores have been much discussed. A debate continues still as to whether their demise or decline was due solely to climate and habitat change, or whether their populations suffered from human hunting and destruction of their habitats, either directly or as an exacerbating factor.

2.4.1 The Woolly Rhinoceros

The Woolly Rhinoceros was a popular subject for Palaeolithic (Magdalenian) cave artists, with probably the most famous example being that found in 1901 in a cave at Font de Gaume in the Dordogne department of south-west France (Figure 2.22). The disappearance of the Woolly Rhinoceros appears to have been due to a gradual shrinking of its habitat. The last refugia were in the Trans-Urals and western Siberia.

It should be noted that unlike the Woolly Mammoth which finally became extinct in Arctic regions,

Table 2.1 Examples of molecular taxonomic units based on chloroplast DNA from the two Chukotkan sediment samples, showing the molecular taxonomic units identified and listed in descending order of their frequency of occurrence (percentage of sequences obtained in t 454 runs). For full details see original paper by Sonstebo et al., 2010.

22,960 ± 120 BP	
Bistorta vivipara	47.25
Equisetum arvense/E. fluviatile	24.31
E. sylvaticum	24.31
Salix sp./*Chosenia arbutifolia*	24.31
Populus balsamifera	4.74
Armeria scabra	3.03
Thymus oxyodontus	2.77
Lagotis glauca	2.17
Avenella flexuosa	1.77
Aconogonon alaskanum	1.77
A. ocreatum/A. tripterospermum	1.36
Rumex sp.	1.31
Packera sp./ *Senecio* sp.	0.96
Ranunculus acris/R. subborealis	0.81
R. turneri	0.81
Festuca sp.	0.76
Hulteniella integrifolia	0.66
Saxifraga hirculus	0.55
Trientalis europaea	0.45
Valeriana capitata/V. officinalis agg.	0.35
Myosotis alpestris	0.30
Empetrum sibiricum	0.30
E. subholarcticum	0.30
Anthoxanthum nipponicum	0.25
Crepis chrysanta	0.25
Saxifraga bracteata/S. cernua	0.25
S. hyperborea/S. radiata	0.25
S. rivularis	0.20
Papaver sp.	0.15
Elymus sp./ *Leymus* sp.	0.15
Trollius sp.	0.15
Koeleria asiatica	0.15
Trisetum spicatum	0.15
Pedicularis oederi	0.10
Viola biflora	0.10
Claytonia arctica/C. scammaniana	0.10
Sanguisorba officinalis	0.10
Vaccinium uliginosum	0.10
Calamagrostis sp.	0.10
Potentilla sp.	0.10
Pulsatilla patens	0.05
Beckmannia syzigachne	0.05
Cardamine pratensis	0.05
Trisetum sibiricum	0.05
Vaccinium alaskense/V. myrtillus	0.05
Castilleja elegans/C. hyperborea	0.05
Deschampsia sp.	0.05
Parrya arctica/P. nudicaulis	0.05
Astragalus alpinus/A. umbellatus	0.05
Thalictrum alpinum/T. minus	0.05
T. sparsiflorum	0.05
Caltha arctica/C. palustris	0.05
15,810 ± 75 BP	
Trollius sp.	27.87
Festuca sp.	16.34
Valeriana capitata/V. officinalis agg.	14.91
Salix sp./*Chosenia arbutifolia*	8.90
Populus balsamifera	8.98
Bistorta vivipara	4.95
Alchemilla glomerulans	4.91
Larix cajanderi/L. dahurica	2.99
L. laricina	2.99
Pulsatilla patens	1.72
Koeleria asiatica/Trisetum spicatum	1.48
Hulteniella integrifolia	0.88
Viola biflora	0.29
Vaccinium alaskense	0.11
Gentianopsis barbata/G. detonsa	0.07
Omalotheca norvegica	0.04

Figure 2.19 The arrow shows the location of the site by the Mayn River, a tributary of the Andyr River, from where the permafrost samples were collected at 64°17′N; 171°15′E.

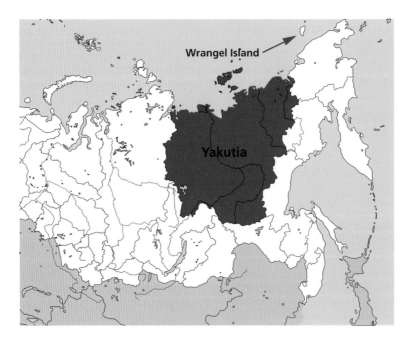

Figure 2.20 Location of Yakutia and Wrangel Island. Politically Yakutia is now the Sakha Republic—one of the ten autonomous Turkic Republics within the Russian Federation.

the final extinction of the rhinoceros took place not in the Arctic but in the temperate zone of the European–Asian border area. Recent radiocarbon dating indicates that populations survived as recently as 10,000 BP in western Siberia (Figure 2.23). However, the accuracy of this date is uncertain, as radiocarbon plateaus exist around this time. The extinction does not fit with the end of the last ice age but does

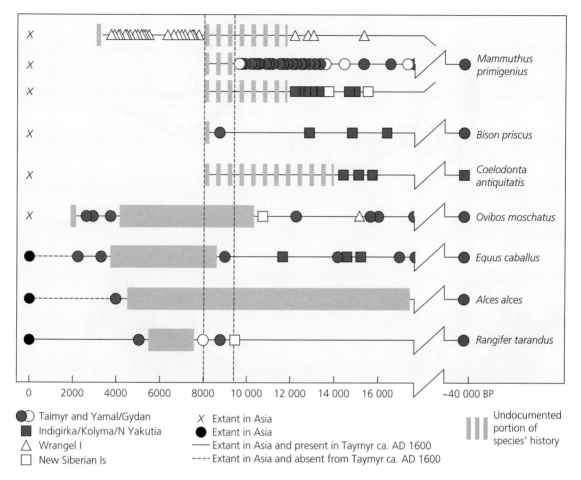

Figure 2.21 Chronometric summary of dates for seven megafaunal taxa from the high arctic periphery of Asia, post-18,000 BP. The number of independent dates per species per area varies greatly. For *Mammuthus primigenius*, the post-18,000 BP dating framework is truly dense only for the Taymyr Peninsula and Wrangel Island, and to a lesser degree for the Kolyma, Indigirka, and northern Yakutia regions. For all other taxa and regions, the number of records is smaller. Date representation is not exhaustive, although it is believed that all currently available 'last occurrence' dates for listed taxa have been included. When late Holocene *Ovibos moschatus* and *Equus caballus* became locally extinct in northern Asia is not known, except that it must have occurred after *ca.* 2,000 BP but before 400 BP (the time of first Russian penetration). (Reproduced with permission from MacPhee *et al.*, 2002).

coincide with the minor yet severe climatic reversal of the Younger Dryas period that lasted for about 1,000–1,250 years (Kuzmin, 2010).

2.4.2 The last of the mammoths

In the Taymyr Peninsula of northern continental Siberia (Figure 2.24) at least some elements of the now-extinct megafauna appear to have survived longer than they did elsewhere on the mainland. The Taymyr Peninsula has long been of special interest to Quaternary palaeontologists for the number of remains of mammoth skeletons that can be found lying on the ground surface. The latest dates for the fossil remains of Woolly Mammoths anywhere in mainland Eurasia are based on specimens from the Taymyr and nearby Yamal Peninsulas (MacPhee *et al.*, 2005). The Taymyr Peninsula has also yielded *last occurrence* dates for several other species including Bison (*Bison priscus*), North Asian Horse (*Equus caballus*—sensu latu—ssp. *leniencies*), and the post-Pleistocene Russian mainland populations of Muskoxen (*Ovibos moschatus*).

Figure 2.22 Cave painting of a Woolly Rhinoceros dating from around 19,000 BP at Font de Gaume, Dordogne department, south-west France. Scratch marks are frequently noticed on the larger horn suggesting that it was used for sweeping away snow to gain access to the vegetation. (Reproduced with permission from the Centre des Monuments Nationaux, Font de Gaume).

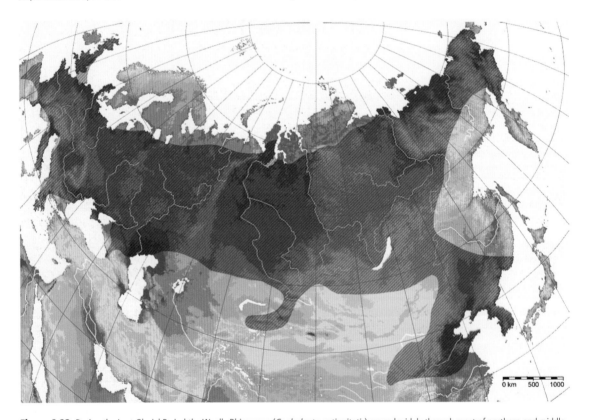

Figure 2.23 During the Last Glacial Period the Woolly Rhinoceros (*Coelodonta antiquitatis*) ranged widely through most of northern and middle Eurasia from the Iberian Peninsula, France, and Britain in the west, through Central Europe, Italy, the Balkans, Siberia, Outer Mongolia to the far-east, including southern Korea, and north-eastern China. (Data and image courtesy of Kahlke, 2013).

Wrangel Island, together with the Pribilof and neighbouring islands in the East Siberian Sea, are generally considered to have been the localities that supported the last remaining populations of Woolly Mammoths (Figure 2.25). Radiocarbon dates fall randomly within a time interval of 8,980–3,685 years BP, with the most recent survival having a radiocarbon date of 3,685 ± 60 years BP (Vartanyan *et al.*, 2008).

Figure 2.24 The Taymyr Peninsula showing location of Lake Taymyr and the adjacent lowland areas which are particularly rich in remains of mammoth bones, often just lying on the surface of the ground.

Fossil remains of mammoth found in caves in the Pribilof Islands, 500 km west of the Alaskan mainland in the Bering Sea, have yielded radiocarbon dates that converge around 5,725 years BP, making these the youngest mammoth remains discovered in the North American section of the Siberian Sea (Figure 2.25).

Persistence of mammoths on the Pribilofs is most easily explained by their isolation and the lack of human presence. An additional factor may have been the local existence of high-quality forage enriched by nutrients derived from local Holocene tephras. This interpretation is reinforced by stable carbon and nitrogen isotope remains. The endpoint of mammoth survival in the Pribilofs is unknown but could have been contemporaneous with the arrival of polar bears, whose remains in the cave date to the Neoglacial (renewed) cold period of 5,725 BP (Enk *et al.*, 2009).

In north-western America the extinction of the mammoth had been placed at 15,000–13,000 BP, based on dating surveys of macrofossils (bones and teeth). This date has, however, had to be revised with the advent of DNA dating.

2.4.3 DNA dating of mammoth survival

Macrofossils, due to their rarity, cannot be relied upon to determine when a species actually became extinct. An alternative approach is to detect *'ghost ranges'* of dwindling populations based on recovery of ancient DNA from perennially frozen soil where securely dated sediments indicate the continued presence of species. Such studies have revealed more recent dates for survival than those suggested from macrofossils.

Modern elephants pass about 50 litres of urine a day, which gives an impression of what might be expected of mammoths. Such a level of urination would leave widespread deposits of DNA. By recovering this ancient DNA from perennially frozen and securely dated sediments it is possible to reveal from molecular evidence the presence of species that have been supposed to have been absent on the basis of the macrofossil record. In this way it has been possible to show that the Woolly mammoth and horse persisted in interior Alaska until at least 10,500 BP, several thousands of years later than indicated from macrofossil surveys.

The cause of the ultimate extinction of the mammoth is not altogether clear. On Wrangel Island it would appear that there was a higher mammoth density in the area during the Holocene than during the Pleistocene, when Wrangel Island was still connected to the mainland. A gap in radiocarbon dates between 12,000 and 9,000 BP indicates a period when mammoths were either rare or absent from the area (Vartanyan *et al.*, 2008). It is still unclear what caused the final extinction of mammoths on Wrangel Island. The two main hypotheses for their

Figure 2.25 Coastlines past and present of Beringia showing the regions that were the last refugia of the mammoth (see text). Ancient coastline (dashed lines) and glaciated regions (shaded areas) at ~14,000 ^{14}C BP, shortly after the Last Glacial Maximum, when the sea level was ~100 m lower than today. The modern coastlines are also marked, including Wrangel Island and the Pribilof Islands. (Adapted from Enk et al., 2009).

demise and that of the other members of the Pleistocene megafauna on the Eurasian mainland have been environmental change and human impact. Neither of these causes is readily attributable to Wrangel Island.

A study of the fossil genetic data points towards a relatively large Holocene population on the island, which is also congruent with the high density of mammoth fossils dating to this period (Vartanyan et al., 2008). This suggests that the final extinction was caused by a relatively sudden, rather than a gradual, change in the mammoths' environment.

One possible explanation for such a sudden change could be the arrival of human hunters on Wrangel Island. Human extermination of a large mammal, such as the mammoth, should be readily visible in the archaeological record with wound and butchery marks on the bones. In the case of Wrangel Island, such direct evidence is lacking. The alternative to human predation as a cause for the extinction could include the emergence of a novel disease or a previously undetected short-term change in the climate. The mammoths on Wrangel Island survived the climate change associated with the Pleistocene–Holocene transition, and eventually became extinct during a period of relatively stable climate. If the extinction was a delayed response to deteriorating environmental conditions associated with the Pleistocene–Holocene transition, or a consequence of a reduced habitat-carrying capacity, it would be expected that there would have to have been a gradual loss of genetic variation in the population. However, in spite of ca. 5,000 years of isolation, no evidence has been found for further loss of genetic variation. The relatively high number of mammoth mitochondrial haplotypes on Wrangel Island near the time of their final disappearance (ca. 3,700 BP) suggests a sudden extinction of a rather stable population (Nystrom et al., 2010).

2.5 Pleistocene megafauna survival

Muskoxen (*Ovibus moschatus*; Figure 2.26) Reindeer/Caribou (*Rangifer tarandus*), Elk/Moose (*Cervus canadensis*), and the Horse (*Equus caballus*) are the only large herbivores of arctic regions and the ancient *Tundra-Steppes* to survive today from the Pleistocene megafauna. For a detailed discussion of their present status and survival adaptations see Chapters 6 and 8.

2.5.1 Horses in the Arctic

The ancestors of the modern horse evolved in North America. The northernmost fossil record of a horse in the Arctic is from Ellesmere Island, Canada (78° 33′N). It is only a partial skull of a juvenile, probably dating from the early Pliocene about 3.5–4 million years BP (Hulbert and Harington, 1999). The Yukon horses most probably arose in Beringia 200,000 years ago and have subsequently been able to evolve to withstand cold sufficiently for survival at high latitudes.

Fossils have been found as far north as the northern coast of Alaska, where radiocarbon dates for fossil horse species range from about 31,500–12,300 BP and indicate that horses were already present in eastern Beringia at the cold peak of the last glaciation. It appears that small groups of wild horses were widespread at the end of the Pleistocene but there is nothing to suggest that there was any domestication of the horse at this period. They did, however, provide a regular item of food for early hunter-gatherers.

Eurasian 'caballoid' horses have been assigned to numerous sub-species, which probably represents no more than over-active taxonomic splitting. Molecular evidence now shows that domesticated the Eurasian horse (*E. caballus*—*sensu latu*) incorporates a wide genetic diversity which undermines classical taxonomic attempts at identifying clear sub-species. Nevertheless, in Eurasia two races of wild horses have been generally recognised that have survived into historical times. In Asia this was Przewalski's horse (*E. caballus przewalskii*). With regard to the fossil remains of the small horses found in the Taymyr Peninsula mentioned in section 2.52, they may have connections with extant Central Asian Przewalski's horse. The other ancient wild horse was the Tarpan (*Equus caballus ferus*) which had a more central European distribution. The Tarpan died out in the wild between 1875 and 1890, and the last captive animal died in 1909 in a Russian zoo.

2.5.2 Muskoxen

Muskoxen (*Ovibos moschatus*) had a very wide distribution during the Pleistocene. Until recently it was thought that they had died out in Eurasia about the end of the Pleistocene, and that it was only in North America and Greenland (Figure 2.26) where they managed to survive. However, there is now ample evidence for the presence of Muskoxen in the Taymyr Peninsula well into the Holocene. From the skeletons that have been examined (MacPhee *et al.*, 2005) it appears that they were still present between 3,800 and 2,700 years BP in the Upper Taymyr River Basin (Figure 2.24).

Muskoxen were successfully reintroduced to the Taymyr Peninsula in the 1970s and now number over 6,000 animals (see Chapter 9). As with other members of the Pleistocene megafauna that were residing in the Taymyr Peninsula in the early Holocene, there is a mystery as to what extent their populations have fluctuated during the Holocene. There appears to have been a widespread megafaunal collapse at the Pleistocene–Holocene boundary, which was then followed by a recovery.

Figure 2.26 Muskoxen near Mesters Vig (73°N), north-east Greenland. Note the very fine wool fleece which provides highly effective thermal insulation.

2.6 The Hypsithermal, or Xerothermic, or Climatic Optimum

The warmth that accompanied the beginning of the Holocene did not last. The culmination of this warm period is variously referred to as the Hypsithermal, or Xerothermic, or Climatic Optimum. At high latitudes the summer temperature appears to have been 4°C above present temperatures, while in the tropics it was no more than 1°C. The timing was also variable. In south-western Saskatchewan the Hypsithermal period lay between 6,400 and 4,500 years BP (Porter *et al.*, 1999), while in north-west Montana the equivalent warm dry period lasted from 10,850–4,750 BP (Gerloff *et al.*, 1995). In western Norway, the Hypsithermal is dated to between 8,000 and 6,000 BP (Nesje and Kvamme, 1991) when Hazel (*Corylus avellana*) reached its most northerly Scandinavian extension (Huntley and Birks, 1983) at the Norwegian North Cape (71°N).

The rise in sea level that brought about an expansion of the Arctic Ocean increased the maritime influence on the northern regions of Russia and western Siberia. This increase in oceanicity began to have an effect on the High Arctic Climate about 6,000 BP. This is seen in the initiation of a retreat of tree cover from the shores of the Arctic Ocean. Between 8,000 and 4,500/4,300 BP, *Picea abies* ssp. *obovata* had been growing farther north than at present. Also, various *Larix* spp. were further north between 10,000 and 5,000/4,500 BP than they are now. Tree Birches (*Betula pubescens*) had already begun retreating from the shoreline of the Barents Sea (68°N) and the Taymyr Peninsula (72°N) between 9,000 and 8,000 BP (Figure 2.27). Subsequently, a more general imposition of more oceanic conditions in areas promoted active bog growth across a treeless landscape (Figures 2.27 and 2.28).

2.6.1 Post-Hypsithermal climatic cooling

A marked change in the vegetation began to take place at northern latitudes after the Hypsithermal.

2.6.2 Permafrost persistence

The Pleistocene permanent land-surface ice has now retreated, but at northern latitudes, despite

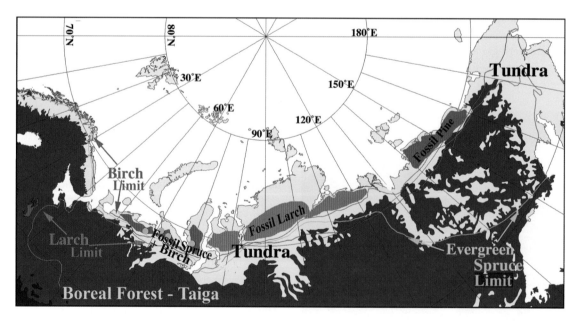

Figure 2.27 Comparison of present and past northern limits for tree survival in northern Siberia. Present-day distribution of the boreal forest (brown) is based on the vegetation map produced by Grid Arendal and published by the World Wildlife Fund for Nature. Mid-Holocene limits to forest trees are regional generalizations from locations of fossil remains (green = evergreen—pine and/or spruce spp.; red = tree birch; purple = larch) based on Kremenetski *et al.* (1999). Modern limits for the northern survival of individual tree species (colours as above) are also taken from Kremenetski *et al.* (1999), as drawn by Callaghan *et al.* (2002).

Figure 2.28 Aerial photograph taken flying north over the West Siberian Lowlands at 75°E between 64° and 66°N showing the ultimate state of paludification of a former ancient forest. (Photo. S. Kirpotin).

the passage of time since the end of the ice age, there still remain large areas of frozen ground. The Arctic accounts for 5% of the total land surface of the Earth, but even now 11,000 years after the end of the last ice age permafrost still underlies vast areas of vegetation and soil. Its effects can be seen not just on the tundra heaths but also on the arctic wetlands and at the Tundra-Taiga Interface. Despite climatic warming, former permafrost zones in the soil can freeze again, especially when trees and dwarf woody vegetation advance into areas from where the permafrost has retreated only from the upper soil layers. The shade created by the colonizing woody vegetation frequently causes the permafrost level in the soil profile to rise, killing the newly established trees and bushes (Figure 2.29). There still remain places in eastern Eurasia where not only has the permafrost from the last ice age persisted, but also the deeper soil layers have remained frozen since the previous ice age.

2.6.3 Ecological effects of permafrost

Biologically, an ice-permeated permafrost zone is more restrictive for plant growth than one without ice. The presence of ice prevents the diffusion of oxygen and therefore creates an anaerobic (oxygen-deficient) environment that is physiologically disadvantageous for most plants. There is also a physical aspect, in that soils with permafrost with ice can suffer cryoturbation. When ice thaws, the volume of terrain supported by ice is drastically reduced, causing slumping and erosion and can result in the release of large quantities of methane, possibly contributing to climatic warming as a result of microbial methanogenic activity in arctic soils.

Arctic methanogenic activity like clathrate hydrate (methane ice) in the oceans is of particular interest in permafrost studies, due to the key role of the Arctic methane cycle and its significance for the global methane budget and climatic warming. As the ancient frozen organic components in arctic soils thaw they are likely to contribute significant amounts

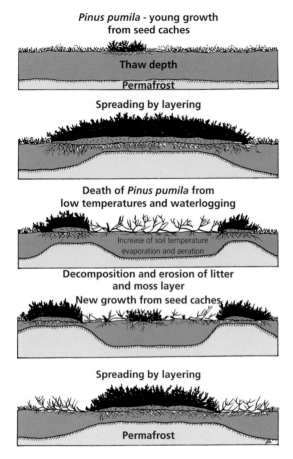

Figure 2.29 Interaction between Dwarf Siberian Pine (*Pinus pumila*) and site conditions in the lesotundra (Tundra-Taiga Interface) in northern Siberia. (Reproduced with permission from Holtmeier, 2003).

of methane to the atmosphere where, even as a trace gas, it can augment the greenhouse effect.

Methanogenesis is the formation of methane by microbes known as methanogens. Organisms capable of producing methane belong to the Archaea, an ancient group of micro-organisms distinct from both eukaryotes and bacteria, many of which live in close association with anaerobic bacteria.

Methanogenic archaea are widespread in nature and highly abundant in extreme environments, being able to tolerate both low and high temperatures. They are also abundant in permafrost soils (Wagner and Liebnerz, 2010). Although they are regarded as strictly anaerobic organisms without the ability to form spores or other resting stages, they are capable of long-term survival in extreme environmental conditions.

Microbial studies on permafrost core samples from the flood-prone lowlands of north-east Siberia have found viable microbes with an age of 3 million years (Shi *et al.*, 1997). Fortunately, the residence time of methane in the atmosphere is relatively short, in the region of 10 years, whereas that for carbon dioxide is measured in centuries.

2.7 Late Holocene climate fluctuations

Until recently, the history of the Holocene climate since the Hypsithermal has generally been one of gradually falling temperatures. There have, however, been periods of warming which have been well documented over large areas of the northern hemisphere and have also been detected at high latitudes. The terms *Roman Warm Period*, *Medieval Warm Period*, and *Little Ice Age* were introduced by pioneering palaeoclimatologists in the 1960s working in the European North Atlantic. Although it has been argued (Jones and Mann, 2004) that these terms cannot be applied to global temperature changes, they are nevertheless relevant to the areas in which they were first used. The particular relevance of these terms to Europe and the North Atlantic regions such as Iceland and Greenland can be clearly seen in the data of global climatologists in figures that show the history of temperature anomalies over the past 2,000 years (Figures 2.30 and 2.31).

A study in Iceland has shown that climate can be modulated by changes in the speed of oceanic deep-water flow. The speed of flow is shown by the increased grain size of the sediments. The data have shown that the *'thermohaline'* circulation (the flow of ocean water caused by changes in density) may modulate European climate. In particular, these studies have noted a coincidence of flow changes with some well-known climate events such as the Little Ice Age and the Medieval Warm Period. It was also suggested that periods comparable to the Little Ice Age and the Medieval Warm Period have been a recurrent feature of earlier parts of Holocene climatic history, with the warm intervals coinciding with faster near-bottom water flow in the south Iceland basin (Bianchi and McCave, 1999).

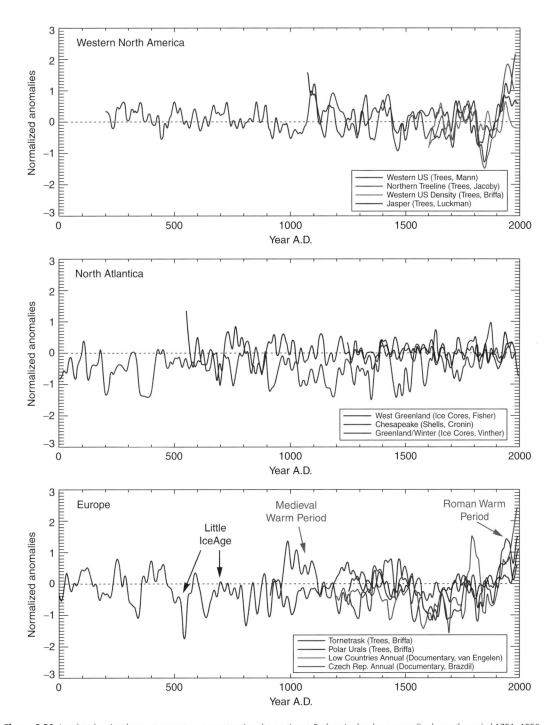

Figure 2.30 Local and regional proxy temperature reconstructions by continent. Each series has been normalized over the period 1751–1950 and then smoothed with a 50-year Gaussian filter. For the decadal resolved data the normalization period is the 20 decades from 1750 to 1949, smoothed using a 5-decade filter. (Adapted with permission from Jones and Mann, 2004).

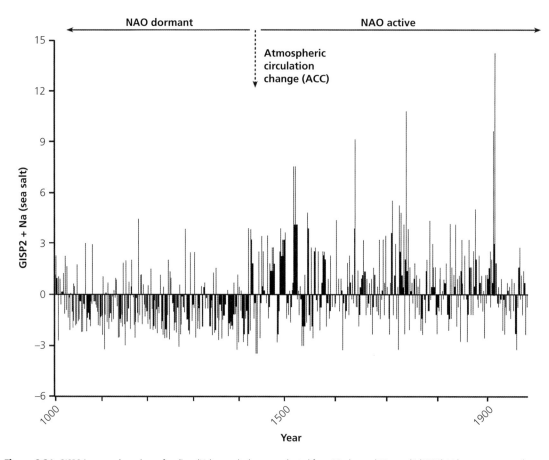

Figure 2.31 GISP2 ice-core chronology of sodium (Na) sea salt changes, adapted from Meeker and Mayewski (2002). Values are expressed as concentration departures in parts per billion (ppb) measured with reference to the long-term average. Also shown is the changing status of North Atlantic Oscillation (NAO) before and after key atmospheric circulation change ca. AD 1420. (Reproduced with permission from Dawson et al., 2007).

Regional studies have now provided additional evidence of periods of warming that had an influence at northern latitudes. Fluctuations in the oxygen isotope $\partial^{18}O$ values of North Atlantic mollusks recovered from near-shore marine cores in northwest Iceland indicate significant variations in seasonal temperatures over the period 360 BC to AD 1660, with warm periods from 230 BC to AD 140 and AD 640–760 (Patterson et al., 2010). This 2,000-year-long dataset also shows a warming trend that occurred after AD 1120 (Patterson et al., 2010). However, by AD 1320 the climate began cooling again, recording record lows for the 2,000-year dataset. Such lows are also seen in Greenland ice cores.

Altogether, these recent regional studies provide evidence for the view that Europe did enjoy a warmer spell around 2,000 BP, the time that is often referred to as the Roman Warm Period, from approximately 250 BC to AD 400. There is therefore scientific evidence to substantiate long-held claims that vines grew in England in Roman times and this warmer weather might even have facilitated Hannibal's elephants in crossing the Alps in 218 BC.

By AD 410 there had been a return to cooler temperatures presaging the onset of a cold and wetter era that marked historically the beginning of the Dark Ages Cold Period between AD 400 and 600.

2.7.1 The Medieval Warm Period

With the passing of the Dark Ages (fifth to eighth centuries AD) a warming trend led to what has been termed the Medieval Warm Period or *Medieval Climatic Optimum*, denoting a period of warm climate lasting from about AD 950–1250 in the North Atlantic region that may also have been related to other climate events around the world during that time, including China and New Zealand.

Some authors refer to the event as the *Medieval Climatic Anomaly* as this term emphasizes that effects other than temperature were important. When viewed on a global basis the Medieval Warm Period can appear a rather minor temperature anomaly (Mann, 2007). However, when examined in relation to its biological impact in northern polar regions, particularly around the North Atlantic, marked effects can be noted. One example is an increased flowering in Iceland of Birch (*Betula pubescens*) as seen in the pollen deposits for the period AD 600–800 which suggests a significant improvement in the climate just before the Norse settlement (Erlendsson and Edwards, 2009).

It appears that the climate changes associated with the Medieval Warm Period took place at a time when the North Atlantic Oscillation (NAO) may have been weak. It is therefore commonly assumed that the Norse expansion across the North Atlantic to Faroe, Iceland, Greenland, and even North America would have been aided by a period of relative climatic tranquillity in the northern seas. There is also evidence that marked changes in atmospheric circulation were coincident with a significant increase in North Atlantic storminess at *ca.* AD 1400–1420 at the time when the Norse settlement of Greenland was in decline (Dawson *et al.*, 2007).

To attribute either the beginning or the end of the Norse settlement of Greenland to climatic factors is, however, a simplification that ignores a number of other important historical political and economic events. These are discussed in Chapter 3.

2.7.2 The Little Ice Age

The most recent climatic event to have had a biological impact at northern latitudes before the present warming trend was the *Little Ice Age*, a name given to a period of world-wide cooling lasting approximately from the fourteenth to the mid-nineteenth centuries. The effect of a general cooling effect can be found in a variety of locations in both hemispheres, but it is the more northern regions of America and Europe that show the greatest evidence of an impact from climatic change that affected human settlements and has consequently left a significant historical record (Grove, 1988).

There is no generally agreed start or end date for the Little Ice Age. Some authors nevertheless confine the term to the period between 1550 and 1850, in which there were three notable cold episodes, beginning about 1650, 1770, and 1850, with each separated by slightly warmer intervals (Figure 2.32).

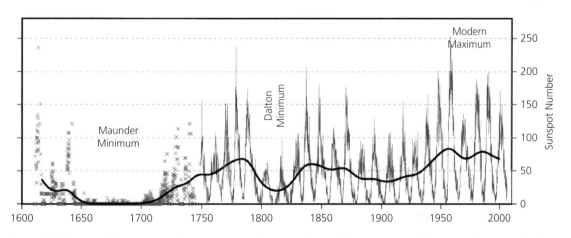

Figure 2.32 Summary of 400 years of regular sunspot number observations. Since ~1749, continuous monthly averages of sunspot activity have been available and are shown here as reported by the Solar Influences Data Analysis Center, World Data Center for the Sunspot Index, at the Royal Observatory of Belgium, and are based on an average of measurements from many different observatories around the world. (Image reproduced from Global Warming Art (http://www.globalwarmingart.com)).

The general cooling and the periods that were exceptionally cold have been widely attributed to a reduction in solar radiation.

A noticeable feature during these periods is the rarity of sunspots (Figures 2.33 and 2.34). Observers at that time noted that they were exceptionally rare between 1645 and 1715, which was one of the colder intervals. This period of low sunspot activity is referred to as the *Maunder Minimum*, after the solar astronomer Edward W. Maunder (1851–1928) who discovered the lack of sunspots from an examination of historical records. During one 30-year period within the Maunder Minimum, astronomers recorded only about 50 sunspots, as opposed to a more typical figure of 40,000–50,000 spots. The Maunder Minimum coincided with the middle and coldest part of the Little Ice Age during which Europe and North America, and many other parts of the world, endured bitterly cold winters.

Whether or not there is a causal connection between low sunspot activity and cold winters is, however, still controversial. During the Maunder Minimum the amount of cosmic radiation reaching the Earth was also reduced. There were, however, other solar phenomena that may have accounted for the lower temperatures on Earth. In relation to climatic variation, the Maunder Minimum does raise the possibility that the time-scale associated with any climatic change may be in the order of hundreds rather than tens of years, which would be the scale for changes directly linked to individual 11- or 22-year cycles.

2.7.3 Solar activity and climate

It has been suggested that a solar-forcing mechanism may underlie at least the Holocene segment of the North Atlantic's '1,500-year' cycle (Bond *et al.*, 2001). The evidence comes from a close correlation between inferred changes in production rates of the cosmogenic nuclides carbon-14 and beryllium-10 and centennial to millennial changes in proxies of drift ice measured in deep-sea sediment cores. Specifically, weaker westerly winds have been observed in winters with a less active sun, for example at the minimum phase of the 11-year sunspot cycle.

There is little variation in solar luminosity in relation to sunspot cycles. It has nevertheless now been reported from satellite measurements that variations in solar ultraviolet irradiance may be larger than previously thought (Ineson *et al.*, 2011). Studies using climate models suggest that peaks in ultraviolet irradiance which are much greater than luminosity variations may be associated with the negative phase of the North Atlantic or Arctic Oscillation. Thus low solar activity, as during years with cold winters in northern Europe and the USA, and mild winters over southern Europe and Canada may have been brought about by a reduction in UV radiation (Ineson *et al.*, 2011). Historically the incidence of cold winters, as documented in harvest failures during the Little Ice Age, may be directly related, for example, to the low sunspot activity, as was noted during the Maunder Minimum.

In terms of ecological effects in the Arctic, a study of peat profiles in Spitsbergen suggests that before or at the beginning of the Little Ice Age a dominance of Purple Saxifrage (*Saxifraga oppositifolia*) indicated a cold and dry climate, followed by a decline of this species and a gradual increase of Polar Willow (*Salix*

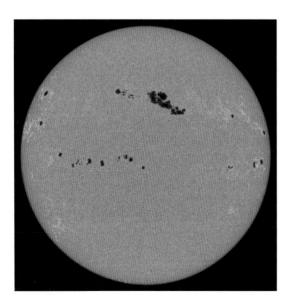

Figure 2.33 Sunspots as observed on 29 March 2001. (Photo Solar and Heliospheric Observatory, European Space Agency, and NASA).

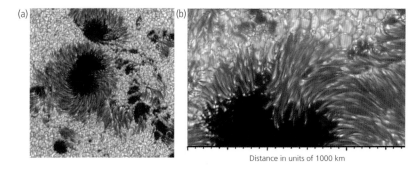

Figure 2.34 (a) Image of an active region with sunspots and pores from the Swedish Solar Telescope on La Palma. (b) Close-up of penumbral structure. (Courtesy of G. Scharmer and the Royal Swedish Academy of Sciences).

Figure 2.35 Polar Willow (*Salix polaris*) showing a female plant growing in north-west Spitsbergen. An increase in the presence of this species has been taken as an indication of a recent onset of milder, wetter conditions (see text).

polaris) (Figure 2.35), indicating a moist and milder climate (Rozema *et al.*, 2006).

In the closing years of the seventeenth century other factors have also had an influence on the climate of north-western Europe. A series of volcanic eruptions—Hekla in Iceland and Serua in Indonesia in 1693 followed by Aboina in Indonesia in 1694—combined to produce a large extrusion of dust into the atmosphere that may have also contributed to the series of harvest failures in Finland, Estonia, and Iceland, with widespread famine leading to large reductions in the human population. The period is remembered in Scotland as the 'seven ill years' (1693–1700) when a succession of disastrous summers caused widespread starvation, soon followed by a necessity for political union with England (Morrison, 1990).

2.7.4 Arctic climate at present

The Arctic, however it is defined, is a region with enormous environmental variation both at present and in the past. Being a marginal area in terms of thermal input, even minor differences between regions or even microsites can have significant ecological consequences. As a large measure of the thermal input to the Arctic comes from ocean currents and winds it is therefore not surprising that geographical location in relation to continentality and oceanicity has in many places a greater effect on the plant and animal communities than latitude. The Arctic is also a place of climatic uncertainty, where plant and animal populations live at the edge of a habitable terrain where minor differences from year to year can result in failure in terms of plant growth or animal reproduction. Consequently, large changes in population size in animals and flowering and seeding in plants frequently take place in response to minor climatic variations. This has always been a feature of high-latitude ecology and will be discussed further in the following chapters.

CHAPTER 3

Human arrival in the Arctic

3.1 Prehistoric arctic peoples

The discovery in 1993 by Russian archaeologists of an early Palaeolithic site on the Yana River at 71°N in eastern Siberia astonished the archaeological world (Figure 3.1). Detailed excavations in 2001–2002 dated the site to 27,000 BP (Pitulko *et al.*, 2004).

The site is generally referred to as the *Yana Rhinoceros Horn Site* due to the magnificent remains of Woolly Rhinoceros (*Coelodonta antiquitatis*) horn that were discovered on digging to a depth of around 30 ft into the banks of the River Yana. The excavation revealed not only an astonishing collection of artifacts but also the oldest known location to date for human occupation within the Arctic Circle. The next oldest widely-accepted settlement discovered before this was a late Palaeolithic site at Berelekh 70°N in Yakutia, commonly referred to as the Mammoth Graveyard and dated to 13,000–14,000 BP (see Figure 3.2).

The Berelekh site comprises two separate entities, a Mammoth Graveyard and a late Palaeolithic site. The deposits of the Berelekh bone bed (or Mammoth Graveyard) accumulated between 14,000 and 11,000 ^{14}C BP, with the most intensive accumulation of bones dating to around 12,300 BP.

The Mammoth Graveyard is described as a natural accumulation of mammoth bones. There is almost no chronological overlap between the 'graveyard' and the 'archaeological late Palaeolithic site', the age of which was estimated to be no older than 11,820 ± 50 years BP. (Pitulko, 2011). The location of the site fits well with the known historical position of the ice sheets (see Figure 2.8). The Novaya Zemlya ice sheet reached its maximum extent *ca.* 39,000 BP. In addition, the Yana Delta, being to the east of Novaya Zemlya, would have been in a rain-shadow area and would probably have had only a minimal amount of ice cover in the late Pleistocene.

Just as remarkable as the age of the Yana Rhinoceros Horn Site are the artifacts that were found. These included shaped stone scrapers and a spear foreshaft carefully carved from a rhinoceros horn. These finds were surrounded by the burnt bones of mammoth, muskoxen, bears, bison, horses, and cave lions. As well as the rare Woolly Rhinoceros spear foreshaft, there were also other shafts from mammoth bones, together with a wide variety of tools and flakes (Figure 3.3). The human race is known for its adaptability and capacity for migration. However, the Yana site has demonstrated a capacity for a human presence 300 km north of the Arctic Circle 27,000 years ago during the late Pleistocene which is truly remarkable, both biologically and technically, for a species that had migrated only relatively recently out of Africa.

3.1.1 Human arrival in North America

The discovery of the Yana Rhinoceros Horn Site not only revised the date for the first human penetration of the coast of Siberia, but also changed the perception of when Palaeolithic hunters reached, and possibly crossed, the Bering Strait into North America (Figure 3.4). Previously it was believed that the harsh climate of the millennia before the Last Glacial Maximum (*ca.* 21,000–18,000 BP) prevented people from crossing into Beringia and thus to the Americas. Such a scenario dated human arrival in North America to the Clovis peoples. After the discovery of several Clovis sites in western North America in the 1930s, the Clovis people came to be regarded as the first human inhabitants of the North

Tundra-Taiga Biology. First Edition. R.M.M. Crawford © R.M.M. Crawford 2013.
© Published 2013 by Oxford University Press.

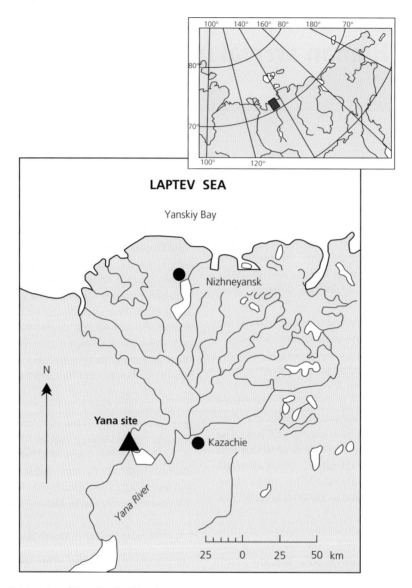

Figure 3.1 Location of the early Palaeolithic Rhinoceros Horn Site on the Yana River. (Adapted from Pitulko et al., 2004).

American continent. However, with the Yana Rhinoceros Horn Site being only 2500 km from the Bering Land Bridge it appears likely that people living near this icy gateway to the New World may have been the ancestors of the first human populations to migrate into North America.

There is now increasing evidence to suggest an earlier, pre-Clovis settling of the Americas. The discovery of the tip of a projectile point made of Mastodon bone embedded in a rib of a single disarticulated Mastodon at the Manis site in the state of Washington together with radiocarbon dating and DNA analysis showed that the rib is associated with other remains and dates to 13,800 BP. This bone projectile, which is common to the Beringian Upper Palaeolithic and Clovis, was made and used during pre-Clovis times in North America. The Manis site has been claimed, together with evidence

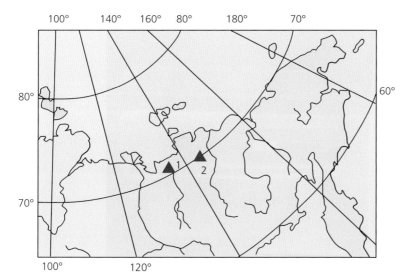

Figure 3.2 Locations of (1) the Yana early Palaeolithic site and (2) the Berelekh site with the Mammoth Graveyard. (Adapted from Pitulko *et al.*, 2011).

of mammoth hunting at sites in Wisconsin, as showing that people were hunting proboscideans at least two millennia before the arrival of the Clovis peoples (Waters *et al.*, 2011).

More recent investigations in the Paisley Caves in Oregon have recorded the oldest directly dated human DNA remains in the western hemisphere. More than 100 high-precision radiocarbon dates found well-stratified deposits containing artifacts and coprolites ranging in age from 12,450–2,295 C^{-14} BP. Projectile points were recovered in deposits dated to 11,070–11,340 BP, a time contemporaneous with or preceding the Clovis technology. The coprolites contained human DNA of pre-Clovis age. Overall there was no evidence of diagnostic Clovis technology at the Paisley Caves site. The artifacts showed evidence of two distinct technologies with parallel developments but no indication of linear evolution (Jenkins *et al.*, 2012).

The study of the peopling of North America is now passing through a period of reassessment accompanied by an extensive debate as to the number of founding human immigrations and their timing and routes. In the past it had been supposed that the main immigration route would have been through northern Canada and therefore would have been feasible only once the mass of Pleistocene ice in that region had melted. The few scattered sites that have been studied in the far north leave the subject of the early history of the human arrival in that region still open to archaeological deliberation.

Recent archaeological discoveries when combined with molecular genetic studies of Siberian and Native American populations now suggest that the initial migration of ancestral Amerindians originated in south-central Siberia and entered the New World between 20,000 and 14,000 BP (Figure 3.5). The movements across Beringia to North America and ultimately South America can probably now be summarized as an appearance of arctic peoples on the west coast of the Bering Strait about 27,000–22,000 BP. Later Beringian populations moving into northern North America after the Last Glacial Maximum (LGM) gave rise to Aleuts, Eskimos, and Na-Dene Indians (Schurr, 2004). The archaeological records of Siberia and Beringia generally support these findings, as do archaeological sites in Oregon now dated to as early as 14,000 years ago (Gilbert *et al.*, 2008) and also in Chile (Dillehay *et al.*, 2008). If this is the time of the first colonization, geological data from western Canada suggest that early human migrants would have moved along the recently deglaciated Pacific coastline (Goebel *et al.*, 2008). A second migration may also have followed later from the same Siberian region and entered the Americas possibly using an interior route, and contributing

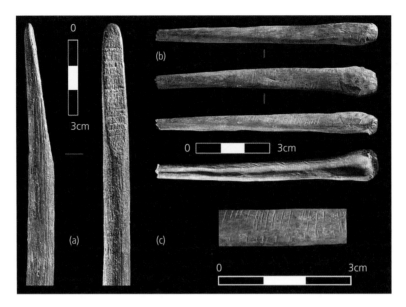

Figure 3.3 Organic artifacts from the Yana Rhinoceros Horn Site: (a) rhino horn spear foreshaft (proximal end); (b) wolf metatarsal bone (a broken awl?) with multiple cut marks; (c) cut marks (enlarged). (Reproduced with permission from Pitulko et al., 2004).

genetically to indigenous populations from North and Central America.

Palaeo-Eskimo human material that can provide usable DNA samples is scarce at high latitudes, despite the relative youth of some of the cultures and the cold preservation conditions offered by the Arctic.

The application of improved investigations into ancient DNA as deduced from archaeological studies aided by evidence gained from analysis of mitochondrial data obtained from ancient hair and nail clippings is gradually shedding more light onto the migration history of these northern peoples. Possible routes for these sequential waves of human migration into the northern extremities of the New World are shown in Figure 3.5.

The coastal movements of the early migrating hunters at the end of the Pleistocene would have been facilitated by the sea level being 120–130 m lower than at present. The terrain newly exposed from the ice would have also provided better routes for human migration than those that exist today. The bogs, thickets, and forest that now cover this ground would at the end of the Pleistocene have been open tundra interspersed with lakes. The warmth that melted the ice allowed a Tundra-Steppe vegetation to develop (see section 2.1) over which the Pleistocene megafauna roamed and provided a ready supply of food for the migrating hunter-gatherer populations. The ease of access for hunting across this open terrain would have also provided an incentive for the exploration of fresh hunting grounds.

Population dispersal is fundamental to survival in the fluctuating environments of polar regions, and these early human migrations demonstrate just how successful our species has been in dispersing throughout the Arctic and adapting to areas that might be considered by modern standards to be marginal for human survival.

3.1.2 The Pleistocene overkill

'C'est la soupe qui fait le soldat'—'an army marches on its stomach'. This famous Napoleonic aphorism in relation to military logistics was probably just as relevant for the Pleistocene men and women hunter-gatherers as it was for 19th century warfare. In the peri-glacial world of the Tundra-Steppe, the Pleistocene megafauna would have given ample human sustenance and the frozen ground would have provided ubiquitous refrigeration facilities.

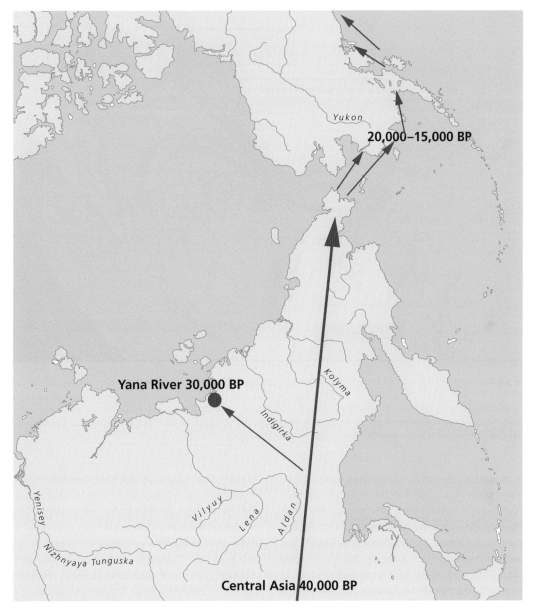

Figure 3.4 Approximate chronology of early human Palaeolithic migration into the Eurasian and North American Arctic—see text for details of controversies concerning the chronology of the migration.

Given the innate human tendency for over-provision, especially in an uncertain environment, it has been suggested that the disappearance of the large mammals was due to early hunters. This possibility that the prime cause of the extinction of the Pleistocene megafauna was due to human hunting in both the Old and the New Worlds was first described as the *Pleistocene overkill* in the 1960s. Indeed, part of the argument for the presence of pre-Clovis hunter-gatherer societies in North America is based on the extinction of the Mastodon (Waters *et al.*, 2011).

It is difficult for us today to have any realistic estimation of just how easily satisfied any group of prehistoric hunters might have been with their kill.

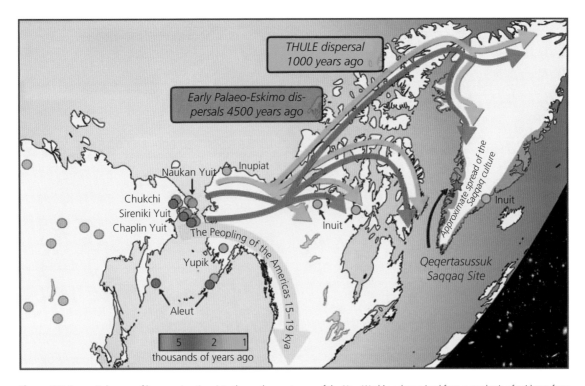

Figure 3.5 Sequential waves of human migrations into the northern extremes of the New World as determined from a synthesis of evidence from archaeological and mitochondrial DNA studies. For clarity, the migration of the later Dorset culture into northern Greenland has not been included. The approximate geographic distribution of contemporary Neo-Eskimo populations is shown together with the location of Qeqertasussuk, the archaeological Palaeo-Eskimo (Saqqaq culture) site, which has yielded usable DNA samples from human hair. The colour codes show the distribution of related extant mitochondria DNA haplo-groups. (Adapted from Gilbert et al., 2008).

Some insight is, however, given from early European travellers who observed the hunting habits of the Northern Indian tribes in the regions between the Hudson Bay and the Mackenzie River. Such a traveller was Samuel Hearne who was one of the earliest explorers in the Canadian Arctic between 1769 and 1772. He travelled for many months, living entirely with the Northern Indians and keeping a careful and detailed diary (Hearne, 1769). He noted that reindeer tongue was a highly valued delicacy and that they would kill far more animals than they otherwise needed in order to enjoy this particular morsel. If this type of hunting was widely practised, then even a small population of hunters might have a significant effect on local populations of their prey.

As already discussed (see section 2.4.3), the application of molecular methods in assessing changes in genetic diversity in ancient DNA suggests that population diminutions as assessed through reductions in diversity are more closely related to environmental variation than predation by pre-historic human hunters. This is also borne out by a correlation in the dramatic contraction in mammoth range *ca.* 12,000 BP, with the extensive spread of trees after the Allerød phase of the Late Glacial Interstadial (14,000–13,000 BP). After this period, known populations of mammoth were confined to northern Siberia (mainly the Taymyr Peninsula and Wrangel Island). The return of open Tundra-Steppe in the Younger Dryas cold phase, *ca.* 11,000–10,000 BP, saw a limited mammoth re-expansion into north-east Europe, followed by retraction and apparent extinction of mainland populations, which matches the marked loss of open habitats in the early Holocene (Stuart, 2005).

When evidence from palaeontology, climatology, archaeology, and ecology are combined it would appear that human hunting was not solely responsible for the similar pattern of extinction everywhere. Instead, it is more probable that it was the interaction of human disturbance along with the impact of pronounced climatic change on the vegetation that brought about their gradual demise (Barnosky *et al.*, 2004). Surprisingly, a dwarf mammoth population inhabited Wrangel Island from 7,700–3,700 BP (Kuz'min *et al.*, 2000; Figures 3.6–8).

Wrangel Island at 72° N lies between the Chukchi and East Siberian Seas and supports a rich and varied vegetation. The island is an important breeding ground for Polar Bears, having the highest density of dens in the world. There are also considerable populations of seals and lemmings. Over 100 species of birds migrate to the island during the summer. Despite being surrounded today by ice for a large part of the year, Wrangel Island was surprisingly largely ice-free during the last glaciation. The dwarf form of Mammoth that survived on Wrangel Island until 3,700 BP possibly owed its survival to being of a diminutive size and therefore able to exist on a limited food supply.

Evidence for prehistoric human occupation of Wrangel Island was uncovered in 1975 at the Chertov Ovrag site on the south coast of the island (Figure 3.8). Site excavations revealed various stone and walrus ivory tools, including a toggling harpoon. Radiocarbon dating shows that the human inhabitation was approximately coeval with the last mammoth survivors on the island *ca.* 3,700 BP. This is the most western site for Palaeo-Eskimo settlement

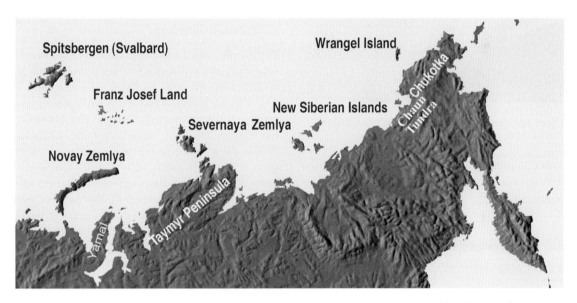

Figure 3.6 Location of sites mentioned in the text. The Yamal and Taymyr Peninsulas and Wrangel Island lie to the north of 70°N.

Figure 3.7 Wrangel Island in summer. (Photograph Professor R.L. Jefferies).

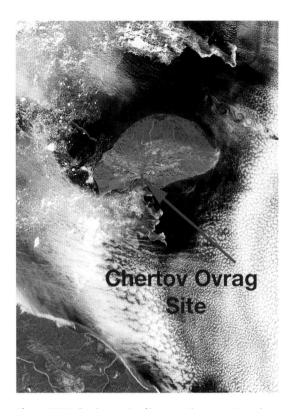

Figure 3.8 Earliest known site of human settlement on Wrangel Island *ca.* 3,700 BP. (Satellite image courtesy of NASA).

based on hunting sea mammals, with no evidence having been found for the terrestrial hunting of mammoth.

3.2 Indigenous peoples of Arctic Eurasia

3.2.1 Saami

Not all arctic peoples are of Asiatic origin. The Northern European Shield, which includes the northernmost parts of Finland, Norway, Sweden, and the Kola Peninsula of Russia, have long been inhabited by an indigenous people commonly referred to in the past as Lapps but now more usually called Saami. This name for these northern people comes from '*Sápmi*', the North Saami word for their territory. The evolutionary origin of the Saami population has long been an enigma. The Saami language suggests an East European background. Even so, as with the Finns, the genes of the Saami are largely West European (Tambets *et al.*, 2004).

Other investigators have detected differences between the western and eastern Saami populations and have suggested that some Saami lineages share a common ancestor with lineages from the Volga-Ural region originating as recently as 2,700 years ago, indicating a more recent immigration of people from the Volga-Ural region to the Saami population (Johansson *et al.*, 2008).

Despite these minor variations, it can be concluded that the Saami mitochondrial gene pool differs substantially from that of the other more eastern arctic peoples. Linguistically, they are even more distinct. The unique position of the Saami on the genetic landscape of Europe suggests that they are an old population in Europe which diverged at an early date from other European populations prior to subsequent linguistic and cultural diversification.

3.2.2 Palaeo-Siberians

The peoples who live to the east of the Kola Peninsula, from the White Sea to the Chukchi Sea, can be generally grouped as Palaeo-Siberians (Table 3.1). Moving progressively eastward, the principal groups are the Samoyeds (Nenets and Ngansan, see Figure 3.9), the Yakuts, and the Chukchi. Within these major groups there are many other groups or tribes often with distinct languages. Some 35 indigenous languages are at present recognized in Siberia. Historically the number was much greater, with about 120 languages being encountered by the Russians when they first colonized Siberia in the 16th century (Forsyth, 1992).

3.2.3 Samoyeds

The term Samoyeds used to be applied to the most western of the Palaeo-Arctic peoples. The word Samoyed is not in any way cognate with Saami. Samoyed has had various interpretations. It has been suggested that it comes from the Russian самоед- (*sam*—self and *ed*—eat), implying an early Russian belief of possible cannibalism. However, this is probably a misinterpretation and it is more likely that it comes from *Samadu*, a name given to the Nenets tribe of the Yenisey by their neighbours (Forsyth, 1992). The name Nenets, meaning a 'man', is officially given to the largest tribe of the

Table 3.1 Principal early peoples in North America and Greenland.

Clovis tradition	Clovis is a pre-historic Palaeo-Indian culture that appeared in North America ca. 13,000 BP and is thought to have lasted between 200 and 800 years. They supported themselves by hunting the remnants of the post-Pleistocene megafauna. The Clovis people were for a time regarded as the first human inhabitants of the New World. This view is now no longer valid. Their disappearance may have been due to a climatic deterioration in the Younger Dryas Period or a decline of the megafauna or merely replacement by subsequent immigrations.
Arctic Small Tools tradition	The Arctic Small Tool Pre-Dorset tradition describes the earliest human settlers to occupy the Canadian Arctic archipelago and northern Greenland, apparently entering those regions from Alaska in a rapid population movement around 4,500 BP. It is characterized by finely made stone tools comprising micro-blades and small scrapers. Hunting was on land with muskoxen their main prey. An onset of colder weather caused a decline in muskoxen and their disappearance by 2,800 BP.
Dorset culture	The Dorset culture occupied the Canadian Arctic and parts of Greenland from approximately 500 years BC until around AD 1000. The Dorset people winter-hunted on sea ice but had no dogs or sledges. With better-constructed houses they were adapted to a climate that was colder than that which had prevailed at the time of the Arctic Small Tools tradition. They also hunted walrus and seals with harpoons.
Norton tradition	The Norton tradition is divided into three cultures (Choris 1000–500 years BC, Norton 500 years BC to AD 800, Ipiutak AD 1–800). It is characterized by extensive all-year-round use of maritime resources. Caribou and small mammal hunting remained important among Choris and Norton peoples. The latter are an Alaskan development from the Choris culture which lasted from 3,000–1,000 BP, and they used flaked stone tools that are similar to those found in the earlier Arctic Small Tool tradition. However, some Norton tradition assemblages also contain pottery and oil lamps. Ipiutak is the most outstanding culture of the Norton tradition, taking its name from a site near Point Hope, Alaska, that contained over 60 semi-subterranean houses and a cemetery.
Thule	The Thule culture (which lasted from ca. AD 1–1600) appears to have replaced the Dorset culture. It is distinguished by the ability to hunt large sea mammals and the use of large skin boats, as well having dogs pulling sledges. The culture developed around the Bering Sea and spread to North America and Greenland by AD 1000. Climatic deterioration in the 13th century is widely credited with causing the Thule people to modify their way of life into that found in the various historic and extant Inuit groups.
Inuit, Aleut, Eskimo	The Inuit are descendants of the Thule people. The names Inuit and Eskimo* are applied to the populations living in the arctic region extending from eastern Siberia to Greenland. Biological, linguistic, and archaeological evidence indicates that the Aleuts who are distinct from all the other aboriginal populations of the Americas and probably derive from a more recent population of Asian origin.

* Eskimo is supposedly the name given by American Indians (Athabascan or Algonquin) to describe a people who ate their meat raw, which was a consequence of living in a land bereft of timber for cooking. The derivation may not be true and other explanations have been given, e.g. snow-shoe netter. In Greenland and eastern Canada the term Eskimo is replaced by Inuit. However, peoples living around the Bering Sea still refer to themselves as Eskimo.

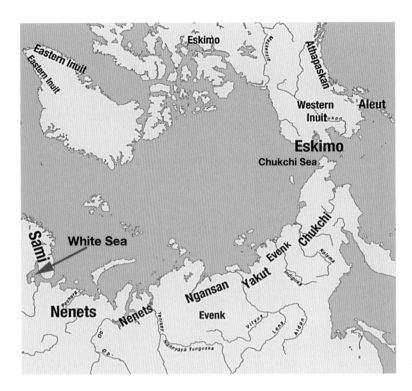

Figure 3.9 Approximate location of major groups of modern arctic peoples. Information based on Krupnik (1993) and various other sources—see text.

western Samoyeds who occupy a large area on both sides of the Urals, from the forest edge to the Arctic coast and into the Kamin, Gydan, and Yamal Peninsulas.

3.2.4 Yakuts

The Yakuts are the most-northern Turkic-speaking group; their language also includes a substantial proportion of Mongolian words, while about 10% of words are of a Tungusic–Manchu origin. According to current views, the early ethnogenesis of the Yakuts was influenced by Scythian and Hun components, whose roots were in a regional culture of early nomads of southern Siberia and central Asia. The ethnogenesis of the Yakuts was generally completed by the early 16th century by which time they had settled in the central basin of the Lena River. The Yakuts are therefore most likely to be descended from a mixture of peoples from the area of Lake Baykal, including Turkish tribes from the steppe and Altai mountains, and indigenous peoples of Siberia, particularly the Evens and Evenks.

When ethnic Russians first arrived in the region of Yakutiya in the 1620s, the Yakuts were living along the Lena and other rivers, functioning in a semi-nomadic, subsistence economy. They are notably distinguished from other arctic peoples by being traditionally horsemen. In the first half of the 17th century the Yakuts occupied a small region, mostly in the central part of modern Yakutia (see Figure 2.20). However, by the early 20th century they had spread over the vast north-eastern region of Siberia and the Yakut population had increased substantially (Khar'kov et al., 2008).

3.2.5 Chukchi

The Chukchi are traditionally divided into the Maritime Chukchi, who had settled on the coast and lived primarily from sea-mammal hunting, and the Reindeer Chukchi, who lead a nomadic existence in the inland tundra as reindeer herders. The Russian name *Chukchi* is derived from the Chukchi word *Chauchu* ('rich in reindeer'), which was used by the

Figure 3.10 Location of the present-day autonomous region of Chukotka in the Russian Federation. (Image from Wikimedia Commons).

'*Reindeer Chukchi*' to distinguish themselves from the '*Maritime Chukchi*'. Chukotka is the most northeasterly region of Russia (Figure 3.10) and, until the sale of Alaska to the USA, was the only part of Russia lying partially in the western hemisphere (east of the 180th meridian).

When in the 17th century the Russians advanced ever further eastward in their quest for furs, they subjugated one native group after another and forced them to pay *Yasak* (a Turkic word for 'tribute'). Russian brutality eventually caused the native groups, who had formerly been enemies, to forge an alliance with one another. Together, the Chukchi, the Koryak, and the Yukagirs had considerable success in resisting Russian domination, even though they had to use desperate measures of resistance. When it came to trying to extract *Yasak* from the Chukchi the Russians found that they would rather die and kill all their women and children than pay *Yasak*. One 18th century Russian writer described the Chukchi as 'the most savage, the most barbarous, the most intractable, the least civilized, and the most rugged and cruel people in all of Siberia' (Bockstoce, 2009).

Although the Russians formally laid claim to Chukotka, the Chukchi were officially listed as 'natives not wholly subjugated'. In 1822 the Chukchi were eventually relieved of all need to pay *Yasak* and other tax payments forever, and to this day Chukotka is an autonomous region of the Russian Federation.

3.3 Indigenous peoples of Arctic America

3.3.1 Inuit

The Inuit are in historic times the most widespread and numerous of the indigenous peoples of North America, covering a terrain that stretches from the shores of the Bering Sea through Alaska to Greenland. They are split into various groups with specific territories, using different names in different regions: Yuit, Yuik, Inipiat, and Inupiak in Alaska, Inuit in Canada, and Kakaallit in Greenland (see Table 3.1). Of all the Arctic peoples, the Inuit have probably been the most studied. As with other present-day native inhabitants of the circumpolar Tundra, even if they are not the direct descendants of the early Arctic peoples, they have nevertheless preserved much of the original way of life that has endured in this region since the late Pleistocene. Consequently, as examples of ancient hunter-gatherer communities they have been extensively studied by anthropologists. In describing the adaptations of these peoples to their environment a distinction has to be made between those who have access to maritime resources, mainly sea mammals, and those who mostly inhabit the inland regions of the Taiga or boreal forest and follow the reindeer in their northward migration into the Tundra in summer, where, if they are within reach of the sea, they can take fish, seals, and whales. However, peoples that live in close proximity to the ocean throughout the year have the advantage of being able to hunt for seals and polar bears on the sea ice in winter and generally have a more dependable all-year-round food supply.

3.3.2 Nunamiut

Until recently the Nunamiut were an isolated inland nomadic group of the Inupiat section of the Inuit who lived inland in northern Alaska and survived principally through hunting caribou, in contrast to other groups that were coastal and hunted marine animals and fish. A decline in caribou populations caused several families to move to coastal villages around 1900 and again in the 1920s. However, in 1938 several families moved back to the region of Killik River in the Brooks Range.

The daily life of such a unique group of inland reindeer hunters was first studied in detail by the Norwegian arctic explorer and anthropologist Helge Ingstad (Figure 3.11). In 1949 Ingstad flew into a remote region in the northern Alaska Anaktuvuk Pass where this inland caribou-hunting group resided. Ingstad was the first westerner to visit the region and to over-winter with the inland Eskimos. The area is known for intense winds and mean winter temperatures of –25°C coupled with a polar night lasting 72 days. He learnt their language, recorded their legends and superstitions, and participated in their caribou hunts and fishing expeditions. Through photographs, film, and audio recordings, Ingstad documented their disappearing nomadic lifestyle which was based almost entirely on caribou hunting (Ingstad, 1951).

The Nunamiut depended on meat more than any other living hunter-gatherer group and their life revolved around the annual migrations of caribou. The main caribou migrations take place in March and April, when caribou move north through the Anaktuvuk Pass to feed on the plains which thaw and become a marshland swarming with black flies and mosquitoes. The caribou hunting is repeated in September and October when the caribou retreat south.

In the short period of time available the hunting strategies have to be well co-ordinated and successful in order to secure sufficient food reserves for survival over winter. A detailed account of the catching, butchering, and preservation of meat by the Inuit in preparation for the winter is given by the American anthropologist Lewis Binford (Binford, 1978). He estimated that each adult ate about one cup and a half of vegetable foods a year, supplemented by the partially digested stomach contents of caribou, as well as some dried fish and sea mammal meat traded from coastal neighbours. These same neighbours also provided them with much vital fat needed for winter cooking and heating. In an arctic environment that has an extremely short growing season, the Nunamiut rely entirely on stored food for eight and a half months of the year, with fresh meat readily available for only two months a year, mainly during the caribou migrations.

The caribou were caught by being driven into the corrals, where they were snared or shot. The hunt continued for days on end for several weeks, with as many as 200–300 animals being taken on a good day. The women skinned the caribou, cutting the meat, drying some of it, and pounding up the rest with fat and berries. They cracked open the leg bones for marrow, and boiled and ate the foetuses, and also the stomach contents. Much of the meat, and sometimes hunting artifacts, were stored at the campsite. Everyone then dispersed into groups of two or three households during the summer months, congregating again in the autumn, when the people again hunted migrating caribou, but this time taking mainly the calves for their skins.

Figure 3.11 Helge Ingstad (1899–2001), the last of the Great Norwegian Arctic adventurers and explorers. As an anthropologist, together with his wife Anne-Stine, he discovered the first European (Norse) settlement in North America at L'Anse aux Meadows, Newfoundland. (Reprinted with permission from Helge Ingstad, *Oppdagelsen av det Nye Land*. © J.M. Stenersen Forlag, Oslo).

Figure 3.12 Ingstad Mountain at the Anaktuvik Pass, Alaska. Officially named in 2006 in honour of the Norwegian Arctic explorer Helge Ingstad. (Reproduced with permission from Grant R. Spearman, Alaska Cooperative Park Studies Unit).

Ingstad's material documenting of the traditional Nunamiut Eskimo culture laid the groundwork for a local museum that opened in 1985 at the Anaktuvik Pass. Today the Inupiaq Nunamiut have settled down and are increasingly connected to the modern world. The book that Ingstad subsequently wrote, *Nunamiut* (Ingstad, 1951), is an engrossing and original work and is a lasting memorial to a way of life that has now largely disappeared. The Nunamiut remember Ingstad's work for preserving their history and have named a mountain in memory of him. They were even conscious of this at the time he was among them. Ingstad recalled the Eskimo elder 'naming' the mountain after him: 'We were sitting in the tent, talking a little bit about my departure. Paneak said, "We will give you the mountain which stands at the beginning of the Giant's Valley. It shall bear your name and we will remember you. Our people remember such things for many generations". When Ingstad passed away in 2001 at the age of 101, a petition was made to the US Geological Survey and permission granted to officially name their most beautiful mountain after Ingstad (Figure 3.12).

3.4 Adapting to unpredictable environments

The unpredictable environmental fluctuations in the Arctic have profound effects on plant, animal, and human populations. Polar animal populations constantly run the gamut from super-abundance to near extinction and back. The human race's unique ability to anticipate the future is particularly relevant in the Arctic. The hunter-gatherer cultures of the far north are obliged to store food for the winter and so have had to develop a large number of social attributes, such as a control of property, scheduling, and co-ordinated labour, to a degree that is not required in tropical or temperate climates (Ingold, 1982).

The acquisition of natural resources in the Arctic has two fundamental aspects that distinguish the exploitation of this habitat from boreal and temperate regions. First, there is the need for rapid harvesting. For the early hunting peoples, short runs in the annual cycle of available game meant that sometimes hunters had only a few weeks or even days in which to provide their families and the entire community with food supplies for possibly a whole year. Second, the uncertainty of the arctic environment makes any prediction of when future resources may become available unreliable. Land hunters do not have the certainty that exists with marine sources (such as annual salmon runs) that next year's hunting will be successful. Many traditional arctic hunting techniques were therefore designed to seize an entire herd or stock of animals when it was encountered so as to obtain a maximum catch. Consequently, some means of food storage was essential in these situations (Ingold, 1982; Krupnik, 1993).

3.5 Eurasian Reindeer herding

The development of herding differentiates Eurasian exploitation of reindeer from that of the North American arctic peoples, at least in historical times. Just when reindeer hunting gave way to reindeer herding is difficult to date with any precision due to a lack of reliable historical resources. A distinction also has to be made between the domestication of the reindeer, which preceded the later development of large-scale intensive herding by several hundred years (Krupnik, 1993).

One clear date is, however, recorded. The Norwegian chieftain Ottar from Halogoland in northern Norway is recorded as having visited the Anglo-Saxon Royal Court of King Alfred in England in *ca.* AD 890 (Meriot, 1984). During his visit Ottar gave King Alfred an account of life in the north of Norway which provided a remarkable insight into Arctic life in the 9th century. Ottar said that he owned a herd of 600 domesticated reindeer. Visible evidence of the wealth of this marginal region of Europe can be seen in the reconstruction of a chieftain's longhouse dwelling dating from *ca.* AD 700 at Borg in the Lofoten Islands (part of Halogoland; Figures 3.13 and 3.14).

It is not until the time of Russian contacts with the northern aboriginal peoples in the 1600s that there is any further written evidence concerning the inhabitants of the Eurasian tundra and the small-scale herding of reindeer as being part of a complex system dominated by hunting and fishing, Such a way of life has long been practised by the Saami of

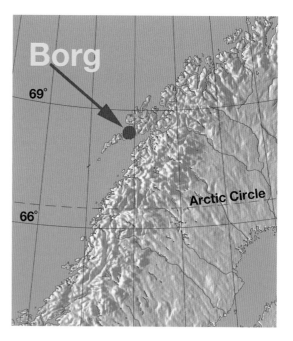

Figure 3.13 Map of northern Norway showing location of Borg in the Lofoten Islands and the site of a reconstructed chieftain's house dating from AD 700 (see Figure 3.14).

the Kola Peninsula in the west, the Nenets of the Taymyr Peninsula, and extended to the nomadic Chukchis at the eastern limits of the Asiatic Arctic.

The Chukchi are the largest group of the northeastern Palaeo-Asiatics and are most probably the descendants of the earliest reindeer breeding nomads to move northward towards the North Pacific coast. Another century was to elapse, however, before there is evidence of extensive reindeer herding. The 1700s saw the beginning of a significant growth in reindeer stocks with the onset of this change taking place virtually simultaneously in the western and eastern fringes of the Eurasian Arctic (Krupnik, 1993). As these domestic stocks grew, the hunting of wild reindeer virtually disappeared.

3.6 Norse settlements across the North Atlantic

One of the most striking movements of human populations into cold subarctic lands at the time of the *Medieval Warm Period* (see section 2.7) was the Norse colonization or *landnam* of the North Atlantic islands from the Faroes to Iceland and Greenland. From

Figure 3.14 Reconstruction at Borg in the Lofoten Islands (68°N) of a chieftain's longhouse dwelling dating from *ca.* AD 700.

here they established an exploratory base in Newfoundland from which expeditions were made to the American mainland. Although the Norse settled in Faroe from the 9th century onwards (Dugmore *et al.*, 2005), they were not the first human arrivals on these islands. Palaeo-botanical and archaeological evidence for an earlier settlement, possibly of Irish anchorites, had been found and given early radiocarbon dates (Jóhansen, 1978, 1985), suggesting AD 600 ± 100, with oats being the first cereal to be cultivated followed by barley. Subsequent research has given more precise datings for the first occurrence of cultivated crops in Faroe. Three locations have been found dated to as early as AD 570 (Hannon and Bradshaw, 2000), which is earlier than that implied from previous archaeological and historical studies but consistent with the findings from earlier palaeo-ecological investigations (Jóhansen, 1971). The earliest known introduction of domestic animals (sheep/goat) was *ca.* AD 700 (calibrated dates). Oat grains (*Avena sativa*) have also been found which has been taken to imply that field crops were being grown possibly by pre-Norse settlers, the Norse being normally associated with barley cultivation. Recently samples of carbonised barley grains have produced 14C dates of two pre-Viking phases within the 4th–6th and late 6th–8th centuries AD which indicates that a re-evaluation is required of the nature, scale and timing of the human colonization of the Faroes and the wider North Atlantic region (Church et al., 2013).

Iceland was initially settled between AD 865 and 930. In the period just before Iceland's initial settlement the climate appears to have been favourably warm for sea voyages but began to deteriorate shortly afterwards (Patterson *et al.*, 2010).

It has been generally considered that the *Medieval Warm Period* in association with a complex of historical, political, and demographic factors facilitated this Norse expansion across the North Atlantic. However, the degree of relative warmth in this Medieval Warm Period compared with temperature conditions today is still under debate (Erlendsson and Edwards, 2009).

Erik Thorvaldsson (AD 950—*ca.* 1003), otherwise known as *Erik the Red*, is remembered in medieval Icelandic saga sources as having founded the first Nordic settlement in Greenland. After being outlawed for murder in Iceland he explored and named Greenland between AD 982 and 985. He then led the first settlers to southern Greenland in 986 and settled in Brattahlið (today's Qassiarsuk) which became the centre of the Eastern Settlement. Between AD 986 and 1000 more settlers arrived from Scandinavia and they gradually

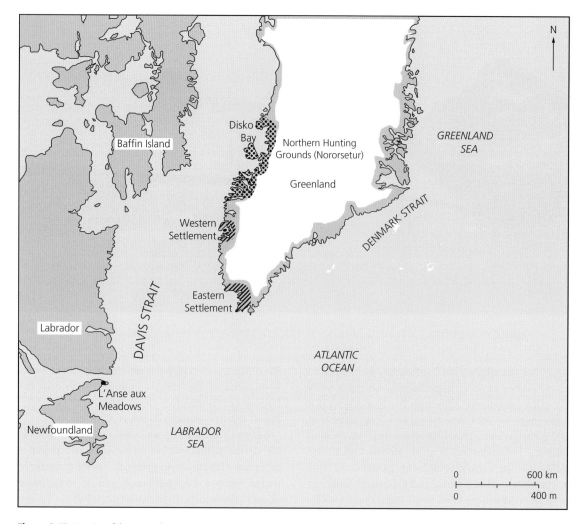

Figure 3.15 Location of the east and west Norse settlements in Greenland and the Newfoundland location at L'Anse aux Meadows. (Reproduced from Dugmore et al., 2005 from Arctic Anthropology with the permission of the Regents of the University of Wisconsin).

settled the south-west coast of Greenland, from Cape Farewell to the Western Settlement (Figure 3.15) further north near the present-day Nuuk.

To this western expansion to Greenland can be added a short-term Norse occupation in America with a settlement in northern Newfoundland shortly before the year AD 1000. The actual site of this settlement, which is mentioned in Vinland sagas (Wallace, 2003), was discovered at the northern tip of Newfoundland at L'Anse aux Meadows by Helge Ingstad and excavated by his wife Anne-Stine in 1960 (Figures 3.15 and 3.16). Subsequent excavation of the site has established a calibrated ^{14}C age range from AD 975–1020 which agrees well with the assumed historical age of ca. AD 1000 (Nydal, 1989). The buildings were permanent structures not seasonal booths. Nevertheless, the occupation of the site was short, probably not more than 10 years. It does not appear to have been an attempt at colonization but served instead as a base for exploration, housing between 60 and 90 persons, comprising mainly men. Botanical remains of Butternuts (*Juglans cinerea*), a North American Walnut, and Riverside Grapes (*Vitis riparia*) indicate that from this base they probably travelled as far south as New Brunswick, which in a straight line is 2,500 km south of their home base in Greenland at Brattahilð (Wallace, 2011).

Figure 3.16 Reconstruction of the Norse longhouse at L'Anse aux Meadows (French—L'Anse aux Meduses—Cove of the Jelly Fish, ~52°N) at the northernmost tip of the island of Newfoundland. The site was originally occupied by Norse settlers from Greenland ca. AD 1000 and discovered and excavated by Helge Ingstad and his wife Anne-Stine in 1960 and subsequently in the 1970s by Parks Canada. It is the only site of a Norse settlement in North America outside Greenland and remains the only widely accepted site of pre-Columban trans-oceanic human contact. It is notable not as an attempt at colonization of North America but as a base for exploration (see text).

3.6.1 The Norse disappearance from Greenland

The 15th century witnessed the disappearance of the Norse. The last document to survive is dated to 1409 with the issuing of a marriage certificate for a wedding that took place in 1408 and witnessed by visiting Icelanders. There is evidence that the couple who married in 1408 emigrated to Iceland in 1410 (Jasinski and Søreide, 2003).

By the 16th century the Norse Greenlanders had vanished in circumstances that are still unresolved. Their disappearance is thought to have involved an interplay between a range of factors, including climatic deterioration and environmental degradation, together with economic marginalization, and conflict with the Inuit (Dugmore et al., 2007). Climatically, the Greenland GISP2 ice core data together with historical documentary information clearly demonstrate that a climate change took place between AD 1270 and 1450 across the North Atlantic region.

It appears that these climate changes took place at a time when the North Atlantic Oscillation (NAO) may have been weak. The results also show that marked changes in atmospheric circulation were coincident with a significant increase in North Atlantic storminess at ca. AD 1400–1420 (Taberlet et al., 2007).

However, to attribute the end of the Norse settlement in Greenland merely to climatic factors is a simplification that ignores a number of other important historical factors. It is also highly probable that periods of unfavourable climatic fluctuations exposed the cultural vulnerabilities of the Norse farming settlers to environmental change to a greater extent than their maritime-based Inuit neighbours (Barlow et al., 1997). Towards the end of the Norse settlement in Greenland there is evidence for an increased emphasis on a marine diet. Samples of bone collagen from the skeletons of 27 Greenland Norse people excavated from churchyard burials from the late 10th to the middle 15th century show from the stable carbon isotopic composition ($\partial^{13}C$) that the diet of the Greenland Norse changed dramatically from predominantly terrestrial food at the time of Eric the Red, around AD 1000, to predominantly marine food toward the end of the settlement period around AD 1450. This evidence from measurements on the remains of the people themselves is more revealing than merely investigating the remains in kitchen-middens (Arneborg et al., 1999).

There has also been recorded a general reduction in coprophilous fungal spores in the period of later Norse settlement through the mid-13th to 14th centuries, which suggests a decrease in domestic livestock. Coprophilous fungal spores typically found on animal dung were abundant during the early phase of settlement but subsequently declined in abundance. This may indicate a likely decrease in grazing intensity or livestock numbers over time, possibly in response to climatic deterioration and/or soil erosion which is expected to have placed increased stress on the pastoral farming system (Schofield and Edwards, 2011).

Much of the initial impetus to colonize Greenland would have come from its importance as a rich source of furs and walrus ivory. After the mid-13th century Hanseatic merchants were trading ivory from Africa, which in European markets would have provided unbeatable competition to walrus ivory. In addition, hemp ropes were now replacing walrus hide for cables. This, combined with increasing operational difficulties in Greenland due to stormier weather and more sea ice, would have reduced the motivation for European merchants to trade with Greenland. The arrival of the Black Death in 1347–1351 together with subsequent plagues when Norway lost 30–50% of its population would have further diminished economic trade for Greenland (Dugmore et al., 2007). The possible maximum size of the Norse population of Greenland, based on the number and size of the farms, has been estimated variously between 3,000 and 6,000. Before ca. 1350 the Western Settlement was abandoned. By 1400 walrus ivory had become unfashionable in Europe, and Greenland's Norse settlers vanished soon thereafter (Roesdal, 2005).

3.7 Northern peoples in recent times

3.7.1 The fur traders

Contact between the indigenous peoples of the Arctic and the European colonizers of adjacent lands has brought both benefits and disasters to the peoples of the North. The Russian colonization of northern Siberia in the 15th century was driven by a demand for the much prized furs of arctic animals and in particular sable (*Martes zibeline*). Similarly, the later European penetration of the Northern Territories of Canada was developed by fur traders such as the French *Voyageurs* and later the British *Hudson Bay Company*.

The Hudson Bay Company, or, as it was simply referred to, *The Company* was founded in 1670 under a charter granted by Charles II. Its formal title was *The Company of Gentleman Adventurers*. 'Adventurers' applies here to the risking of capital and not the lives of their North America agents. In common with other trading adventures this brought many advantages to the northern peoples, not least the opportunity to acquire firearms and other metal goods. However, the spread of European diseases, in particular smallpox and measles, devastated many communities and in some cases caused complete extinctions.

The Hudson Bay Company did what it could to support the native communities through times of famine either caused by local failure of hunting or from a shortage of prey, or due to the disasters that overtook small isolated communities when accidents or disease deprived them of their skilled hunters. Such situations were not infrequent.

In popular imagination the native inhabitants of the High Arctic are perceived as never hostile and not particularly warlike, a view that may have been inherited from the classical imagination of the happy hyperboreans who were imagined as living beyond the source of the north winds (see Chapter 1). Such a view ignores a very real part of life among northern peoples where violence and warfare were common.

An earlier account of an employee of the Hudson Bay Company, Samuel Hearne, has already been referred to (in section 3.1.2). In his journeys between 1769 and 1772 he gives a graphic account of the hostility and brutality of the Northern Indians to the Eskimos if they should come across them on their hunting journeys (Hearne, 1769). This hostility would have been reciprocated given the appropriate circumstances.

An account of the potential for violent encounters with the Eskimos of northern Alaska when they first met with American fur traders in the early 19th century describes the great care that these early traders had to take when approaching groups of Eskimos (Bockstoce, 2009). They were reputed to rob and murder strangers without hesitation especially if they found themselves in the stronger position. The Chukchi to the east of the Bering Strait and the Eskimos on the west lived in a state of eternal enmity with each other, except for a temporary truce period when they gathered for an annual summer fur fair, where the Eskimos traded with the Chuchki who came with their trade goods from Russia and China. The arrival of American ships in the early 1800s was therefore met with much suspicion by the native peoples, as they correctly surmised that they might interfere with their traditional trading habits with north-eastern Asia.

The contact of the Siberian peoples with their Russian colonizers is somewhat different from that of the North American Eskimos. The Russian conquest of Siberia and the demand for furs created a fur fever. Wherever the Russians extended their fur trading they demanded tribute (*Yasak*) in furs, which was a disaster both for the animals and for many of the native arctic peoples. The local people were forced to hand over regularly a set quantity of sables. Hostages were also taken until the pelts were delivered. So great was the Russian demand for furs that they extorted the native peoples mercilessly, especially when tribute was being collected by independent agents working for the local governor. When furs became scarce, money was required and this could force the sale of reindeer herds, depriving the local people of their means of livelihood and independence. The virgin forests of Central Siberia and Yakutia which had been a rich source of furs in the 1630s were largely exhausted within 50 years of the Russian conquest (Forsyth, 1992).

3.7.2 The 20th and 21st centuries

The Russian Revolution in 1917 changed the social order in Siberia. Reindeer brigades were organized and, as in North America, a more European and sedentary lifestyle imposed on the native peoples. Among the Eurasian peoples, however, there is still a section of the population not engaged in the more modern activities of mining and oil extraction who remain dependent on reindeer herding. Reindeer are the only semi-domesticated animals capable of nourishing themselves and putting on body weight on a diet of lichens. Such, however, is the slow growth rate of lichens that migration over considerable distances is a necessary feature of reindeer husbandry. This alone maintains the need for human summer migrations and therefore preserves some vestiges of a way of life that has existed in the Arctic ever since the first hunter-gatherers arrived there in the late Pleistocene. The extent of climatic warming in northern regions together with the disturbance of the Tundra by industrial activity is likely to prove disadvantageous to the health and well-being of the reindeer herds.

Human presence in the Arctic will undoubtedly continue, but it will be the mineral, gas, and oil resources that will dominate the economy. These activities may put the Tundra and adjacent forests at risk from the disturbance that is caused by the movement of equipment and installation of roads and pipelines, as well as the hazards of pollution (see Chapter 10).

CHAPTER 4

Tundra diversity

4.1 Defining the Tundra

The term *Tundra* is currently used to describe arctic lands devoid of forest. Such a blanket description, however, does not do justice to the diversity of plant and animal life that manages to survive in what might be mistaken for a treeless waste. The origin of the word *Tundra* is usually described as absorbed into Russian from the Finnish or Saami word *Tunturi*, meaning a treeless hill. This somewhat philologically vague and superficial explanation does not do justice to the ancestry of the term. Its origin most probably lies in eastern Asia in ancient Turkish and is found today in modern Turkish as *dondurmak*—to freeze and *dondurma*—ice-cream (Stone, 2011). It is also sometimes used in Turkish to describe concrete. As a term for describing frozen ground its ancient Turkish use for the concept of freezing does justice to the permafrost that lies under so much of the Tundra.

As a modern ecological term, Tundra is used to describe the dwarf shrub, herb, and moss vegetation that exists in polar regions that are too cold to support the growth of trees. The first attempt by a scientist to define Tundra was made by Mikhail Vasilyevich Lomonosov (1711–65), the 18th-century Russian polymath and co-founder of Moscow University (Figure 4.1). From an early age, Lomonosov would have been acquainted with the Tundra. His father had lands on the Kola Peninsula and fishing vessels on the Murmansk Bank and he would have seen the vegetation along the White Sea Coast (Figure 4.2). In his work *The Layers of the Earth* (1763), Lomonosov defines Tundra as '*places overgrown with moss and without swamp or forest and covering a large part of the coast of the Northern (Arctic) Ocean*' (Lomonosov, 1763). The presence of moss is still recognized as typical of Tundra vegetation.

Defining Tundra becomes slightly more tedious and pedantic every time it is discussed, possibly due to an over-enthusiastic desire for recognizing minor variations. The result has been a proliferation in the names given to different types of Tundra vegetation. In addition there is confusion as to whether the term Tundra describes a region and its landscape, or whether it refers merely to the vegetation. The Lomonosov

Figure 4.1 Mikhail Vasilyevich Lomonosov—the first scientist to define Tundra, son of a White Sea fisherman, and co-founder of Moscow University. (Reproduced from the Complete Works of Lomonosov, Moscow, 1954).

Tundra-Taiga Biology. First Edition. R.M.M. Crawford © R.M.M. Crawford 2013.
© Published 2013 by Oxford University Press.

Figure 4.2 By the shores of the White Sea at the Tundra-Taiga Interface, a landscape that would have been familiar to Lomonosov when he was a boy.

definition of Tundra based on vegetation, as he originally observed it on the Arctic coast of the White Sea, is used here as this gives the simplest description of the region and its ecological variations.

4.2 Tundra classification

It is not surprising that interpretations differ geographically in relation to describing different kinds of Tundra. Recognition of various arctic and subarctic vegetation zones by ecologists differs between North America and Eurasia due to variations in climate, snow cover, and geomorphology. In the far north of America, the Canadian Arctic archipelago lies in a physically distinct climatic and biotic zone from the mainland Tundra to the south (Bliss, 1988). By contrast the Russian Arctic encompasses a large expanse of continental land together with isolated groups of islands in the Arctic Ocean. Consequently, there is in Eurasia a more progressive transition from subarctic to high arctic conditions in the extensive continental landmass of northern Russia and Siberia. A comparison of the differences between Tundra classification as commonly recognized by North American as compared with Eurasian ecologists is shown in Figure 4.3.

In biology, variation is inherent both in the nature of organisms and the habitats in which they reside. Consequently, attempts to describe the Tundra on a circumpolar basis are merely human concepts based on subjective judgements and influenced to some extent by local opinion. The classification presented in this chapter (Figure 4.4) is based on an international attempt to present an agreed circumpolar vision of the Tundra (Walker et al., 2005).

This map distinguishes five major classes of Tundra:

(1) Polar deserts (or Arctic Barrens)
(2) Prostrate and tall shrub Tundra
(3) Erect shrub Tundras
(4) Graminoid Tundras
(5) Wetlands.

The map (Figure 4.4) further sub-divides these five major types into numerous sub-types, which are described briefly below together with other commonly used terms for polar plant communities.

4.2.1 Polar deserts

These High Arctic regions lack the continuous moss cover that generally typifies the Tundra, as defined

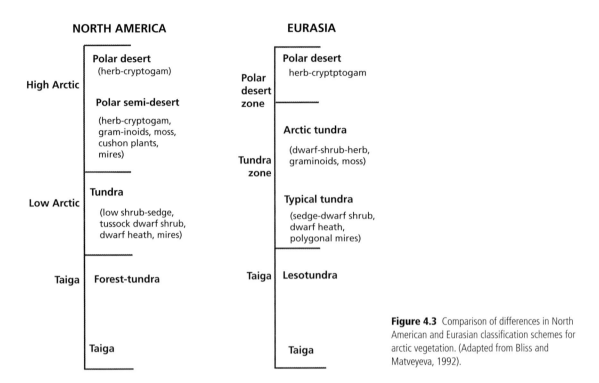

Figure 4.3 Comparison of differences in North American and Eurasian classification schemes for arctic vegetation. (Adapted from Bliss and Matveyeva, 1992).

by Lomonosov; they can nevertheless be considered as the ultimate northern type of Tundra vegetation due to their treeless nature (Figures 4.5 and 4.6).

The prevalence of high atmospheric pressure systems over the North Pole results in polar deserts being areas with annual precipitation frequently less that than 250 mm and mean temperatures during the warmest month of less than 10°C. Polar deserts, although forming what appears on a map as merely the northern fringe of the Tundra, nevertheless constitute a considerable ecological entity as they cover nearly 5 million km^2 including the Arctic islands of Eurasia and most of the northerly Canadian islands, with the exception of the Siberian Wrangel Island. In central Siberia the Taymyr Peninsula reaches sufficiently far north (78°) to be the only place in the world where true polar desert is found on a continental mass.

The typical appearance of the polar deserts is one of gravel plains or exposed bedrock (Figures 4.5 and 4.6). Snow cover is limited to areas where the precipitation is locally more abundant, or where ground is low-lying or sheltered from wind. The open nature of the landscape has led to the frequent postulation that during the Pleistocene glaciation polar deserts may have provided refugia for a number of plant species as the sea level was lower and polar high atmospheric pressures would have severely limited precipitation (see section 2.2).

The polar deserts are also sometimes referred to as *Arctic Barrens* as typically they have a total plant cover of less than 10% and frequently less than 5% (Longton, 1988a). Depending on the relative bareness of the terrain they can sometimes be described as semi-deserts. As a Tundra type they extend as far north as land is available. The slow rate of *pedogenesis* (soil weathering) and constant *cryoturbation* (soil churning) generally inhibit the development of identifiable soil horizons in polar desert soils. The arid soil surface results in the almost total absence of a surface humus layer. In a typical polar desert the summer depth of thawing limits the soil *active layer* to between 20 and 70 mm.

Sand dunes are infrequent and the gravel and clay soils support only a few species of cushion and rosette-forming flowering plants. The moss

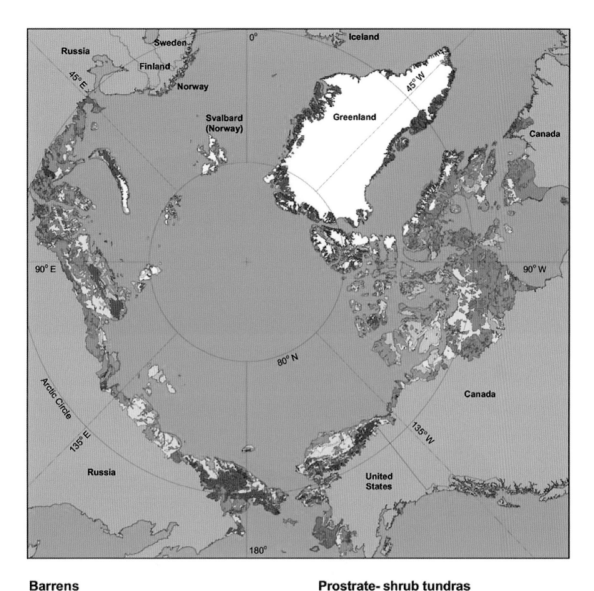

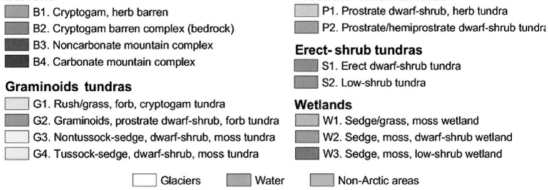

Figure 4.4 Circumpolar vegetation map illustrating the diversity of types of Tundra that can be recognized. When this is attempted on a circumpolar scale regional differences particularly in sub-types can create confusion (see Figure 4.3). (Reproduced with permission from Walker *et al*., 2005).

Figure 4.5 Polar desert as seen at 78°10'N at Fuglafjellen, Nordensköld Land, Spitsbergen.

Figure 4.6 Polar desert as seen on the south side of Kongsfjord, Spitsbergen (79°N).

species that are found here grow as either short tufts or mats. The more common genera include *Ditrichum*, *Hypnum*, and *Racomitrium* (Gudmundson, 2007).

4.2.2 Prostrate shrub Tundra

This is one of the most typical forms of Tundra and is characterized by the presence of dwarf woody plants such as the Arctic Bearberry, the Trailing Azalea (*Kalmia procumbens*), and various species of Dwarf Willow (Figure 4.7). In many areas Purple Saxifrage (*Saxifraga oppositifolia*) can also be widespread. The canopy created by the dwarf woody species can alter the temperature, light, and evapotranspiration regimes in summer, and reduce the degree of exposure in winter. Shrub Tundra can be divided into two categories, namely *Dwarf Tundra* and *Tall Shrub Tundra*. *Dwarf Tundra* contains various species of stunted or creeping shrubs where the overwintering buds are placed close to the ground or in the surface soil, which is usually sandy and poor in

Figure 4.7 Plant species commonly found in polar semi-deserts and prostrate shrub Tundra. (a) Arctic Bearberry (*Arctostaphylos alpinus*) in autumn colours at Mesters Vig (73°N), north-east Greenland; (b) Trailing Azalea (*Kalmia procumbens*), a common arctic alpine species; (c) Crowberry (*Empetrum nigrum*) in Northern Iceland; (d) Purple Saxifrage (*Saxifraga oppositifolia*) at Ny Ålesund, Spitsbergen.

nutrients and, unlike the polar deserts, frequently has a layer of raw humus. Common species include various Dwarf Willows such as *Salix polaris*, *S. reticulata*, and *S. arctica*.

Dwarf-Heath Tundra is a variant that contains ericaceous species including most commonly Arctic Heather (*Cassiope tetragona*), Creeping Azalea (*Kalmia procumbens*), and Arctic Bearberry (*Arctostaphylos alpinus*) (Figure 4.7).

4.2.3 Erect/tall shrub Tundra

Tall shrub Tundra where the height of the woody plants can be greater than a metre does not correspond with what Lomonosov originally described as Tundra. However, woody vegetation does not cease to exist at the treeline and there are favoured spots within the Tundra landscape where if there is some degree of shelter and sufficient

Figure 4.8 An example of substantial growth of woody tall shrub tundra vegetation north of the treeline. View of willow thicket along a lake shore on the Varandei Tundra in the Nenets Autonomous Okrug with shrubs up to 2.5 m in height. *Inset*—transverse section of willow stem. (Photos courtesy of Professor B.C. Forbes).

moisture in the soil, tall shrub vegetation can develop. Such locations are more commonly found in the southern parts of the Tundra, as seen on the Varandei Tundra at 68°N on the Siberian coast just to the west of the Yamal Peninsula (Figure 4.8).

At their northern limits erect shrub plant communities are also found as scrublands nestling below riverbanks where the erosion of the riverbed has created a sheltered habitat. As all the surrounding land is Tundra it can be considered legitimate to refer to the tall shrub vegetation as a component of the Tundra landscape. Typical genera are *Alnus* and *Salix*. In Alaska and north-western Canada tall shrub vegetation is a favoured habitat for Moose (*Alces alces*), permitting them to extend their range northwards. Arctic Hares and Ptarmigan also feed on willow both in summer and winter. The southern tip of Greenland lies just within the treeline where the Norse had their Eastern Settlement (see Figure 3.15). Here tall shrub vegetation is common and Common Birch (*Betula pubescens*) can be found sometimes as high as 3–4 m, with an understory of the grasses *Anthoxanthum odoratum*, *Deschampsia flexuosa*, *Nardus stricta*, and *Poa nemoralis*.

A common feature of woody species such as birch (*Betula* spp.) at the northern limits of their distribution is the abandonment of the single pole-form in favour of the many-stemmed (*polycormic*) bush habit. The latter form has a number of advantages in habitats that are marginal for tree growth. The bush form facilitates regeneration after damage, from climatic adversities such as drought, frost, and heavy snow, as well as from disturbance by grazing or fire. The polycormic form also increases the amount of foliage held near the ground where temperatures are warmer. In species with corticular photosynthesis as in birch, there is a greater area of green tissues under the bark which increases winter photosynthetic activity, when it is facilitated by the absence of shading by leaves (see section 7.6.4—*Stem photosynthesis*). These properties serve to make woody shrubs long-lived and major components of a wide variety of plant communities, as can be seen throughout most of the southern Tundra (see also Chapter 7).

However, this type of vegetation, even though it has been described as tall shrub Tundra, does not have the moss cover that is generally associated with Tundra (Bliss, 1997).

4.2.4 Graminoid Tundras

This group includes the many types of Tundra where sedges are predominant. Frequently they are described as various types of *Sedge Tundra*. In the higher regions of the Arctic, grass species become fewer and are gradually replaced by sedges. The habit of grasses of producing their leaves sequentially is a protection against grazing, but tends to become disadvantageous as the growing season becomes shorter as the brevity of the growing season does not favour this delay in maximizing the production of photosynthetic tissues. Nevertheless, there are some grass species that thrive in the High Arctic Tundra and are found typically in wetland meadows or on lowland coastal plains. Examples of such meadows can be found around the Hudson Bay, the Mackenzie Delta region, and the fjords of Greenland and Spitsbergen. The bulk of the grazing therefore comes from sedges such as *Carex aquatalis*, *C. marina*, *C. membranacea*, *C. magellanica*, *C. rostrata*, *C. rotundata*, *Eriophorum angustifolium*, and *E. scheuchzeri*. Hardy grasses include *Arctagrostis latifolia*, *Arctophila fulva*, and *Dupontia fischeri* which flourish better in the drier regions. There is also continuous ground cover of moss, as befits Lomonosov's view of the Tundra. These Tundra meadows provide grazing for reindeer and muskoxen, with the latter preferring drier ground with dwarf shrubs and especially Arctic Willow (*Salix arctica*).

Geese prefer to graze the wetlands for sedges and especially search out the sedge species with the highest sugar content. Salt pastures enjoying the 24-hour daylight regime provide excellent nourishment for their young goslings. The goose droppings in turn accelerate nutrient recycling which can be a limiting factor in cold arctic soils, which is an example of where both grazer and grazed benefit from mutual association. However, recent increases in goose populations as a result of improved southern winter pasturage and climatic warming are in danger of damaging arctic meadows and salt marshes (Jefferies, 1997).

The availability of Sedge Tundra was incidental in saving the lives of the very first people to survive a winter by accident in Spitsbergen. In 1630 a group of eight British whalers became lost from their mother ship in a storm and had to spend nine months in Bellsund until a British ship rescued them the following year. Food resources were scarce. Fortunately, they were familiar enough with the geography of Spitsbergen to know that on the north side of the peninsula on which they were stranded there was a more fertile (graminoid Tundra) region that was frequented by reindeer. These early sailors also found that this wide valley was largely snow-free in winter (Conway, 1906). To this day it is still one of the best locations for thermophilous vegetation in Spitsbergen (Elvebakk, 2005). They made expeditions across the peninsula (Nordenskold) to this valley which they named Coles Parke, possibly after the name of the whaler who knew about this favoured location (Figure 4.9). In the diary (1630–31) of one of the whalers, a Londoner called Edward Pelham, it was recorded that Coles Parke was 'a fine place for venison'. The valley is now known as Colesdalen (no coal seams reach the surface in this valley). As with most Sedge Tundra, this region is still attractive to reindeer.

With increasing or fluctuating water table levels Sedge Tundra can become dominated by extensive areas of tussock sedges, creating what is aptly called *Tussock Tundra*. This type of Tundra can be so

Figure 4.9 Sedge Tundra at Colesdalen (78°15′N), Spitsbergen, showing a typical Tundra pool bordered by a colony of Arctic Cotton Grass (*Eriophorum scheuchzeri*). As with most Sedge Tundra the vegetation provides good grazing—a fact that was already recorded by Gunner's-mate Edward Pelham (1630–1) as being 'a fine place for venison'. Pelham was the leader of a group of eight British whalers who were probably the first people ever to successfully overwinter (involuntarily) in Spitsbergen, having become accidentally parted from their mother ship (see text).

extensive that it can create an entire landscape based on herbaceous vegetation dominated by tussock-forming sedges. The most characteristic plants are the sedges confusingly referred to as Cotton Grasses in which the Hare's Tail Cotton Grass (*Eriophorum vaginatum*) is one of the most frequently occurring species on moist soils with a risk of flooding (Figure 4.10).

Tussock-forming appears to be a common strategy for plants that live in areas with fluctuating water tables. The tussock has the advantage that it raises the lower leaves to a level where inundation is either less frequent or less prolonged, thus giving the lower leaves greater access to air. This facilitates the diffusion of oxygen from the lower leaves to the submerged organs, which increases flooding tolerance and also aids early recovery from winter flooding. In the snow-covered Tundra, tussocks also have the advantage that the upper part of the tussock emerges first from the snow. This early resumption in growth provides a timely spring bite for grazers such as caribou as well as nesting sites for birds. It should be noted, however, that tussocks or hummocks that are not formed by Cotton Grasses (*Eriophorum* spp.) can be found at much higher latitudes created by physical and biological

Figure 4.10 The tussock-forming Hare's Tail Cotton Grass (*Eriophorum vaginatum*).

processes peculiar to the Arctic (see section 4.5.2 and Figures 4.11 and 4.12).

Figure 4.11 (a) Pingos near Tuktoyaktuk, Northwest Territories, Canada. (Photo Emma Pike, Wikipedia Commons, 2007). (b) View of valley bottom wet meadow formed from earth mounds near Petuniabukta, Spitsbergen (78°44′N), 15 m a.s.l. (Photo courtesy of Dr Tomáš Hájek).

The tussock-forming Cotton Grass species *Eriophorum vaginatum* (Figure 4.10) has a notable northern limit which in North America is sometimes referred to as the *Tussock-Tundra line*. North of the tussock line, dwarf heath species provide the mainstay of the Tundra vegetation. The disappearance of Cotton Grass at increasing latitudes is also followed by the loss of Dwarf Birch (*Betula nana*) and the Labrador Tea Plant species *Rhododendron* (formerly *Ledum*) *groenlandicum* and *R. palustre* (Figure 4.13). The northern limit of the Tussock Tundra—the *Tussock-Tundra Line*—marking as it does the northern limit of shrub Tundra as well as the tussock sedge community, is taken generally as the northern limit of the Low Arctic (Figures 4.3 and 4.4).

4.2.5 Azonal types of Tundra

The primary physical conditions that distinguish azonal vegetation from zonal vegetation types are localized topographic or soil characteristics of either a positive or a negative nature, such as unfavourable soils and poor or excessive drainage. Alternatively, localized regions with more fertile soils can also support azonal plant species associations. Such conditions can create a mosaic of intermediate plant communities determined largely by local intervening regions particularly between typical Low Arctic shrub–tussock Tundra and the polar desert of the High Arctic. Peat swamps which are rich in organic matter occur frequently in Tundra, at the Tundra-Taiga Interface, and in Forest-Steppe zones, and as such can create localized favourable azonal habitats, which are referred to in Russia as *plakor* of *hypoarctic* habitats (Alexandrovna, 1980). Examples are also found in alluvial soils occurring on floodplains under meadow and wetland vegetation. The soil profiles of these azonal regions vary greatly due to layering of alluvial sediments.

Russian ecologists who have studied large expanses of Tundra on level land have stressed the importance of local azonal conditions. These can also be created by the presence or absence of water and its effect in permafrost regions on soil drainage. On these grounds they recognize many variants in Tundra plant communities depending on local soil characteristics. In this connection they make use of the so-called *plakor* or *hypoarctic* concept. In Russia these terms are most frequently employed in an attempt to define non-regional treeless zones with meadow-like vegetation. (Chernov, 1988).

4.3 Arctic bryophytes

The Arctic flora is particularly rich in moss species. The moss flora of Alaska probably extends to 475–500 species. In the cool sites of the northern Ellesmere Islands alone there are at least 166 species. A total arctic bryophyte flora is thought to number 600–700 species compared with an angiosperm flora of approximately 900 species (Longton, 1988b).

The species-richness of the arctic moss flora may be due in part to the high probability that a sizeable proportion of the present flora survived the

Figure 4.12 View of isolated earth mounds at Petuniabukta with sectioned examples revealing typical structural variations. Note differences between a peat-filled hummock (*top right*) and a typical *thufúr* (*bottom right*) showing the effects of cryoturbation. (Photos courtesy of Dr Tomáš Hájek).

Pleistocene glaciations in polar desert refugia lying either to the north or within the area of the ice sheets (Longton, 1988a). The greatest danger to plants from cold is not low temperatures but desiccation (see Chapter 7), and as many mosses are highly tolerant of desiccation this may have favoured their survival in dry, cold polar deserts with minimal ice cover.

Not only are mosses capable of enduring drought, but also they have species that are able to survive prolonged inundation. The Giant Spear-Moss or Arctic Moss *Calliergon giganteum* is an example of a moss that inhabits fenlands and lakes in the High Arctic. Despite the darkness of the arctic winter its leaves can remain green and alive under water for at least 4 years, and shoots are capable of surviving for 7–10 years. Apparently, cold temperatures and low nutrient supply in combination with the short Arctic growing season can account for the low growth rates of this moss which has an unprecedented longevity relative to other submerged macrophyte species (Sand-Jensen *et al.*, 1999).

4.3.1 Mosses as facilitators

Lomonosov's perception of the ubiquity of bryophytes in the Tundra can be a reflection of the positive effect that mosses can exert for the survival of flowering plants. For some angiosperm species mosses can have a role as facilitators, particularly in stressed habitats by ameliorating climatic extremes in a region where other plant cover can often be minimal. Moss cover not only reduces soil temperature fluctuations but also conserves soil moisture, thus acting as a nurse bed for seedlings of flowering plants. Mosses are also of value to the many high-latitude herbivores, ranging from microscopic soil mites to a wide range of arctic grazers from lemmings to reindeer.

Figure 4.13 Inflorescence and distribution of the Marsh Labrador Tea Plant (*Rhododendron palustre* formerly *Ledum groenlandicum*), a circumpolar woody plant of the Low Arctic with numerous sub-species. The disappearance of this species from woody plant communities is an indicator of the boundary between the High and the Low Arctic. (Map adapted from Hultén and Fries, 1986).

Figure 4.14 Moss-dominated terrain below bird cliffs at Steinflåstupet, Brøgger Peninsula (79°N), Spitsbergen. The luxuriant and dominant moss cover has markedly reduced the presence of flowering plants, especially the graminoid species.

Mosses tend to sequester mineral nutrients. This can provide an important source of mineral supplements to the diet of many herbivores. In particular they are beneficial for lactating mammals such as lemmings which require a ready supply of available calcium to rear their large broods. This sequestration of mineral resources from the immediate environment can, however, be inhibitory for more rapidly growing plants such as grasses and sedges (Figure 4.14).

Moss depth manipulation experiments carried out in Spitsbergen with the graminoids *Alopecurus borealis* and *Luzula confusa* have demonstrated both positive and negative responses of flowering plants to the presence of mosses (Gornall *et al.*, 2011). There is generally a negative effect from deep moss with the graminoids but a positive effect with the shrub *Salix polaris* and the herb *Bistorta viviparum*, both of which tend to benefit from a shallow moss layer.

The moss layer and the upper soil layer together provide a favourable environment for arctic flowering plants. As moss keeps the soil moist, this favours the growth of unicellular algae and cyanobacteria.

4.4 Sequestration and refixation of soil carbon

In the High Arctic the primary producers of soil organic matter, particularly in soils where there is extensive bare ground, are not the higher plants, bryophytes, or lichens. It is the cyanobacteria, which initially fix nitrogen and, together with the soil surface algae, sequester both carbon and nitrogen in the soil. The carbon deposits in the soil subsequently contribute to soil respiration and provide in early spring a richer source of carbon dioxide for photosynthesis than that which is found in the atmosphere.

In the short growing season of the High Arctic the flowering plants are metabolically active for a relatively short time and achieve the bulk of their growth and development within a few weeks of snow melt. This episode of early growth coincides with the period when soil respiration releases carbon dioxide, which the dwarf herb and shrub vegetation with their prostrate and cushion growth forms are well adapted to capture.

Measurements made in Spitsbergen show that the carbon dioxide flux from the soil can provide a net carbon dioxide source of as much as $0.3 g\ C\ m^{-2}\ d^{-1}$ in early spring, changing to a mid-summer sink of $-0.39g\ C\ m^{-2}\ d^{-1}$, returning to $0.1g\ C\ m^{-2}\ d^{-1}$ in autumn (Lloyd, 2001). The brevity of the spring and autumn periods as compared with the summer creates a positive gain for the current year's vegetation.

An estimation of the use made by these plants of soil carbon has been made from examination of the stable isotope ratio of the carbon sequestered in the flowering plants. The relative abundance of ^{13}C vs ^{12}C (expressed as $\partial^{13}C‰$) is frequently used as a measure of relative water efficiency over extended periods. It can also be used to distinguish the source of carbon dioxide. Carbon dioxide derived from soil respiration normally has a $\partial^{13}C$ value of around $-19‰$. In a sample of plants of *Saxifraga oppositifolia* growing in Spitsbergen, a mean $\partial^{13}C$ value of $-19.51‰ \pm 0.32$ ($n = 8$) suggests that carbon dioxide derived from soil respiration supplied a considerable portion of their photosynthetic needs (Crawford *et al.*, 1995).

4.5 Disturbance and diversity

Geomorphologically, the Arctic, including the Tundra, is easily scarred, or so it can seem to human eyes. Erosion and cryoturbation are constant features in a region where fragile soils have little protective vegetation cover. Although there is little doubt that the Tundra is physically fragile it does not necessarily follow that it is biologically frail. The flora and fauna of the polar regions have always had to endure such conditions and in many ways the physical instability of arctic landscapes can be ecologically advantageous. The high proportion of clonal plant species, coupled with the individual longevity of individual plants, provides a system that actually benefits from disturbance. Soil movements break up clones and redistribute rhizomes and stolons. This enforced physical dispersion is not necessarily disadvantageous as it promotes dispersal and creates new opportunities for colonization.

4.5.1 Cryoturbation and patterned ground

Cryoturbation occurs to varying degrees in most permafrost soils (gelisols). Repeated freezing and

Figure 4.15 Stone stripes and polygons on gently sloping land at Blomsterstrand Peninsula, Kongsfjorden, Spitsbergen (78°50′N).

thawing of the soil in autumn causes the formation of ice wedges in the more friable parts of the parent rock. The process can continue during the summer when an *active layer* forms in the soil above the permafrost horizon. Eroded material can move upwards and downwards in the soil profile, both from the soil surface downward and from the permafrost table upward. The upper soil material gradually dries out as moisture moves from the warm surface layer to the colder layer at the top of the permafrost. Separation of coarse from fine soil materials produces distinctive patterned ground with different types of soil.

As a consequence of repeated freezing and thawing the groundwater forces larger stones towards the surface as smaller soils flow and settle underneath larger stones. Water-saturated areas with finer sediments have an ability to expand and contract as freezing and thawing occur, leading to lateral forces which ultimately pile larger stones, depending on topography, into clusters, stripes, and polygons.

The larger stones provide a favourable microhabitat for plant colonization as large stones lose heat more slowly than soil. Consequently, snowmelt takes place more readily from stony ground. In addition the centre of the polygon is usually slightly raised in relation to the edges and the shelter that is provided by the slightly sunken edges enhances plant colonization of this microsite. When the ground is sloping downhill, soil creep turns these polygons into stripes (Figures 4.15 and 4.16).

4.5.2 Pingos, earth mounds, and thufúr

Mounds in wetland periglacial environments vary from large permafrost landforms, such as the pingos which can be up to 70 m in height and 600 m in diameter (Figure 4.11a), to small earth mounds or hummocks. *Pingo* is an Inuit word now used generally to describe the large ice-cored mounds in regions of discontinuous permafrost (*hydrolaccoliths*). The word *Pingo* was first recorded

Figure 4.16 Alder shrubs (*Alnus viridis* ssp. *crispa*) colonizing the rims of polygons on the Colville River Delta. The colonizing seeds were dispersed by water movement along the floodplain corridor. It is suggested that this novel habitat may increase across the North Slope of Alaska. (Photo courtesy of M.T. Jorgenson).

in the diary of the Danish-Canadian botanist Alf Erling Porsild (1901–77) when crossing the Mackenzie Delta in April 1927. Porsild had been brought up in Greenland and was already familiar with these features which he described as *mud-volcanoes*. It was Porsild's 1927 diary that contains the first recorded written use of the Greenland Inuit word *Pingo*. Subsequently, Porsild introduced the word *Pingo* through his botanical writings into the English language. In the Eskimo dialect in the region of the Mackenzie Delta these conical hills are called *pi-nok-tja'lui* (Dathan, 2012).

Much smaller earth mounds are found in periglacial regions such as Iceland and Spitsbergen. Recent detailed studies in Spitsbergen at Petuniabukta (78°44′N) by Dr Tomáš Hájek have been able to distinguish two types of small earth mounds. An examination of the structure of the meadow mounds shown in Figure 4.11b demonstrated that the mounds consisted entirely of peat, with no evidence of having a cryogenic origin. By contrast, other isolated earth mounds at the same general location (Figure 4.12) when sectioned revealed they are not built exclusively from peat but have instead a clearly distinguishable mineral base consisting mostly of fine silt, which can be seen in the cross-sectioned hummocks in Figure 4.12. The elevated central parts of the mineral bases indicate that the vegetation was forced upwards by the action of freeze–thaw cycles in the underlying silt.

Such cryogenic mounds are commonly described as thufúr, a term derived from the Icelandic þufur cognate with English tuft. They clearly have a cryogenic origin. There is also the possibility that when the active layer is trapped at the start of winter between the frozen surface and the permafrost below, the resulting pressure distorts the trapped strata. Alternatively, involutions may result from the pressure exerted by expanding ice (Grab, 2005). Other theories have been presented as to the ontogeny of thúfur-dominated landscapes. One view is that the hollows are created by the sinking of the unvegetated soil surface as a result of successive freeze–thaw cycles (Gudmundson, 2007). With time the growth of these Arctic mounds, hummocks, or thufúr may merge into closed Tundra vegetation irrespective of their original origin. It would therefore appear that these small mounds that are such a feature of many arctic wetlands may have diverse polyphyletic origins.

4.6 Tundra animal diversity

Given that the Tundra possess only shallow soils and provides little above-ground cover there is a remarkable range in the physical size of the fauna. Among birds the range is from swans and geese to snow buntings and qheatears. In mammals the size span from Muskoxen to voles is even more remarkable. The frequent fluctuations in population numbers that is such a feature of many arctic mammals and birds might suggest that their survival is precarious. The fact that it has probably always been a feature of arctic demography suggests that it is merely the inevitable outcome of short food chains in an environment with limited resources. (For further discussion see section 8.4, *Predator–prey interactions*).

As animal populations fluctuate, so are plant populations subjected to periods of intense grazing followed by intervals of respite, when the animal populations eventually crash. The capacity of plant populations to recover, often with great rapidity, is another testimony to the resilience of most arctic species to periods of adversity. Periods of overgrazing can be followed by renewed vigorous growth as nutrients have been recycled during the peak grazing period, and the shade of old vegetation no longer prevents the summer sun from melting the frozen soil.

4.6.1 Lemming population cycles

For population biologists the lemming has been one of the most intriguing of arctic herbivores. It has been suggested that Arctic lemmings differ from other microtine rodents by having several features which increase their foraging efficiency under harsh conditions at the cost of reduced agility, and that these features were acquired rapidly at the dawn of the Pleistocene glaciations. It is speculated that when the Arctic Ocean froze, the primary productivity of northernmost Eurasia and North America was reduced, causing a shift from predation-controlled to food-limited dynamics in microtine rodents. This change in population dynamics triggered an extraordinarily rapid change in the characteristics of lemmings and precipitated an intense, sustained lemming–vegetation interaction, as old as the Tundra itself, which has probably played a major role in the evolution of arctic plants in areas where lemmings are numerous (Oksanen *et al.*, 2008).

The causes of the lemming's famous 4-year population cycle in which the population peaks can mark a 100-fold population increase are still the subject of scientific debate. Periodic population fluctuations are not unique to lemmings. Most small rodents of the northern hemisphere exhibit fluctuations in their numbers with similar periodicities. The timing of the fluctuations is not a global phenomenon. A population peak can be found in one area while remaining stable elsewhere (see Chapter 6 for discussion of lemming physiology).

The legend that lemmings commit suicide by mass migrations causing them to jump over cliffs has no substance. They are, however, aggressive animals and at times of peak populations disputes over territory can cause large-scale migrations. However, these are not motivated by any death wish but are merely a quest for fresh territory. The periodic 4-year cycle that has been observed in many localities for lemmings and voles (Figure 4.17) has over the years been variously attributed to a number of possible natural causes.

The various interpretations of the population cycles can be divided into two classes, namely nutrition and predation. Under the nutrition hypothesis, the expansion of the lemming populations to high levels is postulated as exhausting the available food supply. The ability of lemmings to bring about a rapid increase in their numbers is in considerable measure due to their ability to breed in winter in tunnels. With this protection from predation they can feed on accumulated dried vegetation and underground plant reserves in roots and rhizomes. This is a well-adapted strategy for the Arctic as it is the underground plant organs that typically store the bulk of over-wintering carbohydrate. When the winter store of dried vegetation is exhausted, the winter reproduction can be expected to decline and the lemming population numbers will fall (see also Chapter 6, *Lemming nutrition*).

The alternative interpretation of lemming and vole cycles in general is related to predator–prey relationships and may be the seemingly inevitable consequence of living in the world's simplest predator–prey community.

4.6.2 Collapsing vole population cycles

The cyclic population dynamics of voles has long been studied in Fennoscandinavia where regular periodicities have been particularly noted. Since the 1980s, population cycles in voles, grouse, and insects have been falling in Europe. Several lines of evidence now point to climate forcing as the general underlying cause. However, how climate interacts with demography to induce regime shifts in population dynamics is likely to differ between species and ecosystems (Ims *et al.*, 2008). Herbivores with high-amplitude population cycles such as voles, lemmings, and Snowshoe Hares form the heart of the arctic terrestrial food web dynamics. Thus, collapses of these cycles are also expected to imply collapses of important ecosystem functions, such as the pulsed flows of resources and disturbances. Deviations from the '*normal vole cycle*' have been reported in northern Finland (Figure 4.17) over vast areas of the boreal forest zone since the mid-1980s (Ims *et al.*, 2008).

Similar reductions in the size of the population fluctuations in lemmings have, however, prompted a different explanation from that advanced for voles which was based on nutrition. A study carried out over many years in north-east Greenland (Gilg *et al.*, 2009) has monitored the population of the Collared

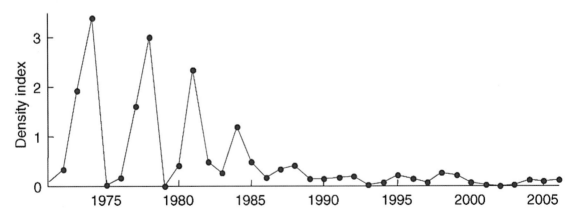

Figure 4.17 Cycle amplitude dampening towards a temporal collapse of the 4-year cycle in boreal Grey-Sided Vole (*Myodes rufocanus*) in northern Sweden. Time series of combined autumn catches from a region covering 10,000 km^2. (Reproduced with permission from Ims *et al.*, 2008).

Figure 4.18 The Collared Lemming (*Dicrstonyx groenlandicus*) feeding on Polar Willow (*Salix polaris*) in north-east Greenland. (Photo courtesy of Dr H. Körner).

Lemming (*Dicrstonyx groenlandicus*) (Figure 4.18). The Collared Lemming is the single main prey of four predators, the Snowy Owl, the Arctic Fox, the Long-Tailed Skua, and the Stoat. Using a 20-year-long time series of population densities for the five species and a dynamic model that has been previously parameterized for the climatic fluctuations taking place in north-east Greenland, it was shown that there was a complex of responses to climate changes that depended on species' life histories.

Climatic warming has been suggested as increasing the length of the lemming population cycle and at the same time decreasing maximum population densities. It has therefore been suggested that these recent diminutions of the magnitude of the population cycle are a consequence of climatic warming (Figure 4.17).

The argument made from modelling studies is that climate change appears to be detrimental to the populations of the predators, which are adapted to make use of the years of the greatest prey abundance. It is therefore possible that climate warming will indirectly reduce the predators' reproductive success and population densities, and may ultimately

Figure 4.19 Lemming predators: (a) long-tailed skua (© H. Körner); (b) snowy owl (© Rich Phalin/iStockphoto); (c) Ermine (© Vladimir Chernyanskiy/iStockphoto); (d) Arctic Fox (© Catharina van den Dikkenberg/iStockphoto).

lead to local extinction of some of the predator species such as long-tailed skua, snowy owls, ermine, and arctic foxes (Figure 4.19).

If this is the case, it might be that recent observations of the lack of expected cyclic lemming dynamics in eastern Greenland may be the first signs of a severe impact of climate change on the lemming–predator communities in Greenland and elsewhere in the High Arctic (Gilg et al., 2009).

These findings were based on populations in the southern part of the range of the Collared Lemming in Jameson Land (eastern Greenland) where lemming cycles were regularly reported up to the 1970s. Since then there have been no reports of recent lemming population surges. Further north at Zackenberg (north-east Greenland) there have been no lemming cycles since 2000. However, still further north at Hochstetter Forland lemming cycles still continue at regular 4-year intervals. It is possibly still too early to be able to unequivocally link the breakdown of population cycles with climatic warming.

4.6.3 Reindeer populations

The adaptations of migrating reindeer to high-latitude habitats are discussed in detail in Chapter 5 and their evolutionary history in Chapter 10. However, the non-migrating reindeer populations of the isolated High Arctic Islands deserve a special mention for their survival in the most northern area of Tundra where they are geographically isolated with no possibility of annual migrations south to more productive or sheltered habitats in winter. Their isolation has been long-standing and probably began about 10,000 years ago with the retreat of the Pleistocene ice.

The Spitsbergen (Svalbard) Reindeer (Figure 4.20) are considered to be a distinct subspecies (*Rangifer tarandus platyrhynchus*) and are the northernmost

Figure 4.20 Svalbard Reindeer (*Rangifer tarandus platyrhynchus*), a sub-species of small stature endemic to Spitsbergen where it has probably lived in isolation from other reindeer populations since the retreat of the ice in the early Holocene (Flagstad and Roed, 2003).

populations of the Eurasian Tundra reindeer (*Rangifer tarandus*). Distinct differences exist between Spitsbergen reindeer and the Novaya Zemlya form. The latter resembles more the Tundra forms of the Eurasian mainland. The Svalbard reindeer, because of their small structure, have been compared in terms of physical appearance with similar northern Greenland populations in Ellesmere and Axel Heiberg Islands, and especially the now extinct reindeer of east Greenland (Hakala *et al.*, 1985). The evolution of the arctic races of reindeer is discussed in Chapter 9 (see section 9.5.3).

4.6.4 Reindeer winter starvation

Reindeer in ice-prone areas such as Spitsbergen are becoming increasingly vulnerable to winter starvation from ice-encasement of the vegetation. Spitsbergen due to its position at the top-end of the North Atlantic Drift is experiencing increased exposure to the phenomenon of rain falling on frozen ground during intermittent warm weather periods in winter (Figures 4.21–3).

Such events encase the vegetation in ice which denies the reindeer access to the vegetation. Recent studies have shown (Figure 4.22) that in Spitsbergen between 2000 and 2010 heavy '*rain or snow*' events were occurring annually, with ground-ice covering from 25–96% of low-altitude habitat. The extent of ground-icing has had a strong negative effect on reindeer population growth rates. These results have important implications for the future viability of the Spitsbergen reindeer (Hansen *et al.*, 2011).

Despite these current risks, it has been noted in the past in Spitsbergen that when reindeer numbers have crashed due to ice-encasement (Figure 4.21), the collapse has been followed by a rapid recovery (Aanes *et al.*, 2000).

4.6.5 Muskoxen

Muskoxen (*Ovibos moschatus*) are remarkable for being the largest herbivores to survive on the Tundra and probably recall the Pleistocene megafauna more than any other extant mammal (Figure 4.24). They are noted for their thick coat and a very fine under-fur (*Qiviut*—an Inuit word for this underwool) which is much prized as possibly the finest wool in the world. It can be collected by gleaning from the branches of woody shrubs or anywhere else where the animals lie and roll. The males emit a strong odour, from which their name derives, which serves to attract females during the mating season.

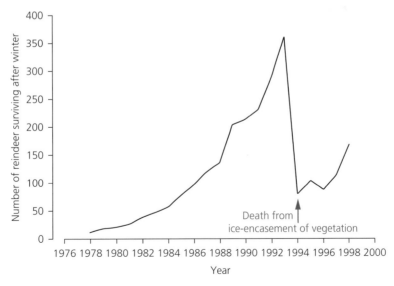

Figure 4.21 The population development of Svalbard reindeer on Brøgger Peninsula since their introduction to the peninsula in 1978. Numbers are from annual counting of reindeer during April. (Reproduced with permission from Aanes *et al.*, 2000).

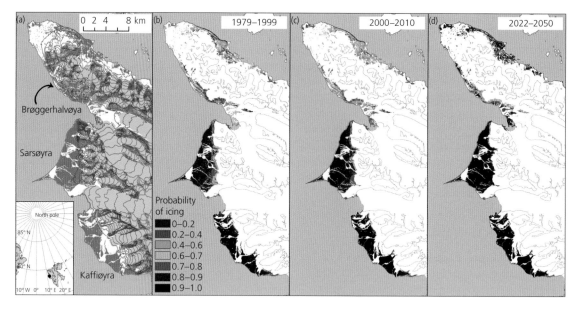

Figure 4.22 (a) The study area on the Brøgger Peninsula and Sarsøyra (78°50′N) on Svalbard, Spitsbergen (*inset*—the black circle shows the study area location). Vegetated habitat is shown in green and glaciers in blue. (b–d) Probabilities of ground-icing in vegetated habitats, 120 m a.s.l. in the average year during (b) the period prior to snowpack sampling, (c) the sampling period, and (d) the climate projection period. Note that vegetated lowlands on Sarsøyra and Kaffiøyra are overall flat, which suppresses surface runoff and generates high icing probabilities compared with the more rugged terrain on the Brøgger Peninsula. All three areas provide winter forage refugia that are aggregated in steep areas above 120 m and thus not covered by the sampling/model. (Reproduced with permission from Hansen *et al.*, 2011).

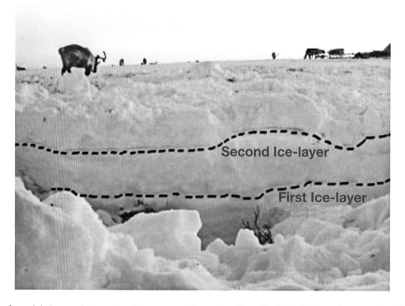

Figure 4.23 Layers formed during a major two-stage icing event on the southern Yamal Peninsula in November 2006. The affected area covered an estimated 60 × 60 km. The same area was affected by an even more extensive icing event a few months later in January 2007, covering approximately 60 × 100 km and causing a large number of animals to perish. (Photo by F. Stammler reproduced with permission from Forbes and Stammler, 2009).

Figure 4.24 Muskoxen (*Ovibos moschatus*) in north-east Greenland in the characteristic grouping that they adopt when threatened by wolves.

As members of the subfamily Caprinae of the family Bovidae, Muskoxen are more closely related to sheep and goats than to oxen. This taxonomic affinity, however, does not make them any less dangerous to arctic travellers if they approach too close. Their massive horned heads are a very significant part of their total body weight and provide a group of Muskoxen *en masse* with a very formidable defence. When launched in a frontal attack they can prove lethal to all oncomers. Muskoxen are native to the Arctic areas of Canada, Greenland, and Alaska. The Alaska population was wiped out in the late 19th or early 20th century, but Muskoxen have since been reintroduced.

The naturally surviving populations are now restricted to western and northern Alaska, the northern regions of the Canadian mainland and its arctic islands, as well as north and north-east Greenland. The US Fish and Wildlife Service introduced Muskox onto Nunavik Island, the second largest island in the Bering Sea, in 1935 as a means of providing sustenance for the indigenous human population. The species has also been reintroduced to northern Europe, including Sweden, the Dovre mountain range of Norway, and Russia with animals from Banks Island, and to eastern Canada with animals from Ellesmere Island. In the province of Quebec, Muskoxen were close to extinction at one point, but have recovered after being protected from hunting. The world population is estimated at between 80,000 and 125,000, with an estimated 68,788 living on Banks Island alone.

The introductions, however, are proving successful. The introduction to Norway made 50 years ago was to a relatively snow-rich alpine environment. Although this was far outside their natural polar distribution it was nevertheless successful. The habitat choice for *Dryas* heath, together with other heath communities, comprising a range of dwarf lichen heath and *Juncus* heath with numerous ericaceous species (*Vaccinium*, *Calluna*, and *Erica*) and shrub willows (*Salix* spp.), appears to have been suitable for providing sufficient winter grazing. In particular the Muskoxen favour the *Dryas* heath on calcareous rock located on very steep slopes (>27°), which offers areas with shallow snow (*ca.* 20 cm) and available graminoid species (Natcheva and Cronberg, 2007).

A similar introduction to Spitsbergen was made in the autumn of 1929, when 17 animals were released in Advent Fjord on west Spitsbergen. Initially, the population seemed to prosper and reached a peak of approximately 50 animals in 1959. By the 1970s the population was declining rapidly and in 1986 only one cow remained from the original introduction. The species is now extinct in Spitsbergen (Myrberget, 1987).

The demise of the Spitsbergen population was probably due to a number of negative factors including competition for grazing with reindeer, which have been increasing in number. Harsh winters with insufficient snow-free slopes would have aggravated the situation in relation to winter fodder. In addition, the young calves tend to show a high mortality in oceanic regions as they are intolerant of wet conditions.

4.7 Conclusions

For both plants and animals, the Tundra can be both a land of plenty and a land of famine. The 24-hour light regime provides ample energy resources for flowering plants as well as the less demanding mosses and lichens. When grazing is taking place, the return of nutrients encourages plant growth. The productivity of salt marshes provides grazing for geese with a herbage quality that repays the cost of their summer migration to high latitudes. Despite the presence of Arctic Foxes they can successfully rear their goslings. Likewise, lemmings recycle nutrients, but when their numbers are low, due to either predation or the exhaustion of winter fodder, there is an interruption in the nutrient cycle. Overgrazing by reindeer is frequently reported, but the fluctuations in their numbers that result from periodic ice-encasement of the Tundra from time to time provides a much needed recovery period for the vegetation, albeit without the continued renewal of nutrients.

Weather is also a feature of the alternations between periods of plenty and times of scarcity. Late springs and cold summers can cause a drastic drop in plant productivity which has an immediate effect not just on growth but also on the ability to reproduce. The life strategy of plants, however, is well adapted to these period fluctuations. Large reserves of carbohydrate in rhizomes and roots together with vegetative reproduction can ensure survival even after several years in succession of sub-optimal conditions. Lemmings and reindeer also show rapid rates of population recovery. The short extent of the food chain and the lack of alternative prey for their predators provide the vegetation with a period of respite from predation. The Tundra may even be described as being uniquely adapted to a *balanced state of non-equilibrium* as populations continually run the gamut from super abundance to near extinction.

CHAPTER 5

Taiga and bog

5.1 Boreal forest diversity

At first sight, the most northerly forests in the world, the Taiga and the boreal forest, can appear to be a monotonous circumpolar belt of coniferous trees with relatively few genera and many similar species. Yet these hardy trees create a forest that dominates a vast and seemingly functionally similar biome across northern Europe, Asia, Alaska, and Canada. Such a simplistic view is, however, misleading. Within the realm of these northern forests there lies a hidden wealth of ecological diversity.

At the northern boundary between the Taiga and the Tundra (Figures 5.1 and 5.2) there exists an interface in a perpetual state of flux as climates alter and the permafrost level rises and falls (Figures 5.3 and 5.4). Cold-adapted forests, mainly of coniferous trees, are found extensively both north and south of the Arctic Circle. The terms *Taiga* and *Boreal Forest* are often used indiscriminately to describe this circumpolar forest which stretches across North America and the northern parts of Eurasia. It is now becoming more common to refer to the northern region of the boreal forest as the Taiga and the more southerly part as the Boreal Forest.

Viewed when travelling in a train or bus, these northern forests can seem to be monotonous tracts of dense coniferous trees. When viewed from the air or on foot, however, it soon becomes apparent just how much of the forest terrain is shared with numerous lakes and bogs, particularly in oceanic areas. There is, in fact, throughout much of the boreal zone, a conflict in terms of succession between forest and bog as to which is the dominant type of vegetation. Studies of forest history in oceanic regions over the past millennia frequently reveal that bog and not forest is the climax vegetation. Once an initial bog is established the process of bog growth (*paludification*) becomes almost irreversible as water tables rise and bog predominates. Periods of warm dry weather can allow trees to advance across bogs, but as the peat still remains, renewed bog growth can resume whenever cooler wetter weather is re-established.

In North America the *Tundra-Taiga Interface* is commonly referred to as the *Forest-Tundra Ecotone*, while in Russia, where this biome is particularly extensive latitudinally, it is called the *lesotundra* (les = Russian—*forest*).

The Boreal Forests are a relatively recent development in circumboreal vegetation and therefore it is not surprising that there is a noticeable taxonomic affinity across the polar continents between the dominant tree species. Although many genera are common to most regions of this circumpolar belt, they are represented in different areas by a series of vicarious species that replace one another in different locations (Latin *vicarious*—a substitute or representative). The principal genera of trees with vicarious species are Spruce (*Picea*), Pine (*Pinus*), Larch (*Larix*), and Fir (*Abies*) in the conifers, and Birch (*Betula*), Aspen (*Populus*), Mountain Ash (*Sorbus*), and Alder (*Alnus*) in the deciduous broad-leaved trees (see Table 5.1). Similarly, the boreal fauna have strong similarities from east to west with examples of vicarious species among deer, beavers, squirrels, voles, and rodents together with numerous other examples in birds and insects.

An understanding of population dynamics in the taiga and the boreal forest needs to take into account the unique combination of edaphic and climatic factors that operate in this region. The natural hazards for woody scrub plants and forest even at high latitudes can include lack of snow-cover, leading to

Tundra-Taiga Biology. First Edition. R.M.M. Crawford © R.M.M. Crawford 2013.
Published 2013 by Oxford University Press.

Figure 5.1 The Tundra-Taiga Interface at Åbisko (68°N), Swedish Lapland. Mountain Birch (*Betula pubescens* ssp. *czerepanovii*) in the foreground and middle distance, and in the far distance the Lapp Gate mountain pass, the traditional entry point for the Lapp reindeer herds to their summer pastures on the Tundra.

frost damage followed by drought stress in the spring, wind and ice abrasion, permafrost fluctuations, flooding, and fire. Insect attacks also cause frequent damaging events. All these factors combine in varying proportions to impose a degree of natural disturbance, which although it causes periods of forest decline, nevertheless also creates a diverse dynamic ecosystem that is continually undergoing destruction, renewal, and ecological readjustment. Despite these numerous hazards it is astonishing how far north some of the boreal conifers can be found (see Figure 1.2).

5.2 The Tundra-Taiga Interface

Given the geographically indeterminate manner in which Tundra replaces Taiga with increasing latitude, it is customary to refer to the transition zone between Tundra and Taiga as the *Tundra-Taiga Interface*. This boundary region has a somewhat schizophrenic existence where patches of forest and Tundra exist side by side, alternating one with the other as the level of the permafrost layer in the soil fluctuates. In such marginal areas, the thermal insulation that develops as tree cover increases causes the upper permafrost level in the soil profile to rise, which can eventually kill the trees and bring about a localized reversion to Tundra.

The Tundra-Taiga Interface is one of the most extensive boundary zones in the world, stretching for 13,400 km around the northern hemisphere. In its northern extremities it is reduced to patches of trees on the Tundra, while within the Taiga zone in areas with higher elevation the reverse situation can exist, with extensive areas of Tundra surviving surrounded by forest. It is also a very dynamic region supporting a mosaic of plant and animal communities that exist in close proximity to each other and where the pattern of distribution is constantly changing in response to fires, floods, soil upheaval (cryoturbation), and climatic fluctuations. The latitudinal depth of the zone also varies. In Russia, the *Tundra-Forest Interface zone* (the *lesotundra*) can measure as much as 600 km (≈5° latitude) from north to south (Figure 5.3), while in other regions such as in North America it can be just a few kilometres.

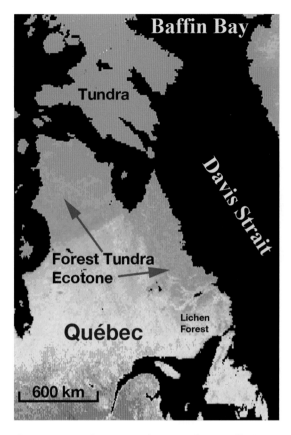

Figure 5.2 View of eastern Canadian as seen from a satellite image recording the normalized vegetation index in May 1988 (a proxy indicator of photosynthetic activity). Colour scale blue indicates snow-covered ground with no emergent evergreen forest. Increasing colour from green to red shows increasing photosynthetic activity. Note the more abrupt zonation between snow and emergent forest in Canada as compared with Russia. (Images were prepared from 8-km resolution Pathfinder dataset—US Geological Survey, EROS Data Center, Sioux Falls, South Dakota).

Figure 5.3 Details of Russian lesotundra zonation (Tundra-Taiga Interface) as seen from a satellite image recording the normalized vegetation index in May 1998. Note the diffuse zonation between evergreen trees (light green) and snow-covered ground (blue), indicating a more open tree cover in this typical Russian lesotundra. The greater extent of permafrost in Russia is associated with the more transitory condition of the Russian lesotundra as compared with the North American Forest-Tundra Ecotone. (Images were prepared from 8-km resolution Pathfinder dataset—US Geological Survey, EROS Data Center, Sioux Falls, South Dakota).

5.2.1 Krummholz and Krüppelholz

Two words emanating from German forestry are used to describe stunted, twisted trees. *Krummholz* (*krumm*—twisted, *holz*—wood) is the term most commonly used to describe the misshapen trees at the timberline. The second term *Krüppelholz* derives its adjectival prefix from the German *Krüppel* meaning crippled. Although *Krummholz* has long been used to cover all twisted dwarf trees as found at both altitudinal and latitudinal treelines, it has been pointed out that this word covers both trees that are genetically dwarfed to keep the twisted form, irrespective of the environmental conditions, and trees that have grown in a twisted form as a phenotypic response to suboptimal environment. For the latter, it was proposed over a century ago that the term *Krüppelholz* should be used (Kiklman, 1892; Schöter, 1898).

There arises, however, the problem as to where and when this distinction can practically be achieved in the field. Only during periods of climatic improvement for the growth at the treeline is it possible to

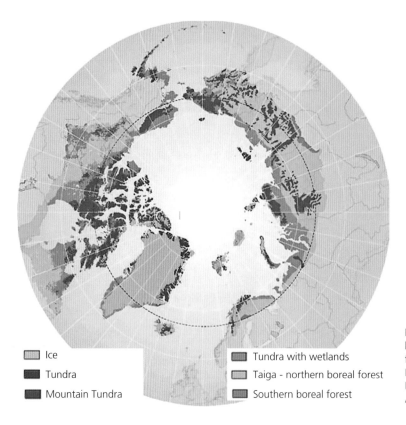

	Ice		Tundra with wetlands
	Tundra		Taiga - northern boreal forest
	Mountain Tundra		Southern boreal forest

Figure 5.4 Distribution of tundra and boreal forest zones. (Modified and adapted from the Arctic Circle Map adapted by Philippe Rekacewicz, United Nations Environmental Programme UNEP/GRID, Arendal, Norway).

Table 5.1 Examples of boreal forest vicarious tree species. N.B. Local floras may differ in their botanical nomenclature; species names are standardized to follow the International Plant Names Index.

Genus	North American species	Eurasian species
Abies	*balsamea*	*sibirica*
Alnus	*rubra*	*incana, glutinosa*
Betula	*glandulosa, papyrifera, cordifolia*	*nana, pubescens, pendula*
Larix	*laricina, lyalli, occidentalis*	*decidua, gmeliniae, sibirica*
Picea	*glauca, mariana*	*abies, obovata*
Pinus	*contorta, strobus*	*sylvestris, sibirica, pumila*
Sorbus	*americana, sitchensis*	*aucuparius, sibirica*

see which individual trees are capable of change to the pole form. To the east of the Hudson Bay, recent climatic amelioration has been observed to improve tree growth in the Black Spruce (*Picea mariana*) with the production of upright trunks (Lescop-Sinclair and Payette, 1995). In this case it might appear to be a more exact terminology to describe the trees as *Krüppelholz* when they are in a dwarfed form but with a potential for upright growth. It is however probably simpler to adhere to the term *Krummholz* and avoid making unsubstantiated statements, unless a particular author has chosen for well-defined reasons to use *Krüppelholz*.

There are tree species such as *Pinus mugo*, *Pinus pumila*, and *Alnus viridis* in which the low (prostrate) growth form is genetically fixed and whatever the environmental conditions they never succeed in producing upright trunks.

5.3 Fire in the boreal forest

Fire is a constant feature of the boreal forest and is a hazard that foresters constantly have had to endure. As a biological phenomenon in northern regions, evidence of fires can be found dating back through the millennia to before there was any significant human disturbance. In Alaska the remains of ancient fires can be traced back 9,000 BP (calibrated) in forests consisting of Birch (*Betula*), White Spruce (*Picea glauca*), and Alder (*Alnus*). By the mid-Holocene (6,500–5,600 BP) fire frequency began to increase as fire-prone Black Spruce (*Picea mariana*) replaced the more fire-resistant White Spruce (*P. glauca*) as the dominant species (Lynch *et al.*, 2002).

Paradoxically, the proximity of the boreal forests to extensive wetlands can in many areas increase the risk of fire. This greater predisposition to fire has been linked to the warmer conditions of the forest increasing evapo-transpiration from the adjacent bog and thus increasing the risk of thunderstorms and lightning-induced fires (Valentini *et al.*, 2000). In this connection it is relevant to note that in Siberia and North America there can be a high frequency of thunderstorms along the Tundra-Taiga boundary which is particularly noticeable as occurring in the month of July. The frequency of fire is noticeably greater in the more continental areas (Figure 5.5). These natural forest fires can be sufficient to delay forest regeneration, as can also insect attacks (Figure 5.6).

Studies of the history of fire along the eastern coast of the Hudson Bay at 58°N (Figures 5.7 and 5.8) have recorded in detail the incidence of fire in eastern Canada (Payette *et al.*, 2008). Between 8,000 and 5,800 calendar BP, the climatic and ecological conditions were less conducive to fire events than after this date. After 5,800 calendar BP, the number of fires increased per 1,000 years, reaching a maximum at *ca*. 3,400 calendar BP. The patchy incidence of forest destruction by fire appears to have induced a rise in the water table during the past 4,000 years.

Currently, each fire event is followed by a drastic reduction of spruce cover. Thus, contrary to a widespread belief of a general northwards migration of the boreal forest due to global warming, there is no

Figure 5.5 Lichen–woodland at its most southern location in North America at 47°N in Quebec. The southern extension of this particular type of boreal forest is due to the cold currents in the neighbouring Davis Strait (see Figure 5.3).

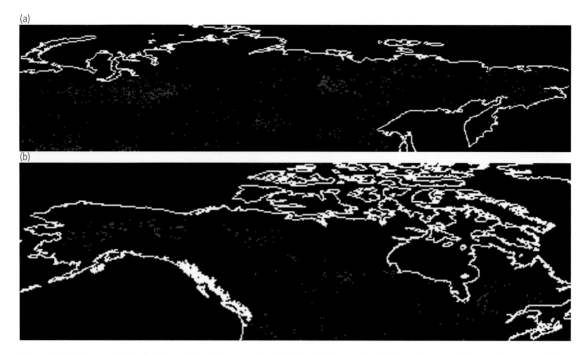

Figure 5.6 Incidence of fire in the Taiga and boreal forest regions of (*a*) Eurasia (May 2011) and (*b*) North America (July 2007), as recorded from satellite detection of night-time thermal hot spots. (Source: *World Fire Atlas*, European Space Agency).

indication that this is imminent either in Quebec or Labrador. In the Hudson Bay region there are extensive areas where forest cover is at present limited either to the remains of ancient trees or else extensive Krummholz (Figures 5.9 and 5.10). The present situation in this region is in sharp contrast to the active tree regeneration that took place here more than 1,000 years ago (Payette *et al.*, 2008). It is therefore probable that for this northern site, currently projected global climatic changes may lead to further expansion of boreal wetlands rather than forest.

Forest flammability is not just dependent on the dominant tree species. Experiments with mixtures of contrasting plant materials have shown that the nature of the fire in terms of spread rate and duration can be affected by the mixture of species present. In particular, the presence of Crowberry (*Empetrum nigrum*) and Glittering Wood Moss (*Hylocomium splendens*) can increase flame speed and time to extinction. Furthermore, such observations suggest that ecological invasions by highly flammable species may have effects on ground-fire dynamics well out of proportion to their biomass (van Altena *et al.*, 2012).

5.4 Taiga regeneration and the seed bank

Regeneration from a buried seed bank after a fire depends both on the species composition of the forest and the nature of the fire. Fire in the Taiga, like other forests, varies depending on whether the fire is top or bottom burning. Surface fires that burn the bottom regions of the forest destroy the understory of the forest as well as a significant proportion of the forest litter. The forest canopy can also suffer to varying degrees. However, certain tree species are able to survive bottom burning fires, especially if they pass over the ground rapidly and do not generate too much heat.

Several species of conifers, notably pines, have evolved a thick bark that insulates and protects the cambium from heat-damage from low-intensity fires. In some cases this is combined with a natural shedding of the lower branches which reduces the probability of the fire spreading to the forest canopy. This feature in the pines is combined with the possession of an aerial seed bank. Pine cones are

TAIGA AND BOG 99

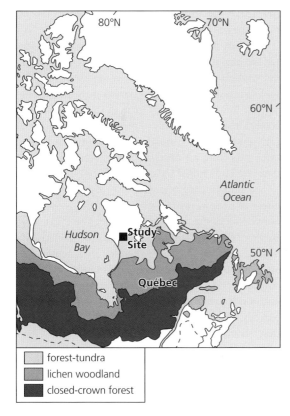

Figure 5.7 Location of forest-tundra study in the vicinity of the Hudson Bay in north-east Canada by Payette *et al.* (2008) (see also Figures 5.8–5.10).

Figure 5.8 Incidence of fire in forests of eastern Canada for the month of July 2007. (Source: *World Fire Atlas*, European Space Agency).

Figure 5.9 Old-growth in an ancient (more than 2,000 years old) lichen–spruce woodland on the east coast of the Hudson Bay with all sizes and all ages of dead and living spruce trees. (Reproduced with permission from Payette *et al.*, 2008).

Figure 5.10 Ancient (more than 2,000 years old) woody site of Krummholz on the east side of the Hudson Bay. Dead stems on the ground date from the 15th–16th centuries when growth conditions were more favourable than at present. (Reproduced with permission from Payette *et al.*, 2008).

described botanically as being *serotinus* (Latin *serotinus*—late-coming) in that they retain their seeds so that they are not dispersed or able to germinate until after an acute disturbance such as fire. After the passage of fire, the aerial seed bank remains in the serotinus cones. In Lodgepole Pine (*Pinus contorta*), the trees can shed an accumulation of more than two decades of seed production onto the forest floor which is now cleared of litter and therefore in a very suitable condition for the establishment of young pine seedlings.

Although crown fires that burn over large areas dominate fire regimes in the boreal forest, these fires rarely burn with the same intensity throughout the entire forest. Consequently, the fire by burning the

Figure 5.11 Spruce roots (*Picea abies*) that had grown in a peat soil and were killed by a bottom burning fire that lasted from June to October and entirely removed the soil peat layer.

vegetation to varying degrees creates an ecological mosaic and can therefore have an important role in restoring biodiversity.

A study assessing regeneration over 6 or 7 years in relation to crown fire severity in Jack Pine (*Pinus banksiana*) and Black Spruce (*Picea mariana*) stands showed that seedlings established more frequently and grew better where the thickness of residual organic matter was lowest. In the long-term ecological view, and ignoring the loss of commercial timber, crown fires, provided there is a viable soil seed bank, may be good promoters of regeneration in coniferous forests (Jayen *et al.*, 2006).

Nevertheless, where the boreal forest and Taiga are growing on soils with deep peat layers these can smoulder for months, particularly during a dry season, and are only extinguished eventually when winter rain and snow arrive, by which time the seed bank will have been destroyed (Figure 5.11).

5.5 Northern mires

Due to the low evaporation and transpiration at high latitudes, extensive areas in the Arctic and subartic boreal regions can only be described in general terms as wetlands. Water-saturated soils are depleted of oxygen and in the north this is further compounded by periods of acute and prolonged anaerobiosis as the soils become frozen in winter. When encased in ice there is no access to atmospheric oxygen until the ice melts, which in the far north may not be until the following June.

Not only are arctic soils oxygen-deficient in winter, but also due to the nature of the topography and the underlying permafrost large areas can remain waterlogged for the greater part of the growing season. At low temperatures growth rates are low, but so also is decomposition of organic matter and peat can gradually accumulate and eventually spread outwards from an initial wet site, creating what is described as a *raised bog* or *Hochmoor* (see Table 5.2).

The outstanding character of raised bogs is the effect of peat in raising the level of the water table above that of the surrounding land. For this reason this hydrological situation is aptly described as a *perched water table*. Consequently, on reaching this stage of development, the nutrition of the bog vegetation is no longer influenced by the geology of the surrounding land as water arrives at the bog surface directly from the clouds. The bog can thus be described as *ombrogenous* (Greek *ombros*—a shower of rain) as opposed to *rheotrophic* (Greek *rheos*—a stream or flowing water). Initially the wetland would have been a *rheotrophic* fenland with water that had gained mineral nutrients as it passed through rocks and soil. This pattern of bog development from rheotrophic to ombrotrophic is widespread and occurs in most humid areas, including arctic and boreal wetlands.

5.5.1 Paludification of the north

In the early Holocene trees advanced northwards relatively rapidly, eventually reaching the Siberian shores of the Arctic Ocean (see Chapter 2). When sea levels began to rise *ca.* 6,000 BP, bog growth accelerated, not just because the climate was more oceanic but because waterlogging was facilitated by low temperatures causing the departure of the forest. Trees through their transpiration capacity can play an important role in controlling the level of the water table in humid climates. The removal of trees, whether by fire or any other agency (e.g. insect attack), nearly always results in rising water tables, which then become a major factor in hindering forest regeneration. Fire, apart from causing water tables to rise, can also aid bog formation as particles of ash and carbon deposited into the soil profile can

Table 5.2 A selection of some of the terms used in describing wetlands. The terms vary in different countries and are arranged in alphabetical order without regard to the circumstances of their formation. (From various sources (see Gore, 1983)).

Aapa mire	(Finnish—open mire) A cold-climate mire with ridge and pool surfaces common in boreal regions
Blandmyr	(Swedish) A mixed mire, an aapa mire with ridges high enough to become ombrotrophic
Blanket bog	An extensive mire that is not confined to depressions and is found mainly in oceanic environments
Bog	*See* Hochmoor
Carr	Fen with scrub or woodland
Fen	(Old English) A eutrophic mire, usually with an accumulation of peat with a winter water table at or above ground level
Flark	(Swedish) An area of exposed peat with an algal film, often having a sparse covering of sedges
Hochmoor	(German) Raised bog, the classic peat bog of north-western Europe with a water table usually perched above the surrounding landscape
Marsh	An ecosysssstem with more or less continuously waterlogged soil dominated by emersed plants and without a surface accumulation of peat
Mire	All wetland types that receive their water from the clouds, i.e. *ombrotrophic*
Mòine	(Gaelic) Blanket bog
Moss	(Scottish) Raised bog, *see* Hochmoor
Niedermoor	(German) A fen, a minerotrophic mire with a wider range of species than a bog
Palsa	A mound of peat with a permanently frozen core, often 3–6 or even 7m high
Schwingmoor	(German) Mire formation floating on water, frequently developed along lake shores
Strangmoor	Patterned mire or string bog with a hummock and hollow relief caused by an abundant local flow of water. Ridges of string often elongated in a ladder pattern at right angles to the downslope direction
Swamp	A wet area normally permanently covered with water and not drying out in summer
Tryasina	Quaking bog or quagmire (Russian - трясина)

reduce drainage and therefore help to initiate peat growth (Mallik *et al.*, 1984). The combination of bog growth and low temperatures therefore exacerbates the loss of tree cover.

Paludification (the growth of bogs) has also been a feature in Labrador and Quebec but is not such a significant factor in the boreal forests to the west of the Hudson Bay. In Canada there are comparable expanses of wetlands in the Mackenzie River basin, while the Western Siberian Lowlands are outstanding as the world's largest high-latitude wetlands and cover over 900,000 km^2.

Once established, unfertile boreal wetlands with their perched water tables, pools, and waterlogged soils inhibit the survival of large trees as they are prone to wind-throw due to shallow rooting.

The slow rate of decomposition of plant debris in raised bogs allows a peat layer to develop which, depending on local conditions, can reach a depth in some areas of as much as 30m. Depending on the humidity and the topography, there can be a distinct upper layer (the *acrotelm*). During the growing season, a warm, drier atmosphere can restore aeration to the soil atmosphere sufficiently to promote root growth in this upper layer. The permanently waterlogged, unaerated (anaerobic) layer below is referred to as the *catotelm* (Figure 5.12).

The terms *acrotelm* and *catotelm* were adopted into English from the work of the Russian mire hydrologist K.E. Ivanov who suggested a two-fold division into an upper active layer (деятельный слой or активный слой) and a lower inert layer (инертный слой). These terms could not be translated directly into English as *active* and *inactive layers* in relation to bogs as they were already in use describing cryosoils in relation to the position of the permafrost zone. In addition, *inert layer* was unsatisfactory as the lower layer can contain live roots. Hence the

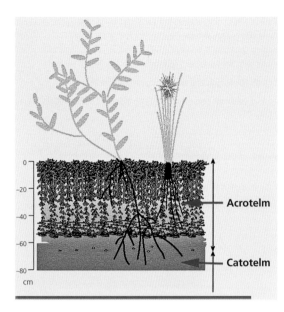

Figure 5.12 Classification of soil profile layers in relation to waterlogging. The *catotelm* and *acrotelm* represent two distinct soil layers in undisturbed peat bogs, first defined by Ingram (1978) in relation to the hydrological regime. The *catotelm* is the bottom layer of peat that is permanently below the water table where, under anaerobic conditions, peat decomposition is slow and root penetration is minimal. The upper layer—the *acrotelm*—is aerated during the growing season and allows active root penetration. (Adapted from Payette and Rochefort, 2001).

terms *acrotelm* (Greek *acro*—upper; *cato*—lower; *telm*—marsh) were proposed (Ingram, 1978).

The Arctic also imposes its own mark on bog structure due to the seasonal effects of freezing and cryoturbation. The very varied nature of boreal wetlands has long been recognized by northern peoples and there is a rich vocabulary of words for bogs in Germanic, Celtic, and Finnish languages. Some of these have been adopted scientifically to describe varying ecological conditions (see Table 5.2). Russian, however, is an exception. Despite Siberia having the largest expanse of northern bogs and wetlands in the world, Russian has the poorest vocabulary for their description. In Russian, the usual word for a wetland is *bolota* (болота, cognate with Latin *palus*—a marsh, swamp, bog, or fen). There is just one other word, *Tryasina* (трясина), which denotes a quagmire and comes from the Russian word for shaking, i.e. a quaking bog.

5.5.2 Sedge mires

Sedge mires (also called sedge meadows) are one of the commonest wetlands in the Arctic. They are generally restricted to lowland sites and broad open valleys that remain wet throughout the growing season. They are often intermingled and border with the graminoid tundra communities, as described in Chapter 4 (see Figure 4.9). Sedge mires can be found in the Canadian Arctic archipelago as far as 80°N in Ellesmere Island where their vegetation is dominated by moss and accompanied by the graminoids (*Eriophorum scheuchzeri*, *Alopecurus alpinus*, *Pleuropogon sabinei*, *Dupontia fisheri*) and some dwarf shrub species (e.g. *Dryas integrifolia*, *Vaccinum uliginosum*, and *Cassiope tetragona*). These meadows are the most productive of High Arctic communities and are important forage communities for resident and migratory herbivores, such as Muskox (*Ovibos moschatus*), Caribou (*Rangifer tarandus*), Arctic Hare (*Lepus arcticus*), and Snow Geese (*Chen caeruluscens*) (Henry, 1998).

5.5.3 Aapa mires

Aapa mire is a Finnish term meaning *open mire* as these mires usually extend across large open spaces with elongated stretches of open water. In Europe, Finland is particularly noted for its range of bog types and with its north–south orientation presents an opportunity to relate their form to latitude, and the factors that bring about the formation of aapa mires The very pronounced structure of the aapa mire is the product of certain physical constraints that are exerted on soligenous mires when exposed to the contrary forces of frost and water-induced movements. Seen from an aerial view, the string-and-pool configuration of the aapa mires is one of the most spectacular natural features of the Taiga (Figures 5.13 and 5.14). In North America they are more commonly referred to as *String Bogs*

The first person to attempt to classify Finnish mires and to use the term *aapa mire* scientifically was Aimo Cajander, Professor of Forestry in Helsinki and ultimately prime minister of Finland up to the the onset of the Winter War in 1939. He gave the name of *aapo miresmto* to this characteristic northern mire (Cajander, 1913).

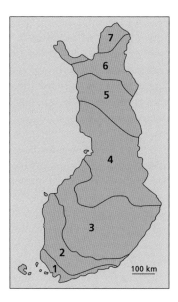

Figure 5.13 Mire zones of Finland. 1: Plateau bogs, 2: concentric bogs, 3: eccentric bogs, 4: southern aapa mires, 5: main aapa mire zone, 6: northern aapa mires, 7: palsa mires. (Adapted from Koutaniemi, 1999).

Aapa mires are normally found on a slight incline with open pools of water running down the length of the incline. The mire is therefore subject to downslope movements from water flow. There is also a converse upslope force from the expansion of ice (Laitinen *et al.*, 2007). Water occupies the linear depressions between the ridges, hence the origin of the alternative name of *string bog* or *strangmoor* (Figure 5.14). They are largely circumpolar in distribution and are particularly common in Fennoscandinavia and boreal Canada. The aapa mire is in effect the extreme form of patterned bog. Further south the surface of raised bogs with their varying amounts of hummocks and pools can show various patterns, varying from plateau to concentric and eccentric depending on slope and the patterns made by surface water pools.

5.5.4 Palsas

Palsa mires are sub-arctic mire complexes with permanently frozen peat hummocks and are common in parts of Fennoscandinavia, Russia, Canada, and Alaska. They are similar to *pingos* (see 4.5.2) which are large earth-covered ice intrusions fed by high levels of water pressure under the ice. The palsas are smaller and do not need water pressure for their formation. Palsas usually occur in areas with discontinuous permafrost and serve as reliable indicators that there is a presence of permafrost. Palsas need large quantities of water for the formation of their ice lenses, and for this reason they occur particularly in very wet bogs.

Wetlands with palsa bogs provide a biologically heterogeneous environment with a rich variety of bird species, and are listed as a priority habitat type by the European Union. In recent decades there has

Figure 5.14 An aapa mire or string bog, Labrador. (Photo C.W. Woodley, Ottawa).

been a noticeable decline of palsa mires in northern Europe, probably due to climatic warming. The degradation of these bogs threatens some of the most important breeding sites for migratory birds in northern Europe.

5.5.5 Long-term forest paludification

Despite an overall improvement recently in the growth of boreal forests in Eurasia, probably as a result of rising temperatures (Briffa *et al.*, 2008), there has also been noted a decline in certain localities. A study of Black Spruce (*Picea mariana*) in eastern Canada has suggested that a consequence of the accumulation of organic matter leading to the paludification of boreal soils is causing a decline in forest productivity (Simard *et al.*, 2007). A lengthy debate has also taken place, particularly among Russian ecologists, as to whether in certain locations climatic warming, instead of causing an advance of the treeline, will result in a continuation of the retreat that began after the Hypsithermal (Callaghan *et al.*, 2002). Those who believe that climatic warming causes an advance of the treeline northwards subscribe to the eventual over-riding influence of climatic factors. There is, however, in the North Siberian Lowlands a powerful argument that climatic warming can cause a retreat of the treeline based on the belief that the presence or absence of trees is due to edaphic factors influenced by the proximity of the Arctic Ocean and the long-term persistence of permafrost.

The open cold water in summer of the Arctic Ocean has a strong climatic effect on the climate of the coastal areas of northern Siberia. High humidity and low temperatures dominate the lowland Siberian Tundra and contribute to the creation of extensive bogs (Figure 2.28). As a result, this treeless terrain supports an impoverished flora and fauna (Sher, 1996). In such an environment any advance of tree vegetation as a result of climatic warming is limited to better-drained sites along river banks.

It is commonly presumed that forest is the natural climax in vegetation succession, and therefore with climatic warming the northern limits to tree growth will move north. However, in areas subject to maritime climatic influences, and these can extend inland considerable distances from the sea, there is always the possibility that the natural climax vegetation may be bog rather than forest (Klinger, 1996). In oceanic environments mineral soil impoverishment is frequently aggravated, as soil leaching accelerates nutrient removal and results in iron pan formation. Waterlogging then follows and impedes mineralization and nitrogen fixation, leading to the process of paludification and establishment of bogs.

Paludification, however, is not just limited to the coastal fringe areas of Western Europe. Extensive regions of Alaska, Newfoundland, Quebec, northwest Europe, and northern Siberia have all seen forest displaced by extensive bogs during the past 6,000 years (see Crawford, 2000). Whether the Tundra-Taiga Interface is moving north or south or remains stationary is at present uncertain and requires further investigation, particularly in areas where cool moist summers may accelerate the rate of peat formation as the growing season becomes longer and facilitates bog growth.

5.5.6 Arctic wetlands and climatic warming

The Arctic is warming faster than the rest of the planet. Notwithstanding the increases in forest wetlands (paludification) in oceanic areas, there are other more continental regions that may be becoming drier. A sign that this is taking place is the reduction in the number and size of some arctic lakes since the 1980s. A satellite image survey carried out across half a million square kilometres of Siberia and covering more than 10,000 large lakes showed that after an interval of three decades of rising air temperatures, there has been a widespread decline in lake abundance despite slight precipitation increases. The spatial pattern of lake disappearance strongly suggests that thawing of permafrost in regions where it is discontinuous is the likely cause of the observed losses (Smith *et al.*, 2005). The situation is, however, complex. Nevertheless, numerous studies have described the opposite effect, with increased surface ponding in warming permafrost environments, driven primarily by slumping and collapsed terrain (thermokarst) that subsequently fills with water (Kirpotin *et al.*, 2009). Where the permafrost is continuous, the total lake area can increase. This trend for net lake growth in continuous permafrost stands is in sharp contrast with more southerly zones where the permafrost is sporadic. In these areas there has been a notable decline in the number of lakes. The declines in the south have apparently outpaced lake gains in the north, leading to an overall loss for Siberia as a whole.

The collapse of peatlands with the increase in thermokarst as a result of climatic warming is therefore causing great concern in Siberia. Healthy peatlands absorb carbon dioxide as new vegetation grows. But as peatlands break down, both methane and carbon dioxide are released into the atmosphere. Calculations have shown that over the long term, Siberian peatlands have absorbed more greenhouse gasses through plant growth and storage than they have released (Smith *et al.*, 2005). This long-term trend may be reduced or even reversed with climatic warming.

The relationship, or rather conflict, between bogs and forests for dominance of the landscape is complex. Trees survive on bogs in many northern areas, and where they do so it is usually in regions where the soil is frozen to a considerable depth throughout a long winter period, when root metabolic activity will be minimal and a period of enforced oxygen deprivation may not be a serious metabolic hazard. The Black Spruce (*Picea mariana*) and Tamarack (*Larix laricina*) are examples of trees that are notable for their ability to grow on wet peatlands in the continental cold-winter regions of North America. When spring does arrive there is a rapid transition from frozen soils to a warm growing season. Consequently, during the summer growing season the aerobic active layer of the soil horizon, the *acrotelm*, will normally be suitable for root growth. However, when late spring flooding occurs on these forested peatlands it can be detrimental to growth for both *P. mariana* and *L. laricina* (Roy *et al.*, 1999; Girardin *et al.*, 2001).

Studies from 1994–98 on the Tanana Flats in central Alaska (Figure 5.15) reveal that permafrost

Figure 5.15 Aerial view of dying trees along the margin of lowland birch forests resulting from thermokarsting as a result of rapid permafrost degradation on the Tanana Flats, central Alaska. (Photo M.T. Jorgenson).

degradation is widespread. Fine-grained soils under the birch forest are ice-rich and thaw settlement can cause a 1–2.5 m collapse. The collapsed areas are rapidly colonized by aquatic herbaceous plants, leading to the development of a thick, floating organic mat. It has been estimated that 83% of the degradation occurred before 1949 (Figure 5.15). The evidence indicates that this permafrost degradation began in the mid-1700s and is associated with intermittent periods of relatively warm climate during the mid–late 1700s and 1900s. If current conditions persist, it is estimated that the remaining lowland birch forests of this region will eventually be eliminated (Jorgenson *et al.*, 2001).

The accumulation of organic matter as a result of paludification in boreal soils has been suggested as a possible cause of reduced forest productivity. A decline in Black Spruce productivity of 50–80% with increased paludification, particularly during the first centuries after fire, has been observed in Quebec. Paludification increases bryophyte abundance and soil moisture, while reducing soil temperature and nutrient availability. The ensuing limitation to tree rooting depth makes the trees unstable.

At the landscape scale, it is evident that the fire regime (frequency and severity) can increase paludification and reduce forest productivity merely through its effect on soil organic layers (Simard *et al.*, 2007). Low-severity soil burns can be more injurious than high-severity fires in accelerating paludification and reducing forest productivity. This successional path to paludification contrasts with edaphic paludification, where topography and drainage primarily control the extent and rate of paludification.

In northern Quebec, the post-fire degradation and disappearance of the conifer stands from the peatlands has created extensive open landscapes which are then subject to a notable cooling due to the proximity of cold ocean currents. This results in summer frosts becoming frequent and causing a stunting of tree growth and further expansion of open spaces. The cold conditions bring about the frequent occurrence of Krummholz (Payette and Delwaide, 2004). Under these conditions any surviving trees are likely to suffer dieback of supranival stems. This scenario appears to explain the southern position of the treeline in Quebec and the anomalous occurrence there of lichen–spruce forest at a latitude of only 47°N (Figure 5.5). In the more oceanic regions the trees may finally succumb to drowning in permafrost-induced ponds. The same sequence of events will probably not occur in non-oceanic areas with lower levels of precipitation.

In more continental sub-arctic Finland it has been found that fire-induced deforestation increases wind velocity by 60%, thus adding to the climatic adversity for trees in an area which is already at the boundary for tree survival (Holtmeier and Broll, 2010). Fire also has mixed effects on the bird and mammal populations. Small mammals benefit in the early stages of recovery from fire when the ground vegetation provides more food and cover, while birds benefit more from the sapling stage onwards as tree cover develops.

Even without human interference the natural disturbances to which Taiga and boreal forests are exposed ensure constant disturbance and renewal. Consequently the tree cover across a wide range of habitat types will be influenced by geographical location and whether or not their climate is continental or subject to oceanic influences. Climate change is not new to this young biome. The Holocene has witnessed many noticeable climatic changes that have had marked effects on the vegetation of northern regions. It is possible that the boreal forests with their bogs will prove increasingly inhospitable to trees in areas with drainage problems unless drainage can be facilitated. In areas prone to paludification both the plant and animal biota may be about to undergo significant alterations in distribution and species composition as a result of current climatic warming.

5.6 Forest grazing

The dynamic nature of the Taiga and boreal forest creates a patchwork with open spaces that facilitates grazing by ungulates. The process is two-way as grazing can further extend open patches for regeneration (Vera, 2000). In the boreal forest the impact of fire alone is sufficient to create clearings which encourage the growth of younger brushwood and provide more palatable fodder for ungulates. The grass and sedge meadows that surround many boreal bogs also provide nutritious forage. The dominant grazing ungulates are various species of deer (*Cervidae*).

5.6.1 Reindeer/caribou migration

Seasonal migration is the outstanding behaviour pattern of the reindeer that graze the tundra in summer. The island populations, such as those that exist in Spitsbergen, are the exception as they are perforce restricted to their respective archipelagoes by geographical insularity. In continental North America the reindeer have never been even semi-domesticated and the Tundra populations migrate spontaneously over vast distances. Some populations of the North American caribou migrate further than any other terrestrial mammal, travelling up to 5,000 km (3,100 miles) a year, and ranging over an area of 1,000,000 km^2 (390,000 sq miles) in their annual migration from their winter refuge in the northern forests to the adjacent Tundra in summer. Presumably the native North American peoples found it easier to wait for the migrations passing through their lands than to follow the herds throughout the year, as became the habit in Scandinavia and Russia.

In Eurasia, forest reindeer are widespread across the boreal zone, migrating into the Tundra in summer. Their effects on vegetation and soils vary depending on local factors such as altitude, exposure, snow-depth, substrate, moisture, prevailing vegetation type, and animal density. Reindeer populations have always fluctuated depending on climate, predation, and human disturbance. At present, the number of Old World reindeer is estimated at under 2.5 million, with the most productive semi-domesticated herds occurring in Fennoscandinavia and the Nenets regions of north-west Russia where they straddle the Ural Mountains (Forbes and Kumpula, 2009). In Fennoscandinavia, the highly managed state of the forest together with relatively high numbers of animals has in many areas reduced the availability of lichens for winter-feeding as well as depleting the ground vegetation of small shrubs. Commercial forestry in many places has also led to a steady reduction of arboreal lichen forests and resulted in a need for supplementary feeding in winter.

In addition, ground lichens are easily damaged in areas where there is intensive grazing pressure. In particular, the practice of double grazing where animals visit the same territory in summer and winter has been particularly injurious. An example of this damage, which is even visible from remote sensing satellites, is at Jauristunturit (Figure 5.16) in the border zone shared by Norway and Finland; here the main vegetation type is lichen-dominated Tundra heath with dwarf shrubs. The difference in whiteness is due to the lichen coverage. The national border between Finland and Norway has a reindeer fence which visibly divides the area. The Norwegian side to the north is grazed only in late winter when snow cover protects the lichen cover from trampling and where there is no grazing or trampling in summer. The Finnish area is used in early summer, when vulnerable lichen mats have been repeatedly trampled over several decades (Forbes and Stammler, 2009).

At present in many areas across their circumpolar range, caribou and reindeer herds are declining. Their reduction appears to be due to increasing arctic temperatures and precipitation, in addition to anthropogenic landscape change. Changes in insect phenology—earlier emergence and increased abundance—indirectly interfere with summer foraging by reducing the amount of time for undisturbed grazing. The increased activity of Warble

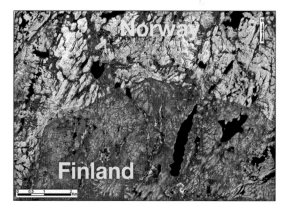

Figure 5.16 Satellite image of northern Norway and Finland. The image of Jauristunturit at the border boundary between Norway and Finland is clearly seen by differences in whiteness due to differences in lichen coverage. (Reproduced with permission from Forbes and Kumpula, 2009).

Flies (*Hypoderma tarandi*) and Nose-bot Flies (*Cephenemyia trompe*) is positively correlated with ambient temperature. Larvae of these insects are endoparasites of *Rangifer*, and heavy parasite loads are associated with poor body condition. Mosquito (*Aedes* spp. and other genera) activity is also correlated positively with air temperature, and mosquito harassment also results in poor body condition. Reindeer calves need to exploit high-quality summer forage to grow sufficiently to achieve a size that ensures survival through the winter. Females, burdened by lactation and demands of gestation, must regain sufficient body mass and fat to conceive the following autumn. The reduction in caribou and reindeer numbers due to various environmental factors is likely to have significant negative socio-economic consequences for northern indigenous peoples (Vors and Boyce, 2009).

The annual treks of the northern populations of caribou to and from their summer and winter ranges are the last remaining large-scale ungulate migrations in the northern hemisphere. It may be that these large-scale migrations will become a thing of the past. The migrations of Bison (*Bison bison*) and Saiga Antelope (*Saiga tatarica*) have long vanished due to human disturbance and habitat destruction.

The extensive distribution of reindeer across the polar regions of the northern hemisphere disguises their vulnerability. When a detailed census is attempted it can be seen that caribou and reindeer can be considered as hovering on the precipice of a major decline because of current changes in plant and insect phenologies, brought about by an increased frequency and intensity of extreme weather events (Vors and Boyce, 2009). Using data from the past 30–50 years, the only site in Eurasia where there has been a population increase is in Spitsbergen (Figure 4.21). This is, however, an exceptional situation as the reindeer populations on this archipelago have been augmented by local introductions to regions from where they have long been absent, as in the Brøgger Peninsula. Although reindeer and caribou populations regularly fluctuate due to periodic climatic adversity such as ice-encasement of their winter forage, there is nevertheless a general decline in most European regions. The Russian Siberian populations have been left out of this particular environmental argument, as recent political changes and expansion of industrial activity with metal and oil extraction are currently the major disruptions affecting the traditional reindeer herding activities in this region.

At the height of the *Little Ice Age* northern summers were marked in many regions by cool moist conditions which would have been ideal for both the reindeer and the growth of their fodder. It is therefore perhaps not surprising that this period saw a sustained increase in the size of domesticated herds (Krupnik, 1993). The 18th century was one of the coldest periods of the Little Ice Age. This would have provided a climate in which reindeer would be expected to flourish. Reindeer are poorly adapted to high summer temperatures. If the temperature rises above 10°C they are noticeably uncomfortable, and when temperatures reach 15°C the animals suffer various physiological disorders. Under such conditions reindeer fail to thrive in the short summer season and do not make the gain in bodyweight that is essential for successful over-wintering. Warm dry summers are also disadvantageous for the lichens and mosses that make up a large part of the reindeer diet (Figure 5.17). The lichens become brittle and in consequence are more easily damaged by reindeer trampling.

5.6.2 Elk or moose (*Alces alces*)

The vernacular names for *Alces alces* in English can be a common cause of confusion. In the Germanic languages of Europe this, the largest of all the ungulate herbivores, is called Elk (English), Elch (German), and Elg (Norwegian). However, in North America, Elk denotes the Wapati (*Cervus canadensis*), a species that resembles the European Red Deer (*Cervus elaphus*). The American name for what is known in northern Europe as Elk is Moose—a Native American Indian (Algonquin) word meaning *twig eater*. Elk/moose (Figure 5.18),) breed in forests throughout the Arctic and sub-arctic but are absent from Greenland, Iceland, and other arctic islands. In Russia they are called *Loss* (Лось) and do not extend east of the River Ob.

Reindeer in forests paw the ground for herbage or pluck lichens from the branches and therefore inflict little damage on trees. However, this is not

Figure 5.17 Northern pine forest (*Pinus sylvestris*) with a ground cover of reindeer/caribou moss. Despite this common name this is not a moss but a lichen. The dominant species in this example is *Cladonia rangiferina* with some *C. arbuscula*. The lichens readily absorb heavy metal and radioactive pollutants which can then be consumed by reindeer grazing, particularly in winter, and subsequently enter the human food chain (see text).

the case with elk. By contrast, these *twig eaters*, when sufficiently numerous, can seriously damage forest regeneration. Moose are substantial consumers of forage. With twig browsing and leaf stripping Alaskan moose consume upwards of 3% of their body weight per day and swallow over 400 kg of woody browse (Rexstad and Kielland, 2006).

Elk grazing can therefore have a considerable effect on forest tree species composition. A study in Swedish Lapland (Åbisko 68°N) of the age structure of a northern stand of the thermophilic Aspen (*Populus tremula*) and its sub-arctic competitor Mountain Birch (*Betula pubescens* ssp. *czerepanovii*) showed that although Aspen grew 45% faster and had seven times higher recruitment numbers than Birch, no Aspen stand expansion was observed. This appears to be due to elk browsing. As the sub-arctic continues to warm, the dynamics between Aspen and Birch in forest ecosystems will probably depend on the number of vertebrate browsers relative to Aspen recruits. The balance could be

Figure 5.18 Elk or North American Moose (*Alces alces*). (Photo courtesy of Istock).

tipped back in favour of Aspen when one of the common major moth attacks on Birch would favour the spread of Aspen by reducing competition from birch (Van Bogaert *et al.*, 2009).

Elk hunting is an active and popular activity in much of Scandinavia and fortunately for some forests has a significant effect in controlling the elk population. In North America, moose populations have more than doubled over the past 30–40 years. In Alaska there are regions such as in the Tanana Flats and the northern hills of the Alaska Range where the population has grown rapidly since the 1950s and they now carry densities of moose as high as 2 moose/km^2. In Alaska, moose have extended their territory as far as the Arctic Ocean. This dispersal from their traditional forest habitat into the Tundra appears to be a relatively recent event, possibly due to climatic warming (Rexstad and Kielland, 2006). It is remarkable that this recent expansion of the moose population in Alaska has been achieved despite the effects of predation and hunting. Moose can live for up to 15 years, but levels of predation mean that the population turnover period is from 4–5 years, with survival limited on average to no more than 20–30% of the potential life span. In part this is aggravated by natural causes. Bear and wolf predation of moose and caribou in the interior of Alaska is estimated to remove annually 20–25% of the total population (Rexstad and Kielland, 2006).

5.6.3 Beavers

Beavers (Figure 5.18) are frequently described as ecosystem engineers, as they are among the few wild animal species that can significantly change the hydrological characteristics and biotic properties of the landscape (Figures 5.19–5.23). As a result of their stream-damming activities significant alterations can be brought about in wetland habitats. Beaver foraging has a considerable impact on the course of ecological succession, species composition, and the structure of plant communities, making them a good example of an ecologically dominant species (e.g. keystone species, see section 5.6.3). (Rosell *et al.*, 2005). The European and American beavers (*Castor fiber* and *C. canadensis*) were once widespread inhabitants of both temperate and northern boreal forests.

Figure 5.19 European beaver (*Castor fiber*). (Photo courtesy of Per Harald Olsen).

Figure 5.20 Beaver dam in the Yukon. (Photo courtesy of M. Pealow).

In North America and Europe their breeding range extends to the northern fringes of the Tundra-Taiga Interface. They are also found north of the Arctic Circle near the Mackenzie Delta. In North America the former abundance of beavers was much reduced by hunting and they now number between 6 and 12 million individuals.

The European species (Figure 5.18) has been hunted almost to extinction, disappearing from Britain in the 16th century and from Finland in 1868 and soon thereafter from Sweden. Beavers were reintroduced in 1922 to Finland and later to Russia in the 1930s (Danilov, 2005). The Finnish introductions were first made with 19 European beavers from Norway, but a later introduction of 17 Canadian beavers was made to central and eastern Finland. It was not realized at that time that they were distinct species. The population of Canadian beavers increased and spread to the north-eastern Karelian territories, where they have penetrated extensively into the former habitat of the European beaver as far north as Archangel (Danilov et al., 2008).

More recently, however, European beavers have returned to south-eastern Finland and Karelia from Russia and now occupy the areas to which Canadian beavers were originally introduced, i.e. the non-native species is being replaced by the indigenous one.

A problem of competition and displacement of one species by another may therefore have arisen. The populations are small and thus chance can play an important role. Nevertheless there are differences between the two species that may indicate decisive ecological features. The Canadian beavers appear more likely to build dams and lodges than the European beaver. They also tend to have a larger litter size.

Figure 5.21 Birches felled by roadside ditch on the road between Petrozavodsk and Vedlozero, Karelia, Russia. (Photo courtesy of Dr F.V. Fyodorov).

Beaver dams can bring about changes that have marked effects on local fish populations. Studies carried out in a small stream in eastern Canada in summer found that growth rates in both length and mass of salmon parr (*Salmo salar*) recaptured in a beaver pond displayed faster summer growth rates in both length and mass than fish that were recaptured immediately above or below the pond (Sigourney *et al.*, 2006).

In terms of invertebrates, beaver ponds harbour more resources than undisturbed waters. The increase in invertebrate abundance in the beaver ponds can enhance teal brood density and reduce mortality as compared with ponds in waters unaffected by beavers (Nummi and Hahtola, 2008).

Whether beavers constitute a benefit or a hindrance for timber production is perhaps another matter. Foresters are known to have to surround the bases of valuable trees with tree-guards to save them from being felled by beavers. Testing the efficacy of these protectors with captive beavers in field pens has shown that protection is possible, but only the most expensive solutions give complete protection.

Most beaver activity takes place in wetlands and the felling of trees obviously has an effect on the riverside forest, bringing about deforestation of the riparian zone. Their effect on many of the deciduous trees they fell is similar to a natural coppicing process where species such as Oak, Rowan, and Willow will send up new shoots from the stump of a felled tree. Aspen also readily regrows, as its main mode of reproduction is vegetative by root suckers or ramets. The presence of beavers therefore tends to encourage the production of young shoots.

Figure 5.22 Beaver lodge in drained pond showing entrance tunnel. (Photo courtesy of M. Pealow).

The flooding caused by dams can nevertheless result in some trees being drowned. However, others Alder flourish from the creation of a suitable wetland habitat. In the natural environment of the boreal forest and Taiga away from the competing interests of farming and commercial intensive forestry, the beaver may have a role as a keystone species by enriching the ecology of stream systems. The disturbance that is caused to trees along the river banks is partially made good by the promoting of Willow, Alder, and Aspen, regeneration which in turn improves forage quantity and quality for grazing animals.

5.7 Boreal insects

5.7.1 Mammal harassment

As every traveller in the north knows, insects flourish in boreal forests and their associated bogs and heathlands. Their presence is one of the less attractive aspects of arctic travel. In order to escape their unwelcome attention in mid-summer it is necessary to travel north to the coldest regions of the Arctic. For reindeer and caribou the northward summer migration serves not just for access to fresh pastures, but also as an escape from insect attack.

The density of insect populations, although an irritant to many animals and human beings, does nevertheless have a positive effect on certain wetland ecosystems. In most humid climates the excess of rainfall over evaporation results in a flow of soluble nutrients from the land into lakes and rivers and then to the sea. However, where insect populations are sufficiently dense the reverse situation can occur. An extreme example of this phenomenon is found in the north of Iceland at Lake Myvatn (Figures 5.24–5.25).

Figure 5.23 Recent spread of Canadian and European beavers across Karelia. (1) Location of introductions of Canadian beavers. (2) Spread of Canadian beaver. (3) Locations of introduction of European beavers. (4) Origins of reintroduction of European beavers. (5) Spread of Canadian beaver. (6) Area of resettlement of European beavers. (Reproduced with permission from Danilov et al., 2008).

The name *Myvatn* is used not just for the lake but for the whole area of surrounding wetlands, which are classed as a nature reserve (the *Mývatn-Laxá Nature Conservation Area*) to protect the large number of birds that nest in this region. Myvatn is a shallow, nutrient-rich lake situated in an area of active volcanism in north-east Iceland (Figure 5.25). The name of the lake comes from the huge numbers of flies (midges) to be found there in the summer (Icelandic *mý*—midge, *vatn*—lake).

During the summer there can be as many as 50,000 larvae m^{-2} in some places in the lakebed. The lake and its surroundings are exceptionally rich in wildfowl, especially ducks, which gorge themselves by feeding on the emerging adult insects lying on the water, as well as diving for their larval stages on the lake bottom. Recent studies have shown that the emergent aquatic insects, which spend their larval stages in lake sediments and emerge as adults to mate over land, can act as vectors of material, energy,

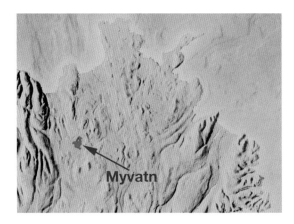

Figure 5.24 North-east Iceland showing the location of Myvatn 65°40′N.

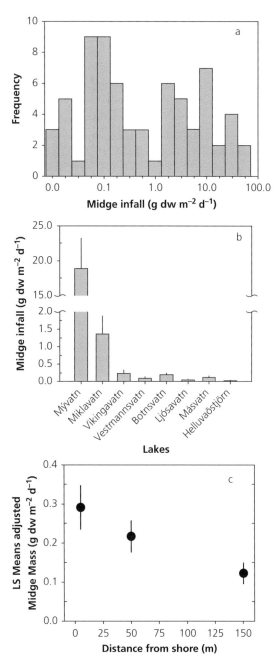

Figure 5.25 (a) Frequency distribution of midge infall captured in midge collectors placed at different distances around eight lakes in northern Iceland. Note log-scale of x-axis. (b) Average midge infall (+ SEM) captured in infall collectors around lakes in northern Iceland over a 5- to 7-day period in early August 2006. (c) Midge infall as a function of distance from shore along transects. (Reproduced from Gratton et al., 2008 with kind permission from Springer Science+Business Media B.V.).

and nutrients from aquatic to terrestrial habitats (Gratton et al., 2008). Annual midge input rates to the lakes in this region vary but can be as high as 1,200–2,500 kg midges ha^{-1} year$^{-1)}$. As midges are approximately 9.2% total nitrogen, this can result in a significant fertilization effect of terrestrial habitats with consequences for plant biomass, quality, and community structure.

In addition it was found from carbon and nitrogen isotope studies that terrestrial arthropods showed increased utilization of aquatic-derived carbon (i.e. midge C) relative to terrestrial sources as midge infall increased. This pattern was particularly pronounced for predators such as spiders and harvest-men (daddy-longlegs—*Opiliones*) and some detritivores (*Collembola*). This study has demonstrated that despite this particular phenomenon being largely ignored, there can be significant aquatic-to-terrestrial habitat nutrient transfer which is the reverse of what is normally expected (Gratton et al., 2008).

Mosquitoes can spread disease in animals as well as human beings. Reindeer in northern Finland were found to have been attacked by a mosquito-borne nematode, *Setaria tundra*, which caused a serious outbreak of peritonitis in 2003–2006 among the reindeer population. The infection has also been found in other deer species including elk. It would appear that climatic warming is a probable factor in the recent northern outbreaks of

Figure 5.26 Myvatn, northern Iceland (65°36′N). (a) View of the shallow lake with a rich fringe vegetation of Archangelica (*Angelica archangelica*); (b) a dense swarm of chironomid midges; (c) green grass growth on the infertile volcanic craters on the lakeshore. The growth of the grass is entirely a result of nutrient input from the chironomid midges. The local farmers call this 'mýgras', i.e. midge-grass. (Photo courtesy of Arni Einarsson at the Myvatn Research Station).

these mosquito-borne infections (Laaksonen *et al.*, 2009).

5.7.2 Insect attacks on trees

In the boreal forest, insect attack as well as being an irritation to grazing animals and human beings in summer can also destroy large areas of forest. The present level of climatic warming, and in particular the milder conditions now prevailing in winter, is increasing the probability, especially in the more oceanic regions, with millions of hectares of forest being killed by outbreaks of phytophagous insects. The insects that prove most dangerous to forests are those that exhibit outbreaks when sudden increases in their number devastate large areas of Birch, Spruce, and Pine. These outbreaks tend to occur in cycles.

Figure 5.27 Example of Mountain Birch (*Betula pubescens* ssp. czerpanovii) growing in the Lofoten Islands, Norway (68°N). The polycormic form can be induced by a variety of causes including flooding, insect attack, and exposure.

Phytophagous insects can be discussed in terms of their taxonomy. However, in a chapter devoted to forest ecology it is more convenient to review their activities in terms of the tree species that they attack.

5.7.3 Birch defoliation

Severe outbreaks of insect defoliation are often cited as one reason for the prevalence of the polycormic (many-stemmed) form in Mountain Birch (Figure 5.27). The most frequent defoliators are *Epirrita autumnata* and *Operophtera brumata*—the autumn and winter moths, which show marked cyclical activity particularly in coastal regions. In the more continental areas, the outbreaks are more irregular but when they do occur can have catastrophic results and leave a mark on the landscape that is visible for decades.

Winter Moth (*Operophtera brumata*) outbreaks are more frequent and usually start on south-facing slopes, as this species is less cold-hardy than *Epirrita autumnata*. When warmer conditions lead to an outbreak in these areas many trees are killed and recovery is slow if grazed by reindeer (Tenow *et al.*, 2001).

5.7.4 Scots Pine defoliation

Scots Pine (*Pinus sylvestris*), as with most trees, can be a host to a variety of insects that can act as defoliators. The Pine Beauty Moth (*Panolis flammea*) has a widespread distribution throughout Europe and Asia. Pine Beauty Moth caterpillars are at their most harmful when they feed on young buds and one caterpillar can consume 200 pine needles during its life. The Pine Looper Moth (*Bupalus piniarius*) also occurs in coniferous woodlands throughout Europe and northern Asia. It is mainly associated with pine trees, especially Scots Pine (*Pinus sylvestris*). It can also affect European Black Pine (*Pinus nigra*) and Norway Spruce (*Picea abies*). The infestations take place in cycles and can cause extensive defoliation.

5.7.5 Spruce defoliation

Spruce Budworm (*Choristoneura fumiferana*) is one of the most destructive native insects in the northern spruce and fir forests of the eastern USA and Canada. Outbreaks occur every few years and Balsam Fir (*Abies balsamea*) is the species most severely damaged by the budworm. These outbreaks have resulted in the loss of millions of acres of spruce and fir. The newly hatched larvae feed on needles or expanding buds. New foliage, which is normally the preferred food, is usually entirely consumed or destroyed before larvae will feed on older needles. Larvae become full grown usually in early July at about 30–40 days after leaving their over-wintering sites. They then pupate in webs of silk they have spun either at the last feeding site or elsewhere on the tree. The pupal stage usually lasts about 10 days. Severe damage to these structures causes the tree to defoliate and die. In Alaska, White Spruce (*Picea glauca*) that are defoliated for several consecutive years suffer from terminal shoot and upper crown mortality and become vulnerable to attack by beetles.

5.7.6 Bark beetles

Bark Beetles are a diverse group of weevils in the subfamily Scolytinae, comprising approximately 220 genera and 6,000 species. The Spruce Bark Beetle (*Dendroctonus rufipennis*) has become a scourge of the spruce forests of the south-west Yukon and has devastated some of the spruce forests of south-central Alaska. In one area of the Chugach National Forest, 51% of the Lutz Spruce (*Picea glauca* x *lutzii*), a hybrid between *Picea glauca* and *P. sitchensis* found only on the Kenai Peninsula of southern Alaska, was killed by Spruce Beetles during a 16-year period (Holsten *et al.*, 1995).

Although these figures suggest that this Spruce Beetle is a dangerous member of the boreal forest community, ecological opinion varies in relation to its effects on the environment. Spruce Bark Beetles occur naturally in Alaskan forests and are important in initiating the decomposition of spruce trees. Under normal forest conditions, Spruce Bark Beetles feed only on wind-felled trees or trees with lowered vitality, and outbreaks occur when there is a sudden increase in prime Bark Beetle habitat, such as an abundance of newly fallen trees, or when climatic conditions are favourable for a rapid increase in beetle reproduction. The natural rejuvenation and biodiversity of the boreal forest depends in large measure on the saprophytic organisms that clear the debris and restore nutrient cycling and make room for natural processes of succession. Thus whether the activities of Bark Beetles in general are to be regarded as positive or negative depends on how the forest is to be viewed. A judgement may have to be made as to whether any particular forest should be managed as an entirely human-controlled timber production unit or else as a biodiverse and naturally regenerating ecosystem that can still provide valuable timber.

5.8 Boreal forest carnivores

The boreal forest and Taiga are home to a variety of carnivores of which the main groups are the bears, various species of dogs and cats, and the weasel family (the Musteidae). The dogs include wolves, coyotes, and red and white foxes. The cats are the Canadian and European lynx. The weasel family comprises wolverine, ermine, stoats, and weasels.

5.8.1 Brown Bears

The greatest concentration of European Brown Bears (*Ursus arctos*) is found in Russia with a population estimated at over 100,000. The Eurasian Brown Bears are smaller than those of North America and are associated with the Taiga and boreal forest. Molecular studies of mitochondrial DNA (Korsten *et al.*, 2009) suggest that *Ursus arctos* colonized America from Asia. The Old World Brown Bears are more diverse in appearance and behaviour than their descendants in the USA and Canada, which reflects their genetic diversity. In North America, Brown Bears prefer open landscapes, while in Eurasia, although formerly more widespread than they are now, they are associated mainly with the Taiga and boreal forest. It has been suggested that this difference may be due to the fact that the particular population that invaded America thousands of years ago did so via Beringia from the Chukotka Peninsula and was therefore pre-adapted to the Tundra environment (Korsten *et al.*, 2009).

5.8.2 Wolves

The Grey Wolf (*Canis lupus*) of the boreal forest and Taiga and its arctic sub-species *Canis lupus arctos*, also called the Polar Wolf or White Wolf, can both be described as glacial relicts. The wolf once had the largest distribution of any mammal except man. However, the loss of its natural habitat, namely forest, due to human settlement and persecution over many centuries has greatly reduced its natural occurrence to the relatively uninhabited northern forests and the arctic Tundra. As with the Brown Bear, there are many sub-species which vary in both size and colour. The largest individuals tend to occur in the northern forests of North America. Like the human inhabitants of these northern lands the wolves are dependent on the herds of reindeer or caribou. They also face the constant risk of human predation as wolf hunting is still widely pursued. However, even small numbers of wolves appear to have an ability to survive as it is usually only with determined attempts to free an area of wolves by poisoning that they disappear entirely.

The wolf is a resourceful and adaptable animal and is maintaining its numbers relatively successfully, and now that it is less feared it is recovering some of its recently lost territory, as in northern Norway. Some concern therefore has to be given to its effects on the wild animals on which it preys, especially if their numbers are low. Wolf predation on domestic animals does cause concern for the individual farmers in some remote areas and, although this loss creates considerable local annoyance, it can be regarded as a sustainable expenditure in the cause of biodiversity, for which the farmers might receive compensation.

Of greater ecological concern, however, is the predation by the wolves of wild animals that exist only at low densities which are also shot, either for food or by people pursuing fire-arm-dependent recreation. Population studies of moose have concluded that yield and density remain low in much of interior Alaska and Yukon, Canada, despite high moose reproductive rates, because of predation from Grizzly Bears (*Ursus arctos horribilis*), Black Bears (*U. americanus*), and wolves (*Canis lupus*). In interior Alaska a 6-year wolf control programme helped to reverse a decline in the moose population, and moose numbers continued a 28-year, seven-fold increase (Boertje *et al.*, 2009).

Such is the reputation of the wolf that its activities as a predator can be over-estimated. Between the years 1930 and 1996, 206 elk (*Alces alces*) and 646 wild reindeer (*Rangifer tarandus*) deaths were recorded in the Lapland Biosphere Nature Reserve located above the Arctic Circle on Russia's Kola Peninsula. Among the deaths where the cause was identifiable, bears (*Ursus arctos*) were responsible for most of the elk and reindeer deaths: 68% and 30% of deaths, respectively. By comparison, wolves (*Canis lupus*) caused 8% of elk and 17% of reindeer deaths and wolverines (*Gulo gulo*) caused 1% of moose and 10% of reindeer deaths. In the area surrounding the Reserve, illegal hunting accounted for 6% of moose mortality and 18% of reindeer mortality, while road kills were responsible for 3% of all elk deaths and 1% of all reindeer deaths. Bears were by far the greatest danger to moose and reindeer in Lapland, while wolves tended to prey primarily on reindeer. Wolverine most frequently targeted weak or sick animals, though they have been known to occasionally attack adult moose (Guiliazov, 1998).

5.9 Conclusions

The Taiga-Boreal Forest complex and its associated bogs and wetlands constitute the world's largest biome and contain 29% of the world's forest cover. The immensity and heterogeneity of this vast region harbours a wealth of biodiversity that has both the space and biological variation to adjust to environmental fluctuations, whether they be from natural causes or human disturbance. The flora and fauna of this region has therefore the possibiity to migrate, hybridize, and evolve and thus respond positively to whatever the future may hold in terms of change. The vitality of the biome, as it now exists, owes its capacity for regeneration to an ability to recover from fire, flooding, freezing, and thawing. This last process due to permafrost degradation and thermokarsting is, however, now occurring in some parts of the Taiga to such an extent that paludification is already occurring and creating extensive swamps and bogs. In the long term this may prove beneficial for the preservation of this biome. Paludification will hinder human over-exploitation and such protection will give time for raised bogs to develop, upon which trees will eventually be able to grow once there is an adequate rooting zone above the level of the perched water table. Although the circumpolar development of the Taiga-Boreal Forest is only a recent post-Pleistocene development, it is likely in a changing world to outlive other more delicate biomes.

CHAPTER 6

Arctic survival in mammals, birds and Insects

6.1 Advantages and disadvantages of arctic habitats

Warm-blooded animals might be thought to have been suitably pre-adapted for the onset of the Pleistocene glaciations. They have the ability, within certain limits, to maintain an active metabolism despite daily and seasonal fluctuations in ambient temperatures. There are nevertheless many risks both to mammal and bird life at high latitudes.

The Arctic has a very variable climate and even in summer it is never free of frost. Cold spells of weather can also interfere with both feeding and reproduction. The fact that the Arctic has its own distinctive mammal and bird species, as well as indigenous human populations, does suggest that for suitably adapted species the polar regions have something to offer that is not found elsewhere (Figure 6.1). For plants, high latitudes with their 24-hour light regime provide an unrivalled potential for round-the-clock photosynthesis. Such a regime produces foliage with high sugar content which makes the vegetation frost tolerant. It also has the effect of making the vegetation attractive to herbivores.

Arctic salt marshes provide an outstanding example of a reciprocal relationship between grazers and the grazed. The summer growth of the sedges and grasses, with their high sugar levels, make the marshes attractive grazing grounds for large populations of migrating geese. The geese in return fertilize the marshes which provide the vegetation with the capacity to nourish the geese. This mutual arrangement nourishes both the vegetation and the geese. There is therefore in the Arctic a state of mutualism between plants and herbivores that compensates for the lack of a sufficiently active soil microflora to maintain the nutrient cycle as it would in warmer climates.

One of the costs for migrant arctic geese is the need to gain enough body weight to support the 1,000-mile or more migration southwards in autumn. For goslings hatched in July this requires a high level of nutrition if they are to be ready to begin the long flight south by September. Such is the quality of the nourishment provided by the salt marshes that the rate of growth of arctic goslings can exceed that attained by highly fed domestic geese at southern latitudes.

The Arctic also has a further attraction in that it can provide a degree of freedom from disturbance and predation that is not found elsewhere. Lemmings can breed under snow and thus escape the attacks of Snowy Owls and Skuas. Geese can manage to defend and fledge their young by either nesting in inaccessible sites (Figure 6.2) or grouping together, as do the various species such as Terns (*Terna* spp.) which often nest on islands and promontories where their collective presence provides defence against marauding Arctic Foxes and Skuas.

6.2 Low temperature survival

Survival at sub-zero temperatures, as with most environmental stresses, has two possible routes, tolerance or avoidance. These routes are not incompatible and many species exhibit adaptations for both avoidance and tolerance. In all cases energy conservation is paramount, and here both birds and mammals use a variety of morphological and physiological adaptations that minimize heat loss. The manner in which energy conservation is achieved has never-

Figure 6.1 Lush vegetation in summer below cliffs on Prince Charles Foreland, Spitsbergen. Note bottom right, reindeer proceeding to snow patch where it retired to keep cool while ruminating (see inset).

theless evolved in different ways depending on the ecological life strategy of the species, sub-species, or even local populations.

For birds a seasonal migration to a warmer climate may seem a simple strategy but it has its disadvantages in terms of energy requirements, as well as risks from bad weather and predation while migrating. There is also the need for careful timing of the return journey so as to arrive in good condition for breeding at the same time as feeding becomes available near the nesting site after the arctic winter (see Chapter 8).

In the case of reindeer (caribou), migration does not necessarily lead to a warmer climate. The continental Boreal Forest can be colder than the more oceanic parts of the Tundra (see Chapter 7). Reindeer migration to the Taiga is primarily a quest for access to the lichens that hang from trees, as the usual source in the Tundra can become inaccessible when the Tundra is frozen.

For arctic mammals and birds, being warm-blooded (endothermic; see Table 6.1) demands that the animals have to be able to maintain their core body temperature, otherwise death will rapidly ensue. The ability to remain warm-blooded during the cold months of the polar night requires the securing in advance of adequate over-wintering fat reserves combined with a high degree of thermal insulation in order to achieve energy conservation. Only by successfully using a range of adaptations can the dangers of irreversible hypothermia be avoided.

Figure 6.2 Barnacle Goose (*Branta leucopsis*) on a predation-secure nest site in north-east Greenland. (Photo courtesy of Dr H. Körner).

6.2.1 Huddling and burrowing

Behaviour can also aid cold survival in reducing heat loss in both individuals and groups. This can be seen in the species that make use of burrowing and huddling. This is in effect a seasonal adaptation which is very common in small mammals such as lemmings, voles, and mice which rely on activity and agility in their search for food. By huddling in a collective nest they are assuming a collective body size which is thermally advantageous. Clustering lessens heat loss in euthermic animals (see Table 6.1) by decreasing the exposed surface area. However, clustering is also used during heterothermic periods such as hibernation, when heat loss should be minimal while body temperature is reduced (see section 6.24).

Rodents living in cold climates employ both behavioural and physiological mechanisms to achieve thermoregulation. Huddling and nest building both help to reduce heat loss especially for young rodents that have not yet developed their fur. When more than one mode of thermoregulation is employed, as when huddling is combined with hibernation, significant energy savings can be made. Burrowing and nest building and especially community nests also owe much of their efficiency to the insulation provided by the burrow in addition to the increased total body mass of the individuals. Nests of Taiga Voles (*Microtus xanthognathus*) occupied by five to ten individuals can be between 7 and 12°C warmer than ground temperatures and when the ground is covered by snow temperatures up to 25°C higher than those prevailing above the snow. The nest is normally never left empty and therefore foraging animals are able to return to a warm refuge (Wolff, 1980).

In the past, *huddling* was also practised by the native dwellers in the Arctic. The traditional custom

Table 6.1 Terms generally used to describe animal body temperatures in relation to the environment.

Ectotherm	Any animal whose source of body heat is primarily external
Endotherm	Animals that generate their own body heat
Euthermic	Organisms that can adapt to a wide range of temperatures and can maintain an optimal temperature in a varying external environment
Homoiotherm	Animals that maintain a more or less constant body temperature regardless of external temperatures
Hypothermia	A condition in which core temperature drops below that required for normal metabolism and body functions and leads to death if external heat is not applied
Poikilotherm	Body temperature varies with the surrounding medium. In popular terminology *cold-blooded*
Regional heterothermy	Organisms that are able to maintain different temperature 'zones' in different regions of the body. This usually occurs in the limbs, and is made possible through the use of counter-current heat exchangers. Common in arctic birds and some mammals
Temporal heterothermy	Animals that are poikilothermic or homeothermic for a portion of the day, or year, e.g. bats and humming birds tend to be homoeothermic when active and poikilothermic when at rest

of the Inuit to lie naked together on a sleeping platform under reindeer or polar bear skins is just another example of the benefits of social aggregation in winter for securing significant increases in ambient temperature.

The use of lodges by family groups of beavers (Figure 6.3) provides these animals with year-round access to an equitable microclimate. A seasonal study from June to the following March of a number of beaver lodges recorded external air temperatures that ranged from –41 to +32°C, while temperatures recorded from within the chambers of occupied lodges varied only from 0–36°C. Surprisingly, despite the large metabolic mass of resident animals, a high level of lodge occupancy resulted in only a limited disturbance to the respiratory gas concentrations inside the lodges. The carbon dioxide levels of occupied lodges ranged from 0.03–1.8%, with no evidence of significant seasonal variation in either CO_2 accumulation or O_2 depletion (Dyck and Macarthur, 1993).

6.2.2 Form and thermal regulation

The often-cited Bergmann's rule postulates that *in many birds and mammals, individuals in colder environments grow more slowly and are larger as adults than their temperate counterparts*. In addition, another rule, Allen's rule, first published by Joel Asaph Allen in 1877 and still much quoted in relation to temperature and body size, states that '*endothermic animals with the same volume may have differing surface areas, which will aid or impede their temperature regulation*'. Basically this means that endothermic animals will have shorter limbs in cold climates than their counterparts in warmer regions.

Figure 6.3 Beaver lodge in winter in Karelia. These structures prove to be highly efficient in protecting beavers from fluctuating external temperatures. (Photo courtesy of Professor F.V. Fyodorov).

Both these rules are merely generalizations. Examples of Bergmann's rule include Polar and Grizzly Bears, and Ravens in northern regions which are larger than their southern congeners. Also the Spitsbergen race of reindeer has shorter limbs than the more southern forms found on the Eurasian continent (see section 9.5.3). Eskimos and Inuit compared with more southern peoples have a lower ratio of surface area to body volume and consequently are better adapted to conserve body heat.

Ravens, which are nearly twice the size of crows, are resident all year round in the Arctic (Figure 6.4) where they feed on small mammals, birds, eggs, insects, worms, and berries. The raven is also a scavenger which enables it to feed in the arctic winter by stealing food from other animals and by following Polar Bears, eating any meat that they have left behind.

When global maps of avian body size are examined, although there is a general pattern of larger body sizes at high latitudes, conforming to Bergmann's rule, there are also exceptions. It would appear, however, that median body size within species assemblages is systematically larger on islands and smaller in species-rich areas (Olson *et al.*, 2009). Once again, although spatial models show that temperature is the single strongest environmental correlate of body size, there are secondary correlations in relation to resource availability and a strong pattern of decreasing body size with increasing species richness. Thus, although prediction based on physiological scaling is remarkably accurate, it is far from the full picture. What might be called lifestyle is largely the cause of many of the exceptions. Whether an animal is agile and nimble or placid and sedate can outweigh the basic concepts in both Bergmann's and Allen's rules.

The Arctic Fox (*Alopex lagopus*) is smaller than the more southern Red Fox (*Vulpes vulpes*). Its survival depends on agility and speed of movement in rocky terrains where a large body weight would be disadvantageous. Voles depend on being able to avoid predators by retreating into holes. Among the many predators of voles are weasels and ermine, and the

Figure 6.4 Contrasting plumage pigmentation in all-year-round resident arctic birds which has evolved as a consequence of different life strategies (see text). *Left* Snowy Owl (*Bubo scandiacus*) (© Rich Phalin/iStockphoto). *Right* Raven (*Corvus corvex*) (photo courtesy of Dr H. Körner).

smaller the vole and its burrow the more difficult is it for the weasel to catch the vole.

In Fennoscandinavia, the Least Weasel (*Mustela nivalis nivalis*) is a very effective vole predator, even capable of entering vole burrows and eating the young from the nest. In winter, the Least Weasel, unlike the Snowy Owl, can still hunt voles under snow. Much has been written about cyclic vole populations and their possible causes (see Chapter 4). One noticeable feature stands out and is known as the *Chitty effect*, named after the first person to have observed that when vole populations are in decline, as well as the population becoming smaller, the size of the voles also decreases (Oli, 1999).

As a reward to successful predators, predation pressure should be expected to be greater on larger voles (Sundell and Norrdahl, 2002). In these conditions it will be the smallest voles that will survive as they can retreat into small holes reinforced by ice against predator excavation. Despite the appealing logic of this argument for the decline in vole size with population predation, there are other possible explanations of the decrease in vole size in relation to nutrition and reproductive effort. Thus, even though the Chitty effect is considered to be a ubiquitous feature of cyclic vole populations, there exists no unanimity of agreement among biologists regarding its causes. When vole populations are in decline, larger predators such as Snowy Owls and Skuas can no longer find sufficient prey and tend to move to new hunting grounds.

In relation to body size voles may seem not to be entirely relevant to the topic of cold adaptation. Nevertheless, it is appropriate to take into account other considerations apart from thermal efficiency that can affect animal size. Evidently, comparative morphology is still a subject for research in relation to cold adaptation.

6.2.3 Pigmentation in the Arctic

The Raven, being black, stands out among the other resident birds of the Arctic such as the Snowy Owl (Figure 6.4) and the Ptarmigan. It could be argued that a bird given to scavenging rather than hunting has no need of camouflage and the dark plumage will absorb heat. However, this is an exception to another ancient rule with regard to avian plumage. This is Gloger's rule, originally based on birds, which stated that 'in humid habitats the plumage tends to be darker than in birds from arid zones'. The rule was extended to animals where there is a tendency for skin colour to be darker in equatorial and tropical species than in polar relatives.

The adaptive nature of pigmentation in skin, fur, or plumage is, however, highly varied. Feathers in humid areas are prone to attack from microbial growth and dark feathers break down less readily than light.

At low latitudes and high altitudes ultraviolet (UV) radiation carries the risk of human skin cancer and can be reduced by pigmentation. At high latitudes darker skin colour can likewise be maladaptive with the absorption of UV to such an extent that a vitamin D deficiency ensues. There are, however, notable exceptions. Among polar peoples such as the Inuit who feed regularly on polar bear, seals, and fish, the high level of vitamin D in their diet means that a darker skin colour is not a disadvantage. In any case, given their effective cold-climate clothing their skin is only exposed to sunlight for short periods in summer (see section 9.6.1).

Fur and plumage colour in the Arctic can also be related to camouflage. The Polar Bears, Arctic Foxes, Ermine, and Snowy Owls seek minimum visibility as they pursue their prey. Ptarmigan and Arctic Hares also seek invisibility in relation to their predators. Thus, even among the few birds that inhabit the Arctic all year round, neither the Black Raven nor the Snowy Owl are disadvantaged by having opposing strategies in relation to external pigmentation.

6.2.4 Hibernation

Hibernation is the popular term to describe winter sleep in animals. The general use of this word, however, means that it is physiologically imprecise. Animals that undergo a true hibernation with deep sleep and reduced metabolic rates are relatively few and include mainly some rodents, hedgehogs, bats, and other small insectivores. Such obligate hibernators spontaneously enter hibernation which is a state of torpor as a result of a circa-annual behaviour rhythm. In the Arctic this deep winter sleep is uncommon. The energy required for arousal from torpor is considerable and in very cold habitats there is always the risk of hypothermic death before arousal takes place. More usual in the Arctic is *facultative hibernation*, either when food supplies become scarce or when temperatures fall. With larger animals, such as bears, and badgers, this can be described best as a state of lethargy from which they may be aroused at any time. In ectoderms, such as reptiles, amphibians, and some fish, a state of torpor is more usual.

6.2.5 Acclimatization and metabolic rates

Hibernation and torpor in animals are accompanied to varying degrees by metabolic down-regulation. For animal species that enter a period of deep sleep, more correctly termed *torpor*, body temperature can be significantly reduced. When hibernation is merely periods of sleep with intermittent periods of arousal, the reduction in body temperature is less marked. In both cases, the onset of the period of hypothermia and a lower metabolic rate are initiated by a fall in environmental temperatures. The subsequent metabolic response or *basal metabolic rate* (*resting metabolic rate*) refers to the amount of energy expended while at rest in a neutrally temperate environment, including the *post-absorptive state* in which the digestive system is inactive. The release of energy in this state is sufficient only for the functioning of the vital organs, such as the heart, lungs, brain and the rest of the nervous system, liver, kidneys, sex organs, muscles, and skin. Basal metabolic rates are prone to modification by body mass and temperature and therefore differ depending on the species in question and expected exposure to environmental fluctuations.

One of the most astonishing cases of *euthermic adaptation* to seasonal changes in temperature is found in the Arctic Ground Squirrel (*Urocitellus/ Spermophilus parryii*). These squirrels are found in Alaska and have a hibernation period that can last for 6–9 months (Figure 6.5), when their body temperature alternates between levels near 0°C degrees during torpor and 37°C during periods of arousal (Williams et al., 2012). This ability is not restricted to just the arctic species as it is also found in the the European Ground Squirrel (*Spermophilus citellus*) which occupies more temperate regions.

Heat production during hibernation in ground squirrels is provided, in part, by non-shivering thermogenesis (see the following section 6.2.6) fuelled by large deposits of brown adipose tissue (Yan et al., 2006).

6.2.6 Brown fat

Brown fat, or *brown adipose tissue*, has long been a subject of interest in relation to adaptation to cold in warm-blooded animals because of its heat-generating properties. Mammals that possess brown fat can generate heat without resource to shivering (*non-shivering thermogenesis*). The tissue is brown due to the density of mitochondria and the iron that they contain. In relation to its origin, brown fat is distinct from white fat (*white adipose tissue*) as the brown lipocytes (fat-containing cells) originate from muscle tissue (Christian and Parker, 2010). Under resting conditions mammal tissues function in the normal manner and generate adenosine triphosphate (ATP), but when stimulated by norepinephrine (noradrenaline) from the sympathetic nerve endings, the *brown adaptive tissue* (BAT) mitochondria are switched to a different proton route by thermal-uncoupling proteins. An uncoupling protein is a mitochondrial inner membrane protein that can

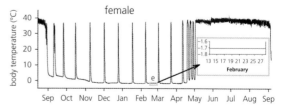

Figure 6.5 Annual patterns of core body temperature of representative male and female Arctic Ground Squirrels. (Adapted from Williams et al., 2012).

release heat from the proton gradient before it can be used to provide the energy for oxidative phosphorylation.

To achieve this, mitochondria operate what is termed a futile cycle in that they catabolize fatty acids from fat and glycerol reserves by β-oxidation. The derived hydrogen is oxidized in the respiratory chain, but the resulting proton-motive force is not used for ATP synthesis but instead simply dissipates as heat. Animals such as hedgehogs and bats when arousing from torpor can regain their active-state temperature by this means without resorting to shivering. Bears, including the Polar Bear, also have brown fat. The rate of heat generation can be two to four times that which is normally associated with vertebrate metabolism.

Brown fat is found in most rodents and small mammals, and some larger hibernating mammals including Brown and Polar Bears. It is located mainly around the neck and large blood vessels of the thorax. Many young mammals have brown fat at birth. Until recently it was thought that brown adipose tissue in human beings was limited to infants, but research now identifies that 20–80% of adults might also have brown adipose tissue provided that they are not obese (Christian and Parker, 2010).

The recent discovery in the north of the Siberian Yamal Peninsula of a female Woolly Mammoth calf (*Mammuthus primigenius*) that died *ca.* 40,000 years ago has shed light on the physiology of these ancient animals. In particular, the presence of brown fat in the body of the baby mammoth has attracted much attention. The mammoth calf, known as *Lyuba*, was about a month old when she died in the Siberian Tundra, where she remained until discovered by a reindeer herder in 2007 (*The Times* 5th October 2009). The calf weighed about 50 kg (110 lb) and was about the size of a large dog. The baby Woolly Mammoth apparently died after being sucked into a muddy riverbed 37,000 years ago. Her body was so well preserved in the permafrost that her stomach retained traces of her mother's milk, and scientists identified sediment in her mouth, trunk, and throat—suggesting that she suffocated while struggling to free herself from the mud. The discovery of brown fat in the baby 1-month-old mammoth calf would indicate the probable importance of this thermogenic tissue for a very young animal born in early spring in the bare, exposed conditions of the Siberian Tundra.

Mammoths were regularly depicted by Palaeolithic cave artists (Figure 6.6). One of the most remarkable sites for these Palaeolithic cave paintings is found at Rouffignac in the Perigord (France). The caves form an extensive labyrinth, with a total length of all the chambers and galleries in three levels of about 10 km. The cave itself has long been known, but the mammoth paintings on one of the lower galleries were only discovered in 1956. One of

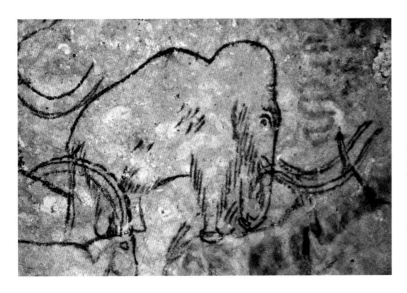

Figure 6.6 An upper Palaeolithic cave painting from Rouffignac, Perigord (France), dating from the Upper Palaeolithic (Magdalenian) (more than 13,000 BP). The picture depicts a mammoth, showing the large hump at the back of the neck which served for the location of brown fat. (Photo courtesy of Grotte de Rouffignac, Perigord, Dordogne, France).

the most famous paintings in the caves is the *Mammoth with eye* which also shows the hump which is probably where brown fat would have been stored. Such fat reserves would have enabled mammoths to withstand both food and water shortage during severe dry cold winters.

6.3 Low temperature protection

6.3.1 Cryoprotection in ectothermic species

Protection against freezing injury (*cryoprotection*) is necessary for all ectothermic animals that over-winter in the Arctic. It is also an essential adaptation for plants, and in particular the trees of the northern boreal forest (see section 7.2).

In both plants and animals, cold tolerance is acquired by acclimatization, with an initial degree of cold-hardiness being acquired after exposure to a mild degree of chilling or freezing. Certain insects, particularly ones that live in higher altitudes or near the Earth's poles, use a state of torpor to survive a drop in temperature. The temporary state of torpor can be combined with an accumulation of glycerol which lowers the freezing point, making insects more cold-tolerant, and provides protection from tissue damage during icy conditions. Glycerol (glycerin), like ethylene glycol (automobile anti-freeze), is a highly effective *cryoprotectant*. Under conditions where it is so cold that there is no free water, survival is only possible provided certain vital organs are protected against cold injury. For plants it is buds and meristematic tissues (see Chapter 7). For insects, as well as other ectothermic animals, vital organs and tissues have to be protected from physical disruption by ice crystals.

At one time it was thought that no adult insects survived the arctic winters and survival was possible only in the larval or pupal stages. Arctic insects were all thought to enter the diapause state in winter. However, examples have now been found in arctic springtails and beetles where the imago enters a state of torpor which is so well protected against frost damage that they can even survive immersion in liquid nitrogen at −196°C. In the Arctic Beetle (*Pterostichus brevicornis*) the adult in its natural environment can endure temperatures below −35°C. Under experimental conditions in the laboratory it has been possible to subject the beetles to −87°C for 5 hours without apparent injury. The waxy coverings on the insect wings also reduce the risk of freezing when temperatures drop. There are also striking examples of adult insects surviving the winter due to a capacity for supercooling (see below).

6.3.2 Supercooling

Supercooling, also known as *undercooling*, takes place when lowering the temperature of a liquid below its freezing point does not cause it to become a solid. Differential scanning calorimetry can determine the temperature to which tissues and organs can be supercooled. Supercooling can be seen by recording the temperature at which the latent heat of freezing by ice formation causes an instant rise in tissue temperatures (see also 7.4.4). When such measurements are made on the larvae of the Flat Bark Beetle not only do they supercool, but also when their water content freezes at temperatures of −58°C it does so only into a glass-like state and the larvae still survive. It was this novel finding that has added *vitrification* to the list of known insect overwintering strategies. Some cryoprotectants have the capacity to make water harden like glass without any crystal formation. *Vitrification* avoids freezing damage to cells as there is no disruption of cell membranes by the formation of ice crystals. It also avoids the dangers of tissue desiccation.

Experiments on the larvae of the freeze-avoiding Flat Bark Beetle (*Cucujus clavipes puniceus*-Coleoptera) in Alaska have found supercooling points in winter between −35 and −42°C, with the lowest supercooling point recorded for an individual of −58°C (Sformo et al., 2010). While over-wintering beneath the bark of fallen trees, the larvae of this species may experience ambient temperatures as low as −40°C (and lower) as the sub-bark microhabitat is uninsulated due to the lack of snow cover. As temperatures decrease in winter they are accompanied by a loss of body water from summer-high levels from near 2.0 mg^{-1} d.wt to winter lows near 0.4 mg^{-1} d.wt. Such changes are accompanied also by increases in glycerol concentrations.

Remarkably, some species of frogs can spend days or weeks with as much as 65% of their total body water as ice. Some amphibians achieve their protection due to the glycerol manufactured by their livers. The sugar glucose, like glycerol is also a cryoprotectant. Arctic frogs have a special form of insulin that accelerates glucose release and absorption into cells as temperatures approach freezing.

6.4 Adaptation case studies

6.4.1 Arctic herbivores

Muskoxen

As already described (see section 4.6.5), Muskoxen (*Ovibos moschatus*) are now left as the largest and hardiest herbivores to inhabit the Tundra (Figure 6.7). They never need to take shelter even in the severest blizzards. This degree of cold-tolerance adaptation obviously excites admiration and curiosity as to how it is achieved. The fleece of this animal is quite unique. It has two layers: an outer coarse layer with underneath a soft fleece of very fine wool that the Inuit term *qiviut*. Wild Muskoxen have qiviut fibres approximately 18 μm in diameter. Qiviut is stronger and eight times warmer than sheep's wool, and softer than cashmere. Unlike sheep's wool, it does not shrink in water at any temperature. The Muskox sheds this layer of wool each spring. Females and young animals have slightly finer wool. This wool is so prized by weavers of fine scarves and hats that student arctic expeditions have been known to finance themselves by collecting qiviut from places where it has been shed by rolling Muskoxen.

Muskoxen are large (> 200 kg adult body mass) which favours stabilization of their core body temperature. Mating takes place in August and September, with most calves born in late April or May when temperatures are still low and there is also the risk of blizzards. Within a few hours of birth, calves are able to follow their mothers back to the protection of the herd. For some time their body weight is less than one-third of the body mass of mature females and therefore they are likely to suffer greater thermal stresses than adults due to a higher surface area to volume ratio (Figure 6.8). Nevertheless, the newborn calves are known to tolerate ambient temperature of −35°C and still maintain a *deep body temperature* of 39.5°C (Blix *et al.*, 1984). Two adaptations provide the necessary thermal protection for young calves. The first is the generation of metabolic heat, i.e. the futile cycle catabolizing of the brown fat with which they are endowed at birth. The second is the prime fur insulation with which they are born.

Remarkably, calves have been found not to lose relatively more heat than larger adults. Coat surface temperatures were only 2–5°C above ambient even

Figure 6.7 Close-up of Muskox head, showing some of the finer under-fur layer (Inuit—*qiviut*).

Figure 6.8 Muskoxen calf in north-east Greenland. Aided by a thick fleece and reserves of heat-generating brown fat they are able to survive low temperatures and blizzards despite a higher surface area to volume ratio. (Photo courtesy of Dr K. Körner).

when air temperature fell to –40°C. Body temperatures recorded deep within the ear canal near the tympanum fluctuated in both cows and calves. Apparently Muskoxen combine peripheral heterothermy and an exceptional winter coat to minimize sensible heat loss in winter (Munn et al., 2009).

The diet of Muskoxen differs with the seasons. In summer they graze preferably on the more nutritious vegetation of sedge meadows and continue with this as long as possible. The nutritious value of sedges and grasses of arctic meadows stems from their own particular adaptations to the arctic winter.

In plants, tolerance of ice-encasement demands an ability to survive enforced deprivation of oxygen. For this to be possible a high over-wintering sugar content is necessary, which makes these tolerant plants, and particularly certain sedges, highly nutritious in late summer and autumn (see section 6.1). When the snow cover becomes too hard for the Muskoxen to break they move to drier ridges where wind-blow reduces snow cover and here they feed on willow twigs. In Greenland this is principally the circumpolar *Salix arctica*, which supports their grazing during the polar night. Food intake is both lower in quantity and protein content in winter.

Rates of ruminal degradation in captive Muskoxen have been observed to increase more than 100% between spring and autumn for cellulose and hemicellulose. This change is accompanied by increased bacterial counts in ruminal fluid (30%). There can also be an increase in the concentration of fermentation acids (16%) and a fall in ruminant pH. These seasonal changes in feeding and fermentation in Muskoxen have been suggested as minimizing winter costs and maximizing nutrients and the gain in energy when grazing on coarse forage (Barboza et al., 2006).

The diet in winter is lower in protein than in summer. However, this is not necessarily disadvantageous, as a lower nitrogen content in the diet reduces renal activity and urination, and the need for water. The consumption by arctic ruminants of cold water and food can increase costs of thermoregulation and potentially impair the bacterial fermentation on which these animals rely for digestion. Experiments carried out on a castrated adult given a consistent diet throughout winter have shown that rumen temperatures were not static but were punctuated by cold shocks down to +26°C in each month. Water turnover rates were high in October (11.1 kg day^{-1}) but declined by January and March (9.8–7.7 kg day^{-1}). The cost of warming ingesta was at times estimated to be as high as 14% of the predicted intake of digestible energy (Crater and Barboza, 2007).

Nutritional deprivation in winter has long been suspected as a cause of population decline in both Caribou and Muskoxen. Quantitative studies have shown that Muskoxen lose a considerable amount of body mass over winter (–6% to –12%) and fat (–22% to –24%), and that pregnant Muskoxen lose a further amount of body protein (–6%) in late gestation. However, indications were also found that amino nitrogen from body protein was reutilized in late winter. It appears, therefore, that female Muskoxen conserve the capital of body protein stores for reproductive investment while using income of dietary protein for maintenance functions (Gustine et al., 2011).

Reindeer

Reindeer are remarkable for their ability to survive on some of the most northerly terrestrial habitats on Earth. By contrast to other arctic mammals, reindeer, due to a specialized gut microflora, are able to digest the carbohydrate content of lichens. The unusual chemical structure of lichen carbohydrates renders them indigestible by other herbivores (Mathiesen et al., 2000). The substances concerned are lichenin and usnic acid. The lichens most commonly eaten by reindeer, namely *Cetraria islandica* and *C. sylvatica*, are rich in the carbohydrate lichenin, a glucan containing 80–200 beta-glucose units. Of these, about 30% are glycosidically linked to the third carbon atom, and the remainder to the fourth carbon atom of the neighbouring sugar units (Figure 6.9).

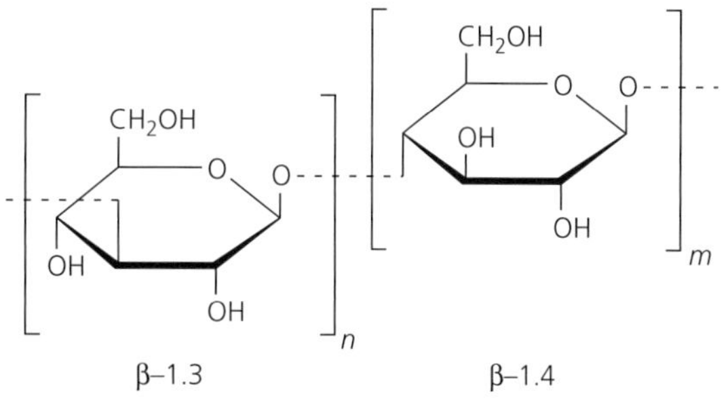

Figure 6.9 Structural formula and linkage of lichenin chain.

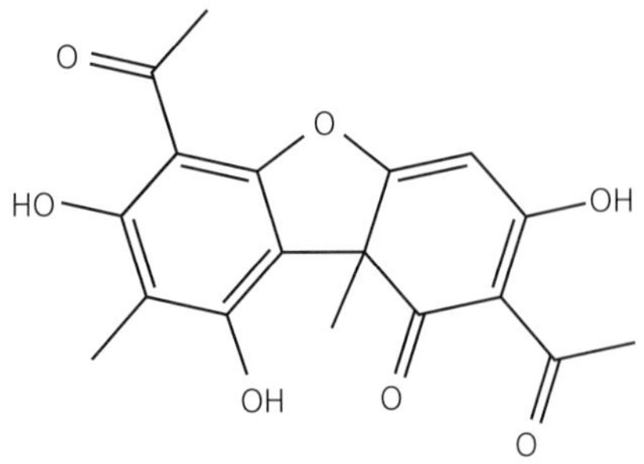

Figure 6.10 Usnic acid.

Lichens also synthesize and accumulate a wide variety of phenolic secondary compounds, such as usnic acid (Figure 6.10), which act as a defence against most herbivores and provide protection against damage by the UV component in solar radiation.

When reindeer are experimentally tested for their ability to digest usnic acid, by first feeding them on vegetation devoid of the acid and then suddenly reintroducing usnic acid into their diet, no traces of the acid or its conjugates are found, either in fresh rumen fluid, urine, or faeces. This suggests that usnic acid is rapidly degraded by rumen microbes, and that consequently it is not absorbed by the animal (Sundset *et al.*, 2010).

The uptake of volatile fatty acids that are produced by bacterial fermentation of forage in the rumen is enhanced by the presence of papillae which greatly increase the surface area of the mucosa of the rumen in several wild ruminant species. The degree of papillation (the *surface enlargement factor*) appears to be related to the level of microbial activity and the rate of production of volatile fatty acids in the rumen. In several species of wild ruminants the surface enlargement factor decreases markedly in winter, apparently in response to a decrease in the quality and availability of forage and also, presumably, to the level of ruminal microbial activity. However, when observations were made on domesticated reindeer in a natural pasture in northern Norway, contrary to expectations, it was found that there was no reduction in the degree of papillation of the rumen or the rate of production of volatile fatty acids (Mathiesen *et al.*, 2000).

This unique ability to graze on lichens finds its maximum benefit in the migrating populations of reindeer. In summer they can graze over wide areas of Tundra when it is snow-free. In winter a retreat to the Tundra-Taiga Interface allows them to access the arboreal lichens that hang from the trees above the cover of winter snow. Lichens are deficient in minerals, proteins, and lipids. Nevertheless, in winter when there is a limited availability of grasses and herbs, lichens provide a useful source of energy for surviving the coldest part of the year.

The Spitsbergen (Svalbard) Reindeer are considered to be a distinct sub-species (*Rangifer tarandus platyrhynchus*) and are the northernmost populations of the Eurasian Tundra Reindeer (*Rangifer tarandus*; see also section 9.5.3). These northernmost populations do not have all-year-round access to lichens that are a major component of the diet of the more southerly populations. One of the most extreme examples of reindeer survival is on Nordaustlandet, the most northerly and second largest island in the Spitsbergen archipelago (Figure 6.11).

Despite the lack of lichens, the polar desert vegetation of Nordaustlandet supports a reindeer population of 400–500 animals. Ecological studies on Nordaustlandet of the reindeer and their summer and winter diets have shown that the reindeer utilize all ice-free areas and adjacent islets, but the preferred plant communities are snow beds and marshes and the vegetation below bird cliffs. At this latitude plant coverage in the polar deserts varies

Figure 6.11 Location of Nordaustlandet in the Spitsbergen archipelago, 79°45′–8°30′N.

from 0–2%, and the dominant species that they consume is *Saxifraga oppositifolia* on dolomite rocks and *Luzula confusa* in other areas. Naturally manured bird cliff sites have high vascular plant cover and diversity and provide important forage for reindeer and geese in an otherwise sparsely vegetated region (Cooper, 2011).

The rumen content of sample animals that were shot for examination contained 45% *S. oppositifolia* herbage, whereas monocotyledons amounted to a similar quantity in rumen samples from animals shot adjacent to bird cliffs. Other important grazing plants that were identified were *Cerastium regelii*, *Papaver dahlianum*, and *Cochlearia officinalis*. Mosses and lichens appeared unimportant in the diet. The results were taken to indicate that at this high latitude there was no pronounced difference in winter and summer habitats of the reindeer, possibly due to the effect of more favourable snow and ice conditions.

When the digestive tract of Svalbard reindeer is compared with that of animals from northern Norway the main difference found is that, relative to their body weight, the Svalbard reindeer have a larger caecum/colon complex than the Norwegian animals. It was also found that in Svalbard reindeer the whole digestive system was larger in animals during winter than in summer. No such seasonal variation has been noted in Norwegian reindeer (Staaland and Punsvik, 1979).

It is suggested that the development of a large caecum/colon complex in Svalbard reindeer (Figure 6.12)

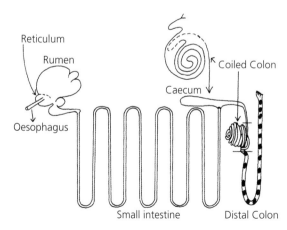

Figure 6.12 Principal features of the reindeer intestines. (Adapted from Staaland and Punsvik, 1979).

could allow for enhanced absorption of water, ions, and products of caecal fermentation when reindeer consume foods of extremely low digestibility.

Lemming nutrition

Fluctuations in lemming populations have already been discussed (see section 4.6.1). For many parts of the Arctic the lemming is the link in the food chain between vegetation productivity and population levels of numerous arctic carnivorous birds and mammals.

In common with other microtine species, lemmings rely on snow to give them sheltered access in winter to the hay-like forage that lies under thick snow cover. Lemmings therefore differ from the larger herbivore species which seek relatively snow-free areas for grazing. As a consequence of this, snow-burrowing mode-of-life lemmings benefit doubly from snow protection as it gives shelter from cold and predation and also provides all-year-round access to fodder. Consequently, the extended breeding season allows their populations to multiply rapidly when conditions are favourable.

Lemmings also differ greatly in their physiology from other herbivores in that they deliberately eat mosses, which due to the phenolic-like compound in their cell walls are indigestible to most mammals. Mosses dominate the Tundra, and the energy they store is only slightly less than that found in flowering plants. They are also reservoirs of inorganic nutrients. As moss is plentiful in the Arctic it is not surprising that it is commonly found in the faeces of many arctic herbivores. A certain amount of moss is normally consumed by Soay Sheep, Reindeer, Barnacle Geese, and some rodents including the Wood Mouse (*Apodeumus sylvaticus*), the Bank Vole (*Myodes glareolus*), and the Field Vole (*Microtus agrestis*).

It is, however, the lemming (*Lemnus* spp.) that consumes moss in greatest abundance (Prins, 1982). The ability to digest fibre with the aid of symbiotic micro-organisms in rodents and lagomorphs (hares, rabbits, and pikas) is facilitated by the morphological development of the hind gut and the coiled colon (Figure 6.13).

The promotion of hind-gut fermentation is dependent on maintaining a positive mineral balance which is also essential for reproduction and

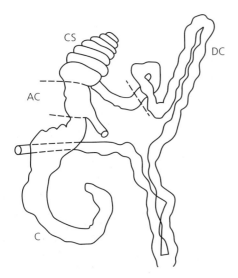

Figure 6.13 Caecum and colon of the Scandinavian lemming (*Lemmus lemmus*). AC, Ampulla coli; C, caecum; CS, colonic spiral; DC, distal colon. Note the relatively large caecum that is a feature of the microtine alimentary system. (Reproduced with permission from Sperber *et al.*, 1983).

lactation (Batzli *et al.*, 1980). A part of the nitrogen and mineral pools in the alimentary tract comes from daily food intake, but a substantial part originates from endogenous secretion into the tract. Nitrogen and minerals are essential to maintain optimal concentrations for the symbiotic digestive processes in the hind gut of rodents and lagomorphs.

6.4.2 Coprophagy

Coprophagy is the consumption of faecal material. For some animals the consumption of their own material serves to retrieve resources that would otherwise be lost. After ingested vegetable matter has served in part for the activity of the gut flora, the minerals and nitrogen (as ammonium) can be recovered by ingestion of faecal pellets (*coprophagy*). Faecal pellets that have been exposed to the external environment become enriched by aerobic microbial activity. However, rabbits and some other small mammals take soft faecal pellets directly from the anus (*caecotrophy*) at night and store them in the stomach to be mixed with food taken during the day. The passing of food through the alimentary canal twice not only increases digestion but also adds metabolites essential for digestion, together with certain vitamins (e.g. vitamin B_{12}) that are produced in the caecum. Rabbits become ill, and some animals will even die, when prevented from eating their faeces. The recovery of minerals from hind-gut digesta and faecal pellets is also essential to maintain the animal in mineral balance. The efficiency of hind-gut absorption is especially important in those species subjected to seasonal declines in food quality, as any loss of endogenous minerals through lowered absorption may not be recoverable in the diet. Coprophily can also be a means of escape from potentially toxic substances in the herbage as well as being an aid to general digestibility.

6.4.3 Grazing-induced digestion inhibitors

Many plant species when actively grazed can develop defences against herbivore activity by means of secondary metabolic products. Two common arctic plant species that are able to counteract intense grazing with a very proficient defence are Stiff Sedge (*Carex bigelowii*) and Common Cotton Grass (*Eriophorum angustifolium*) (Figures 6.14 and 6.15). Both species respond to wounding by producing trypsin inhibitors. The grazing strategy of lemmings and several other species of arctic rodents is to maximize nutrient uptake through rapid fodder ingestion. Such a grazing habit relies on quantity more than quality. Consequently, they are liable to be vulnerable to the induced action of trypsin inhibitors in their fodder, particularly as the vegetation that they eat in winter is often of low nutritive value. This condition can be alleviated by the consumption of even larger quantities of herbage. This then leads to a greater intake of the trypsin enzyme inhibitor which when ingested by the herbivore greatly reduces the nutritional value of its fodder. Neutral extracts of experimentally wounded specimens of these species have been shown to produce an inhibitor of the proteolytic activity of the enzyme trypsin. Increasing trypsin inhibitor (TI) activity is associated with a concomitant decrease in the concentration of soluble plant proteins (TI/SPP-ratio) in the ingesta of the herbivore.

At different phases of a 3-year lemming cycle, the TI/SPP-ratio has been observed to be highest at

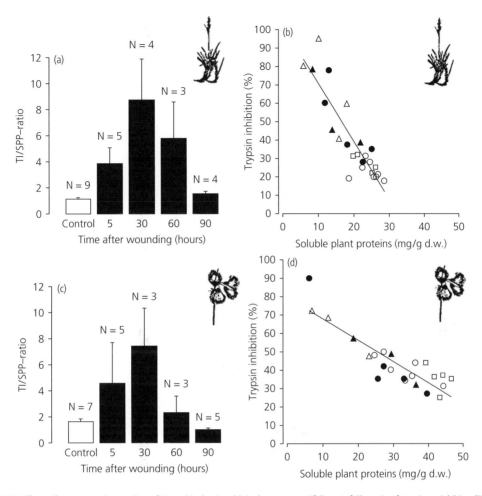

Figure 6.14 Effects of experimental wounding of *Carex bigelowii* and *Eriophorum angustifolium*. *Left* The ratio of trypsinase inhibitor (TI) to soluble plant protein (SPP) in relation to time after wounding; *right* as evidenced in relation to trypsinase inhibition activity as a percentage of activity in the absence of inhibitor in the assay system. (Adapted from Seldal et al., 1994).

peaking and declining densities of the lemming population and lowest when lemming densities were increasing. Moreover, lemmings from a declining population have shown pancreatic hypertrophy, which is a well-documented pathology caused by a prolonged dietary intake of trypsin inhibitors. Thus, a grazing-induced production of trypsin inhibitors can lead to an excessive excretion of undigested dietary proteins from the alimentary tract and can explain the adverse effects typical for the declining phase of cyclic herbivore populations, e.g. retarded growth, compressed breeding season, delayed sexual maturation, low recruitment, low survival rates, and increased dispersal.

Lemmings are monogastric herbivores (non-ruminants) that maximize nutrient uptake through rapid ingestion, as do several small species of arctic rodents. Such a grazing habit relies on quantity more than quality. Consequently, they are vulnerable to the induced action of trypsin inhibitors in their fodder, particularly as the vegetation that they eat in winter is often of low nutritive value and they tend to consume large quantities.

Theories for lemming population cycles are numerous and, depending on the local environment and its vegetation, there may be more than one explanation for the cyclic dynamics of lemming populations (see also section 4.6.1). Some authors

Figure 6.15 Stiff Sedge (*Carex bigelowii*), a species that produces trypsinase inhibitor as a grazing deterrent.

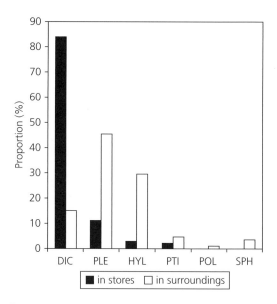

Figure 6.16 Proportion of different moss genera in lemming winter food stores (*n* = 18) and the proportion of the different species in the surrounding vegetation. DIC, *Dicranum*; PLE, *Pleurozium*; HYL, *Hylocomium*; PTI, *Ptilium*; POL, *Polytrichum*; SPH, *Sphagnum*. (Adapted with permission from Eskelinen, 2002).

emphasize arguments that are based on nutrition, while others claim that predator–prey relationships are so strong that the nutritional factor is irrelevant. It should be noted, however, that lemmings have a grazing strategy that is different from most herbivores in that they consume a very large quantity of low quality vegetation. They also have in their gut microflora enzymes that specifically attack the lignon-like phenolic compounds found in moss cell walls. It has been suggested that the all-year-round breeding pattern of the lemming creates a need in this small lactating mammal for adequate supplies of calcium and phosphorus. It could be therefore that it is the demand for these elements that necessitates the consumption of herbage to an extent that is more than is necessary just for energy needs (Batzli *et al.*, 1980).

Lemmings and voles can, despite their bulk grazing habits, be selective grazers with regard to mosses. Studies of the diet of the Wood Lemming (*Myopus schisticolor*) in eastern Finland showed that this species feeds mainly on the moss genera *Dicranum* and *Polytrichum*. The contents of nitrogen and carbon as determined in the moss species correlated significantly with that in faeces. The nitrogen content was highest in the *Dicranum* and *Polytrichum* species (Figure 6.16). This may be one reason for the preference of this moss species rather than the more common *Pleurozium* and *Hylocomium* species (Eskelinen, 2002). The search for one prime cause for the vole population cycles remains elusive. It is possible that in different habitats with varying ecological histories there is a plurality of possible causes. The recent observations that vole cycles are in decline (see Chapter 4) does, however, suggest that population cycles could be merely the consequence of living in one of the world's simplest predator–prey communities, with the simplicity of the food chain having a bearing on nutritional factors.

6.5 Arctic terrestrial carnivores

6.5.1 The Arctic Fox (*Alopex lagopus*)

The Arctic Fox (*Alopex lagopus*), sometimes known as the Polar or White Fox, is the smallest homeothermic carnivore to remain active in the Arctic during the winter. This agile, active animal ranges widely and can survive temperatures as low as −50°C. From this point of view, it can be considered as the hardiest, circumpolar terrestrial mammal. It is found throughout the Arctic and, together with the Polar Bear (*Ursus maritimus*), has been observed close to the North Pole, scavenging remains of Polar Bear kills. To venture so far from land at such high latitudes is remarkable considering the fox's small body size. Fridtjof Nansen (1861–1930) in his voyage through the polar ice (1893–96) recorded seeing the tracks of an Arctic Fox near the North Pole a few days after starting his return journey from his furthermost point north at 86°N (Figure 6.17).

The remarkable degree of cold tolerance shown by the Arctic Fox is due to the excellent insulative properties of its fur and, as mentioned before, it is sometimes referred to as the White Fox due to its white coat, which is predominant in the winter season. With the arrival of spring the white coat tends to become slightly thinner and the legs, back, and tail tend to become brown (Figure 6.18).

Some Arctic Foxes have a bluish coat during the winter season, which also becomes darker in summer. Apart from having the best fur insulation of any arctic animal, the Arctic Fox also has a cold-adapted blood circulation through a counter-current heat exchange system. The paws are by necessity cold, but blood can circulate to bring nutrients to the paws without losing too much heat from the body. For this, the arteries and veins in the limbs are in close proximity to each other which results in a heat exchange, so that as the blood flows down to the paws it becomes cooler and does not lose excessive heat to the snow. As the blood flows back from the paws it regains heat and thus helps to maintain the body core temperature. The fox has a low surface to volume ratio, as is seen in its generally rounded body shape, when compared with the more southern Red Fox (*Vulpes vulpes*). This example of Allen's rule (see section 6.2.2) is a further aid to minimizing heat loss.

Well-insulated paws which have a capillary network of blood vessels prevents freezing when standing on a cold substratum, which enables the Arctic Fox to walk on ice in search of food. This, combined with very acute hearing, allows the fox to accurately locate prey under the snow. When not hunting, the fox seeks shelter in snow lairs or in underground dens.

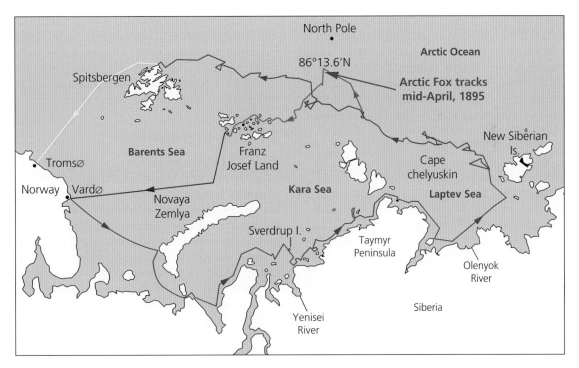

Figure 6.17 Voyage of the *Fram*. Red line—eastward journey from Vardø to the New Siberian Islands (July–September 1893). Blue line—westward journey adrift in the ice to Spitsbergen (1893–96). Green line—Nansen's journey on foot with Johansen reaching their furthest north point (mid-April 1895) where shortly afterwards they saw the tracks of an Arctic Fox. Return to Cape Flora in Franz Josef Land, June 1896. (Image adapted from Wikipedia: http://en.wikipedia.org/wiki/Nansen's_Fram_expedition).

Figure 6.18 Phenotypic plasticity in the Arctic Fox (*Alopex lagopus*). *Left* In summer coat; *right* in winter coat. (Photo *left* courtesy of Dr H. Körner; *right* © Catharina van den Dikkenberg/iStockphoto).

The Arctic Fox copes with seasonal fluctuations in food supply by storing a high level of body fat and caching food items during summer and autumn. Studies in Spitsbergen of post-absorptive resting metabolic rates (RMRs) and body mass after *ad-libitum* food intake have shown an annual cycle in captive Arctic Foxes (Fuglei and Oritsland, 1999). During the light days of summer, food capture was up to 15% higher than the values in the dark days of winter, suggesting a physiological adaptation aiding energy conservation during winter. More significant, however, is the fact that body mass increases in winter despite a reduced food intake (Figure 6.19).

A significant decrease in resting metabolic rate (metabolic depression) is also observed during periods of starvation both in winter and summer, indicating an adaptation to starvation in Arctic Foxes. However, at very low ambient temperatures, Arctic Foxes may require increased heat production which cannot be achieved with below-average rates of metabolism (Fuglei and Oritsland, 1999).

6.6 Arctic terrestrial birds

There are only six species of land birds that regularly over-winter in the Arctic. Among birds of prey (raptors) the Gyrfalcon (*Falco rusticolus*) and the Snowy Owl (*Bubo scandiacus*) are the only two arctic residents. The Raven (*Corvus corax*) is also resident and is omnivorous, preying on small birds and mammals as well as scavenging on carrion.

Rock and Willow Ptarmigan (*Lagopus mutus* and *L. lagopus*) together with the Arctic Redpoll (*Carduelis hornemanii*) survive by feeding on seeds and buds of willow and birch. For the raptors their distribution is dependent on access to prey and here the presence or absence of lemmings and voles is of crucial importance. In Spitsbergen, for example, there are no lemmings and consequently no Snowy Owls, Gyrfalcons, or Long-Tailed Skuas. In terms of living furthest north, the record is held by the Rock

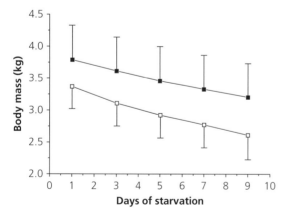

Figure 6.19 Mean values of body mass during starvation in Arctic Foxes at Svalbard in the light period in May (open squares) and dark period in November (filled squares). (Reproduced with permission from Fuglei and Oritsland, 1999).

Ptarmigan which is the only bird to over-winter in Franz Josef Land and is therefore discussed below in detail in relation to its adaptations to this most exacting of winter environments for terrestrial birds.

6.6.1 Gyrfalcon

The Gyrfalcon (*Falco rusticolus*) is the world's largest falcon and the largest resident bird in the Arctic. Its distribution is circumpolar but it is absent from Spitsbergen. It is also the earliest nesting arctic bird and begins egg laying in early May even in the High Arctic (Sale, 1988). Incubation lasts 1 month and the young are able to fly 7 weeks after hatching. When nesting, a favourite site is a ledge on a cliff face (Figure 6.20). Gyrfalcons use the same nest sites over long periods of time, and in the cold dry climate guano and other nest debris decay only slowly. Carbon dating (calibrated C^{14}) of the stratified faecal accumulation in central-west and north-west Greenland estimated the age of the oldest guano levels to *ca.* 2,740–2,360 BP. Isotope discrimination data ($\partial^{13}C$) showed that Gyrfalcons in the more coastal sites had a more marine diet than those nesting further inland, which had more terrestrial sources of subsistence. It has been suggested that the duration of nest-site use by Gyrfalcons is a probable indicator of both the time at which colonization occurred and the palaeo-environmental conditions and patterns of glacial retreat. Nowhere before has such an extreme long-term and continuous use of raptor nest sites been documented (Burnham *et al.*, 2009).

Gyrfalcons have often been seen in the drift ice many kilometres from the shore. But until recently little has been known of their winter movements. However, recent tagging of Gyrfalcons with satellite transmitters in different parts of Greenland have recorded an adult male that travelled 3,137 km over a 38-day period (83 km/day) from northern Ellesmere Island to southern Greenland, and an adult female that travelled 4,234 km from Thule (76°N) to

Figure 6.20 Gyrfalcon (*Falco rusticolus*)—portrait and guarding its eyrie on Baffin Island, 73°N. (Photo courtesy of Clare Kines).

southern Greenland (via eastern Canada) over an 83-day period (51 km/day). In general, return migrations were faster than outward ones. These records highlight the importance of sea ice and fjord regions in south-west Greenland as winter habitats for Gyrfalcons, and provide the first detailed insights into the complex and highly variable movement patterns of this species (Burnham and Newton, 2011).

Gyrfalcons prey on birds of medium size such as ptarmigans, ducks, and sandpipers. Ptarmigan are a significant item. Large birds (geese) and small birds (passerines) become Gyrfalcon's prey less often. In Yakutia (Figure 6.21) Gyrfalcons nest on the Tundra, as well as in the Tundra-Taiga Interface and in the Taiga itself. Long-term studies have shown that the birds remain in central Yakutia and further south in all seasons, including the winter. The Gyrfalcon diet in Yakutia includes 30 species of birds and 6 species of mammals. In the composition of bird species, the share of willow and rock ptarmigan is high everywhere, although in the Lower Kolyma Tundra they give way to ducks. Among mammals, species of open and semi-open habitats are the most preferable. Drastic changes in gyr falcon prey density are one of the most serious causes of changes in their number and ability to reproduce (Labutin and Ellis, 2006).

In the flat plains of the Yamal Peninsula, Gyrfalcons nest mainly on trees, where they take over the former nests of other bird species, using nests ranging in size from the smaller nests of Ravens and Rough-Legged Buzzards to the larger nests of the White-Tailed Eagle. The survival rate of Gyrfalcon nestlings in smaller nests has been noted to be considerably lower than that in the larger nests. A current reduction in brood sizes overall in this region has been attributed to an increase in snow cover and density caused by climate warming which is making the larger nests less accessible to the Gyrfalcon (Mechnikova et al., 2010).

In Iceland, the Gyrfalcon is a resident specialist Ptarmigan predator. A cycle has been noted in juvenile mortality being in excess of adult mortality. This mortality rate appears with a time lag of 2–3 years in relation to Ptarmigan density. It has been suggested that the Gyrfalcon might contribute to the generation of the multiannual cycles of the Ptarmigan in Iceland (Nielsen, 2010). A similar situation is found in Gyrfalcon populations in northern Sweden (66°N), where Rock Ptarmigan (*Lagopus mutus*) are the Gyrfalcons' most important quarry, constituting more than 90% of the prey biomass. A 21-fold difference in Ptarmigan abundance was found across Gyrfalcon breeding territories. However, this great variation in prey availability corresponded to only about a 10% shift in Gyrfalcon diet across territories, suggesting that the Falcons were reluctant or unable to compensate for declining Ptarmigan availability by using alternative prey categories. The Gyrfalcons acted as true specialist predators, and their narrow food niche probably reflects a general lack of suitable alternative prey in the study area (Nystrom et al., 2005).

6.6.2 Snowy Owl

The Snowy Owl (*Bubo scandiacus*) deserves to have the same iconic status for the terrestrial Arctic as that enjoyed by the Polar Bear for maritime regions. Like the Polar Bear, in comparison with its congeners it too is large, second only in size among owls to the Eurasian Eagle Owl (*Bubo bubo*). The Snowy Owl is ubiquitous throughout the Arctic wherever there are lemmings. They have been found at 82°N on Ellesmere Island and as far south as 60°N when nesting in sub-arctic regions such as Shetland.

With regard to territory, the Snowy Owl is a nomadic species with a circumpolar distribution (Figure 6.22). Being an opportunist in its search for lemmings, which it can consume, depending on the

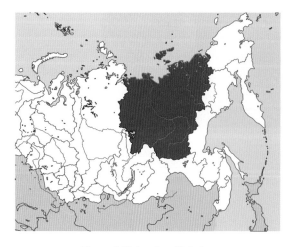

Figure 6.21 Location of Yakutia.

Figure 6.22 Distribution (in green) of the Snowy Owl (*Bubo scandiacus*). (Adapted from Blix, 2005). In Shetland, one pair of Snowy Owls nested on the island of Fetlar from 1967–71 and 1973–75 and raised 20 young (Robinson and Becker, 1986).

seas, from 5–7 per day. There is little geographic variation in form, although plumage is highly variable both between sexes and from year to year in the same individual. An investigation of two mitochondrial genes at different locations throughout the Snowy Owls' breeding range found no phylogeographic structuring between populations. This would indicate an evolutionary history of gene flow in the recent past as well as at present (Marthinsen et al., 2009).

Despite its great size the Snowy Owl has the same characteristics as other owls for silent flight. This is achieved by having serrated fringes along the front vanes of the outer primary feathers which have comb-like structures which serve to reduce noise from flapping. The extremely low frequency of the sound created by owl wing beats is less than 1 kHz. Consequently, when flying they are able to hear the movements of their prey, such as the rustling of voles and lemmings in the undergrowth, which have a much higher frequency of between 6 and 9 kHz (Birkhead, 2012).

Owls have very acute hearing. The breadth of the head increases the stereophonic capacity of their hearing so that they can locate their prey with great accuracy. Sight is less important for accurate detection of prey as their large eyes, although ideal for poor light and crepuscular flying, result in a loss of definition.

Snowy Owls have hunted across the open plains of the Tundra throughout the Pleistocene and have become well adapted to the uncertainties and rigours of this habitat. Their insulation is second to none among arctic birds and exceeds by a factor of four that achieved by Rock Ptarmigan (see section 6.6.3).

Although they feed almost exclusively on lemming and voles they can have difficulties in winter hunting. Satellite tracking of Snowy Owls in Canada that were over-wintering at high latitudes in the eastern Arctic has shown that they can spend up to 14 weeks on sea-ice between December and April. Analysis of high-resolution satellite images of sea-ice indicated that the owls were primarily gathering around open water patches in the ice, which are commonly used by wintering seabirds, a potential prey. Such extensive use of sea-ice by a Tundra predator considered a small mammal specialist was unexpected, and suggests that marine resources may subsidize Snowy Owl populations in winter (Therrien *et al.*, 2011).

In their breeding habit they nest in a shallow scrape in open ground. They are highly opportunistic and adapt to the availability of prey, laying up to 12 eggs when lemmings are abundant. When prey is scarce they may lay only one egg or not nest and abandon a region altogether.

6.6.3 Rock and Willow Ptarmigan

Of the 19 species that make up the grouse family (Tetraonidae) only two inhabit the northern regions where winters are extremely cold and snow cover prolonged, namely the Willow and Rock Ptarmigan (*Lagopus lagopus* and *L. mutus*). The generic name comes from the Greek *lagos* (λαγως)—hare and *pous* (πους)—foot, which describes their feathered feet. The English word *Ptarmigan* is taken from the Scottish Gaelic language *Tarmachan* from *torm*—noise or murmur, which describes the various clucks, croaking, and hissing sounds made by both species. Willow and Rock Ptarmigan are resident over vast expanses of the Arctic and sub-arctic throughout North America and Eurasia. The Rock Ptarmigan is completely circumpolar but the Willow Ptarmigan is absent from Greenland and Spitsbergen. Both species have a geographical range that also extends south in mountain regions.

The genus as a whole is therefore adapted to a wide variety of climatic and biotic conditions. The populations that live in the Arctic are true arctic residents even though there is some relocation in winter to more southerly parts of the Arctic, particularly in the Willow Grouse.

Rock Ptarmigan have a number of sub-species (Figures 6.23–4). One of the hardiest of Rock Ptarmigan populations is the Svalbard Rock Ptarmigan (*Lagopus mutus hyperborea* ssp. *nana*) which is found in Spitsbergen (74–81°N) and Franz Josef Land (80–81.9°N.) The latter group of islands are the most northerly Eurasian islands and the Rock Ptarmigan is the only bird to over-winter at these high latitudes, and as such is a remarkable example of survival on one of the most exacting of winter environments for terrestrial birds.

Throughout its range across the Arctic, the Rock Ptarmigan has to survive extremely low temperatures along with progressive depletion of food resources throughout the winter. As they cannot roost in the snow at –40°C, they shelter by burrowing in shallow snow burrows for up to 21 hours per day in winter. Extensive all-year-round studies of a Ptarmigan population in Spitsbergen have shown a remarkable fluctuation in body weight, which is at a minimum when the snow eventually melts in late June and then almost doubles by the end of November (Figure 6.25). Food intake has not stopped over winter, but the inferior nutritive quality of the diet can fail to maintain body weight.

Figure 6.23 Ptarmigan (*Lagopus mutus*) in summer in Spitsbergen—*left* female; *right* male. (Photos courtesy of Dr H. Körner).

Figure 6.24 Ptarmigan in winter on Baffin Island. (Photo courtesy of Clare Kines).

The Rock Ptarmigan in Spitsbergen prepare for winter by depositing large stores of fat, which can amount to as much as 35% of body mass in autumn (Figure 6.26). This is partly achieved by a marked reduction in their daily energy expenditure as measured by the *basic metabolic rate* in birds in captivity, which permits such fattening to occur despite the fact that food intake is simultaneously reduced (Stokkan, 1992).

To meet daily energy demands on a midwinter day a Ptarmigan needs about 6g of food (dry weight), consisting mostly of willow buds and twigs. They remain active in winter, and their feathered feet not only serve to keep the extremities warm but also act as snowshoes so that they can run about on the surface of the snow. Over winter the nutritional quality of their diet deteriorates as the fibre content increases. Consequently, the birds lose body weight constantly at this time of the year. However, they recover rapidly when a more nutritious diet becomes available after the snow starts to melt.

In Spitsbergen, Ptarmigan have been observed co-feeding with Svalbard Reindeer (*Rangifer tarandus platyrhynchus*) Both pairs and single hens or cocks were seen to make use of the feeding craters excavated by reindeer in search of food, and it has been suggested (Pedersen *et al.*, 2006) that the use of reindeer feeding craters may be important for the Svalbard Rock Ptarmigan during snow-rich events in winter or after terrestrial ice-crust formation resulting from mild spells and rain-on-snow events. This co-feeding strategy may be important for saving energy in periods when territorial defence and preparation for the breeding season make high energy demands on Ptarmigan of both sexes.

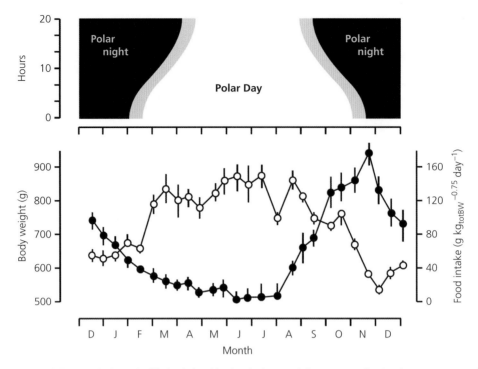

Figure 6.25 Seasonal changes in body weight (filled circles) and food intake (open circles) in captive Svalbard Rock Ptarmigan exposed to natural temperature and light conditions in Svalbard. Upper panel shows periods when the sun is above the horizon (in white), polar night (in black), and civil twilight (in grey). (Adapted from Stokkan et al., 1986).

Birds do not have special heat reserve fats such as the *brown fat* reserves found in some arctic mammals (see section 6.2.6). Nevertheless, the Ptarmigan of the Arctic show an exceptional capacity to put on weight essentially as white fat during the growing season and early autumn (Figures 6.25–6). Their annual cycle of weight gain and loss is an astonishing energy-balancing achievement which appears to operate within set specific limits, as increasing population density or severe and prolonged winters can cause physiological damage which eventually leads to lower reproduction rates within the breeding population (Andreev, 1991).

In this it is revealing to look at the demands that winter makes on their reserves during the period of incipient starvation. Studies have been carried out in Spitsbergen (Lindgard et al., 1992) in which the fat and protein contents of birds over-wintered in captivity were monitored. During the first phase (15 days) of an experimentally imposed starvation imposed on fat birds, body protein fell to a low level,

Figure 6.26 Example of Ptarmigan in Spitsbergen that has exploited to the full the capacity to gain weight when access to food is not restricted. (Photo courtesy of Nicolas Lecomte).

but subsequently in phase II of the enforced starvation it remained stable, and approximately only 9% of further energy demands were covered by breakdown of body protein. By contrast, when lean birds were starved for 5 days there appeared to be no stable phase II, and the energy demand on protein increased to 41% at the end of starvation period and was paralleled by increased plasma uric acid content. Well-fed Spitsbergen Ptarmigan (Figure 6.26) appear to have a much greater capacity to spare body protein than lean birds. Fat birds effectively reduce protein catabolism and maintain this at a low and stable level, whereas starving lean birds increase protein catabolism (Lindgard et al., 1992). This latter condition could account for the adverse effects noted above on population viability when population density increases and winters are prolonged.

The role of external stimuli on winter metabolism has also been studied in Spitsbergen Ptarmigan by following the daily rhythm of melatonin secretion throughout the year. Melatonin, known chemically as N-acetyl-5-methoxy-tryptamine, is a naturally occurring compound found in animals, plants, and microbes. In animals, circulating levels of melatonin vary in a daily cycle, thereby regulating the circadian rhythms of several biological functions. In the Spitsbergen Ptarmigan no daily rhythm has been found in plasma melatonin in May–July, but it varies significantly throughout the day in all other months of the year, with the nocturnal increase reflecting the duration of darkness. The highest mean plasma concentrations were recorded at midnight in March (110.1 ± 16.5 pg/ml) and represented the annual peak in estimated daily production. Around the winter solstice, melatonin levels were significantly reduced at noon but rose during the 18 hours of consecutive darkness that prevailed at this season of the year.

Thus, at the times of the year characterized by light–dark cycles, melatonin may convey information concerning the length of the day and therefore the progression of the seasons. It has been suggested that a flexible circadian system may reflect an important adaptation to life in the Arctic (Reierth et al., 1999).

6.7 Conclusions

To give a detailed account of the adaptations found in every bird and mammal that inhabits the cold climate regions of the Tundra and Taiga is not possible without an exceedingly long text. The examples discussed in this chapter have been chosen to provide an indication of the enormous range of adaptational possibilities that have evolved in the polar biota. The variation of these adaptations in form, function, and behaviour are a reflection of the heterogeneous nature of the arctic environment and the extraordinary facility for the evolution of a diverse range of adaptations, with each one suited to a particular niche or lifestyle. The capacity of the Arctic to provide a habitat for animals that range from Muskoxen to soil mites and for lifestyles that vary from months of torpor, as in insects, to the perpetual activity of lemmings and foxes is a source of wonder in a landscape that at first sight can appear barren and formidable.

The changeable nature of the Arctic climate where times of plenty can oscillate with periods of famine, forcing populations that had reached super-abundance to plummet to near extinction and then recover, is an extraordinary testimony to the tenacity of life in the face of apparent disaster. It is one of the wonders of evolution that such biodiversity can be sustained in such an uncertain environment.

CHAPTER 7

Plant survival in cold habitats

7.1 Plant tolerance of cold

It should not be surprising that plants are able to adapt to low temperatures. The very reactions that trap light energy are independent of temperature. The impact of photons of light in exciting chlorophyll receptor molecules is not affected by the vibrational energy of the receptors. The initial photosynthetic light reactions can therefore be described as having a Q_{10} of 1; i.e. for a 10°C rise in temperature there is no increase in the rate of the primary light reactions. It is the subsequent dark reactions that are temperature dependent. It should therefore not be unexpected that there are numerous cold-climate plants that are capable of carrying out photosynthesis at sub-zero temperatures. Sitka Spruce (*Picea sitchensis*) can still photosynthesize at –5°C (Neilson *et al.*, 1972). Lichens, especially in Antarctica, are also well known for species that remain photosynthetically active below 0°C.

In the Arctic, land plants can be found as far north as there is land on which they can grow. Such is the ability of flowering plants to endure cold that just travelling north in the Arctic does not reveal the ultimate record for low temperature polar survival: as we run out of land before we run out of plants. The ultimate test for High Arctic survival therefore has to be found by searching for maximum altitude plant records in the most northern mountains (Table 7.1).

The highest arctic altitude and latitude so far recorded for a flowering plant is held by Arctic Mouse-Ear Chickweed (*Cerastium arcticum*) (Figure 7.1) at 990 m on Greenland's most northerly mountains in Peary Land (Schwarzenbach, 2000). This record is closely followed by the Arctic Poppy (*Papaver radicatum*) and the Purple Saxifrage (*Saxifraga oppositifolia*) (Figure 7.2).

Records for species of flowering plants nearest the summits of higher mountains further south in Greenland vary with location. Species that have been recorded there on nunataks above 2,400 m include *Papaver radicatum*, *Minuartia stricta*, *M. rubella*, *Saxifraga oppositifolia*, *S. cernua*, and *S. cespitosa* (Halliday, 2013).

What is probably most remarkable in terms of plant adaptations to ultra-short and cold growing seasons is the number of species of arctic plants that not only survive at high latitudes but also manage to flower profusely even at the northern edges of their distribution.

The Purple Flowering Saxifrage (*Saxifraga oppositifolia*) as found at 83°N in the extreme north of Greenland is not a poor under-nourished outlier but a vigorous and flowering plant, even though it is only 778 km (483 miles) from the North Pole (Figure 7.2).

Among woody plants, the Arctic Willow (*Salix arctica*) has the high latitude record for successful establishment, as is illustrated by its colonization of the northernmost regions of the Canadian Arctic archipelago, occurring as far north as Ward Hunt Island (83°N), a small off-shore island to the north of Ellesmere Island (Figure 7.3).

The northern hemisphere low temperature records for plant survival are not found in the Tundra but in the coniferous forests of Siberia and Canada. In the coniferous forests of Siberia and Canada the winter temperatures regularly fall to below –40°C, with –58°C not uncommon. It is, however, in the Canadian forests (Figure 7.4) that the lowest temperatures occur. The Yukon has the distinction of having recorded the lowest-ever northern hemisphere temperature at –62.8°C on the 3rd of February 1947.

Tundra-Taiga Biology. First Edition. R.M.M. Crawford © R.M.M. Crawford 2013.
Published 2013 by Oxford University Press.

Table 7.1 Highest records of vascular plants as recorded in Peary Land (83°N) (F.H. Schwarzenbach, 2000).

Species	Number of records	Highest record (m)
Cerastium arcticum	23	990
Papaver radicatum	25	970
Saxifraga oppositifolia	28	940
Poa abbreviata	12	930
Poa arctica	13	930
Festuca hyperborea	8	910
Luzula confusa	10	910
Minuartia rubella	18	910
Festuca baffinensis	10	880
Stellaria longipes	25	880

Figure 7.2 Purple Saxifrage (*Saxifraga oppositifolia*) flowering vigorously towards the northern limits of its range at 83°N in Peary Land, north Greenland. (Photo courtesy of Dr Jean Balfour).

Figure 7.1 The Arctic Mouse-Ear Chickweed (*Cerastium arcticum*) photographed in Spitsbergen at 79°N. This species holds the arctic high-altitude record of 990 m (3,248 ft) for plant survival in the mountains of northern Greenland.

In the island archipelagos of the Arctic Ocean plants are not exposed to such low winter temperatures as those that commonly occur in the Taiga. In Spitsbergen in the High Arctic zone, only some 800 km (500 miles) south of the geographic North Pole, the January mean temperature is only –16°C. Nevertheless, colder temperatures can be experienced, and the lowest cold temperature so far recorded in Spitsbergen was –49.2°C on the 28th of March 1917.

The island of Nordaustlandet at 80°N in the Spitsbergen archipelago (Figure 7.5) is an example of an extreme polar habitat being briefly free of snow and ice only along its coastal fringes in mid-summer. Investigations at Kinnvika at Murchisonfjord in Nordaustlandet found 28 species of flowering plants, with 15 of the species also recorded as seedlings, indicating an impressive level of recruitment in this ultra-exposed and snow-covered habitat (Cooper, 2011).

7.2 Cold climate adaptations

Plants react to cold in a manner that is different from that found in warm-blooded animals. For plants, the threat to survival in the Arctic does not come from directly low temperatures damaging cellular metabolism. Cold-induced reduction of metabolic rates is rarely a cause of death in plants. Instead, it is the effect of cold in denying access to essential resources such as water, oxygen, light, and nutrients that has to be overcome to ensure prolonged survival under frozen conditions. The ultimate cause of death from cold in plants is more commonly due to a water shortage or, in the case of ice-encasement, the denial of access to oxygen. Tissue disruption from ice crystal formation is less common but remains a possibility in any non-hardened plants (see section 7.4.2)

Cryoprotection in plants differs therefore entirely from cryoprotection in warm-blooded animals as it principally centres on desiccation tolerance. The adaptations used to this end in plants are many and varied and can be used in a variety of combinations depending on the particular lifestyle of the species.

For plants, morphological plasticity provides a pathway to survival that is developed greatly

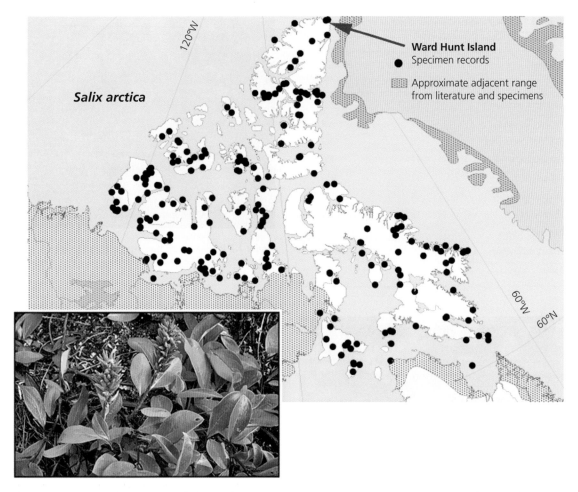

Figure 7.3 Distribution of the Arctic Willow (*Salix arctica*) in the Canadian Arctic archipelago. This woody plant reaches 83°N on Ward Hunt Island (Aiken *et al.*, 1999 (onwards)). Inset: *S. arctica*. (Photo courtesy of Dr H. Körner).

Figure 7.4 Winter in the forests by the banks of Paradise River, Canada. (Photo courtesy of W.C. Woodley, Ottawa).

beyond that usually available to mammals and birds. A diminutive size is one of the commonest responses used by plants in minimizing the need for resources. The ability not to grow should therefore be regarded as a positive rather than a negative response to a resource-limited frozen habitat. Too frequently, physiological assessments of plant survival in adverse environments look for growth as a sign of success. This is logical when selecting for crop plants. However, when considering survival of

Figure 7.5 Location of Kinnvika on the north shore of Murchisonfjord at 80°N on Nordaustlandet, Spitsbergen.

native plants in polar regions, growth is not in itself always a priority. For these plants, avoiding the risls of desiccation and carbon starvation are of greater importance.

Slow-growing plants can of course be shaded out in warmer habitats by competitors where more demanding species have access to sufficient resources (Sala et al., 2010). This can arise in the boreal forest but is less likely to occur in the Tundra, where light is not usually a limiting factor and low growth rates are therefore not a disadvantage. Plants, as compared with animals, are particularly adept at using a minimalist strategy in relation to survival. This is evident in the many minute plant species and ecotypes that are found in the Arctic. Even in exposed situations many diminutive plants can flower and produce at least some fertile seed and complete their life cycles irrespective of their size. In some cases this reduction in growth may be merely a phenotypic response due to a lack of resources. However, it may also be governed by genetic changes.

Furthermore, plants have an advantage over animals in relation to cold tolerance in that they can survive for long periods without sexual reproduction. Many plant species, particularly in the Arctic, multiply vegetatively and consequently survive as individual clones for centuries or even millennia. In the Siberian Arctic, estimates of over 3,000 years have been made for the longevity of clones of *Carex ensifolia* ssp. *arctisiberica* (Jonsdottir et al., 2000).

More examples are still coming to light as molecular markers present new and remarkable insights into the longevity of clonal species.

7.2.1 Seed survival in the Arctic

During periods of adverse conditions plants have a capacity to enter a state of prolonged dormancy that exceeds by far the winter torpor that is found in some animals. In many species, seeds (embryo plants endowed with food) can lie dormant in the soil for decades and in some cases even centuries. Brought to the surface they can then germinate and wake up, like Rip Van Winkle (who slept only for 20 years), into a world that due to environmental change may have been altered from that known to their parents. Plants emerging from long-buried seeds can thus have both post-emergent phenotypic and genotypic differences from modern populations.

An outstanding example of a species with seeds capable of long-term survival in arctic soils is the Stiff Sedge (*Carex bigelowii*) (Figure 6.14) where seeds have been successfully germinated after having been buried for up to ca. 200 years (Vavrek et al., 1991). Seeds of *Luzula parviflora*, excavated from under a solifluction lobe at Eagle Creek, Alaska, have also been carbon dated from their seed coats and found to have been capable of germinating after 175 years of burial (Bennington et al., 1991). These examples are not just isolated or freak occurrences. Whole populations can lie dormant for decades, or even centuries, buried in the soil seed bank. A study in Spitsbergen found that 71 of the 161 species indigenous to Spitsbergen had the capability of persisting in the soil seed bank (Cooper et al., 2004).

Anaerobic conditions generally enhance seed longevity. The Chinese Water Lotus (*Nelumbo nucifera*) has frequently been reported as yielding viable seeds after centuries of burial in anaerobic mud. Similarly, in the Arctic, seeds buried in cold or frozen peat will be protected from the effects of lipid peroxidation of their membranes which is the principal cause of loss of seed viability (Murthy et al., 2003).

7.3 Metabolism and low temperatures

Two types of metabolic adaptation can be discerned in the adaptations that fit species for life in cold climates. The first is *capacity adaptation* in which morphological and physiological properties allow species to increase their capacity to carry out common metabolic processes under sub-optimal conditions, such as low temperatures and short growing seasons. The second is *functional adjustment* which describes changes in behaviour (e.g. phenology) or the adoption of different physiological mechanisms (biochemical adaptations) as a response to alterations in environment.

7.3.1 Capacity adaptation

Low temperatures both limit growing season length and impose temperature regimes that challenge metabolic activity. For many species these limitations can be overcome by having the ability to increase low temperature metabolic rates by means of increased amounts of active enzymes, an adaptation that is referred to as *metabolic* or *capacity adaptation*. It is in essence a quantitative metabolic response and as a means of adaptation it is found in cold-water fish as well as in plants.

Arctic plants are notable not only for having high levels of ribulose bisphosphate carboxylase (RUBP) but also for having enhanced rates of mitochondrial respiration (Friend and Woodward, 1990). This can also be combined with a capacity for the down-regulation of respiration when temperatures rise.

The Glacial Buttercup (*Ranunculus glacialis*) (Figure 7.6) holds the high-altitude record for flowering plants in Scandinavia at 2,370 m on Galdhøpigen (Lid and Lid, 1994) and at 4,725 m on the Finsterahorn in Switzerland. Experiments carried out on this highly successful arctic-alpine species have shown a capacity for phenotypic thermal acclimation to warmer temperatures, as indicated by similar dark respiration rates measured when grown at 8°C in cold-grown plants and at 18°C in warm-grown plants. This ability to avoid a temperature-induced respiration rate increase conserves carbohydrate reserves, which are essential for survival for a plant that normally has to endure a low temperature environment.

Physiologically, the 24-hour light regime makes it economic for arctic plants to have high levels of the carbon dioxide fixing enzyme ribulose bisphosphate carboxylase. This enzyme has a dual function as a carboxylase and an oxygenase and can therefore fix or liberate carbon dioxide. In warm climates this functional duality can cause a significant loss of fixed carbon through light respiration. However, in the Arctic temperatures are so low that the oxygenase activity is not significant and the 24-hour usage

Figure 7.6 The Glacial Buttercup (*Ranunculus glacialis*) maximizing utilization of heat. *Left* Pre-fertilization state where the petals are white and reflect light and heat radiation to the centre of the flower; *right* post-fertilization stage when the petals close, turn red, absorb heat, and protect the developing seeds. This species holds the high-altitude record in Scandinavia (2,370 m, Galdhøpigen, Norway).

of the high investment in carbon dioxide fixing capacity is amply repaid.

Plants in cold habitats are frequently noted for their thicker leaves as compared with related species from lower altitudes (Körner *et al.*, 1989). This is also a form of capacity adaptation, as thicker leaves, with a greater development of palisade tissues and a higher internal volume, will facilitate greater assimilation of carbon dioxide.

7.3.2 Functional adaptation

Functional adjustment differs from capacity adaptation in that it is a qualitative metabolic response to environmental alteration. It can take a number of forms in flowering plants but is most frequently associated with phenotypic plasticity (the ability of an individual to change during its lifetime), as opposed to genetic variation (heritable variation) which varies both between species and within populations of the same species. The crucial difference in terms of plant responses to the environment is that phenotypic plasticity provides an immediate reaction to environmental stress, while genetic variation only allows change from one generation to another as a response to selection.

7.4 Cold tolerance adaptations

7.4.1 Acclimation vs acclimatization

These two terms are often confused and used indiscriminately. Despite their linguistic similarity they have nevertheless evolved their own specific scientific meanings. Unfortunately, this distinction is not always followed even in biological texts.

In the more pedantic scientific circles it has become customary to restrict the term *acclimation* to phenotypic responses and *acclimatization* to genotypic adaptation. It is, however, difficult to be doctrinaire about the respective meanings of acclimation as opposed to acclimatization as usage alters in different parts of the English-speaking world and even phenotypic responses will have a genetic basis.

The term *acclimatization* was first used botanically in 1788 on the creation at Orotava on the island of Tenerife (Canary Islands) by order of King Carlos III of Spain, a *jardin d'acclimatation* to serve the needs of selecting for cultivation suitable strains of new species of plants being introduced to temperate climates from the tropics. Subsequently, a *jardin d'acclimatation* was created in Paris in 1860. This latter garden displayed a luxuriant flora in the winter garden hothouses as well as some exotic animals. The purpose here, however, was more educational than the more practical garden in Orotava.

Acclimatization denotes a genotypic adjustment of physiology and morphology and refers to changes observed in a species or population over a number of generations, which implies natural selection and genetic change. The terms *accommodation* and *habituation* have been used to describe this same response.

Acclimation if used in its strict sense describes the ability of many plants to survive freezing temperatures if the ambient temperature drops gradually over a period of days or weeks. A reduction in day length in temperate climates is usually accompanied by lower temperature. For plants, shorter days and longer nights elicit a photoperiodic response that induces bud dormancy and with it an increase in frost tolerance. This same drop could be lethal if it occurred suddenly during the peak of the growing season. This process of induced cold adaptation is often referred to as hardening in horticulture and involves several changes, such as a decrease in water status or an increase in plant sugar content, and the consequent lowering of the freezing point of sap.

Changes in the lipid content of cell membranes can also occur with the onset of low temperatures even before the plant is subjected to freezing temperatures.

A functional change that is associated with cooling (as opposed to freezing) is the ability to change the state of saturation of membrane lipids. Lipids exhibit temperature-dependent phase transitions, from being a gel at low temperatures to a fluid at higher temperatures. In the gel state, membrane integrity and function are damaged. The exact mixture of lipids present in a membrane determines the temperature at which the phase change takes place between a gel and a fluid. Membranes of cold-adapted plants that have been through a period of acclimation have a higher proportion of fatty acids with short chain lengths and unsaturated double bonds than cold-sensitive plants. This leads to more fluid membrane structures, which can pack more closely together than fully saturated lipids which have straight hydrocarbon chains.

Plants acclimating to low temperatures (see further in this section) shift the lipid composition of their membranes so as to lower the temperature at which the lipid phase change from a fluid to a gel state takes place. Such an adaptation demonstrates the phenotypic and reversible nature of this aspect of functional adaptation in that the membranes are constantly turning over, with all the components being renewed on a regular basis (Bowsher *et al.*, 2008).

Under natural conditions, acclimating to low temperatures in woody plants takes place in two distinct stages. The first hardening (stage 1) is induced in the early autumn when the onset of short days and non-freezing but chilling temperatures (between +5° and 0°C) combine to trigger the synthesis of abscisic acid (ABA). It appears that ABA can substitute for the low temperature stimulus provided there is also an adequate supply of sugars. Nevertheless, it is unlikely that ABA regulates all the genes associated with cold acclimation; however, it clearly regulates many of the genes associated with an increase in freezing tolerance, including the cessation of growth and the withdrawal of water from the xylem vessels (Gusta *et al.*, 2005). Cells in stage 1 of acclimation can survive temperatures well below 0°C. The degree of hardening, however, is not uniform throughout the plant.

The tissues that are essential for plant survival, namely the apical buds and the bark, are preferentially protected against low temperatures. This freezing point depression type is also found in some animals, including small mammals (hedgehogs), insects, and beetles, often associated with an increase in cell glycerol.

In hardening stage 2, direct exposure to freezing is the stimulus. A period at sub-zero temperatures at −2 to −5°C can generate subsequent tolerance of temperatures as low as −18 to −25°C.

The mechanisms involved in this second stage of hardening appear to be complex and are as yet not fully understood. Genetically controlled changes in proteins appear, however, to have a significant role.

7.4.2 Freezing tolerance

Plants have cell walls and animals do not. As a result, plants react differently from animals in their response to freezing. Plants that are hardened to live in habitats where it is impossible to avoid being frozen exhibit the phenomenon known as *extracellular freezing*. In the frost-hardened plant the cell membranes allow water to seep out through the cell wall into the spaces between the cells, where they can freeze without harming the contents of the cell. As the water exits from the protoplasm leaving behind the greater part of the cellular solutes, the cell contents become more concentrated which reduces the freezing point of the contents of the cytoplasm. Equally, as the cell membrane retains the osmotically active substances within the cell there is no freezing point depression in the extracellular water. Once frozen, the extracellular water vapour pressure gradient across the cell wall increases progressively, removing ever more water from the cell, thus decreasing the danger of freezing within the cytoplasm as the osmotic content of the cytoplasm and vacuoles continues to increase. In the most extreme case all free water can be removed. What water that remains in the cell is associated with proteins and other complex molecules as water of hydration and thus is prevented from freezing. In this condition some plant organs are capable of surviving immersion in liquid nitrogen at −196°C. Here again, plants exhibit similarities between drought and frost tolerance.

Not all plants can achieve or withstand this reduction in water content in their cells. In these cases some cell water remains free and not attached to macromolecules for mutual protection. For these plants, the longer and more severe the frost, the greater the risk of desiccation injury as water is progressively extracted from the cell, and finally the free water that remains in the cell freezes and produces potentially lethal internal ice crystals.

7.4.3 Dangers of thawing

Thawing for plants can be even more dangerous a process than freezing. In the Arctic, plants face two possible sources of cell injury when ice melts. The first is the release of water. If thawing is rapid, water re-entry to the cell can take place at a rate that causes severe disruption of cell plastids. The second danger is when ice melts and plants that have been under ice and deprived of access to air are suddenly exposed to oxygen, which can cause post-anoxic injury (see section 7.5). In an environment where frosts are regularly interspersed with warm periods, repeated oscillations between the frozen and the unfrozen state can be damaging, especially if they take place rapidly. Most at risk are plants exposed to full sunlight during the warm part of the day. Every time the temperature rises above the freezing point there is always a metabolic drain of sugar reserves, which further reduces the resistance of exposed buds to frost injury.

7.4.4 Supercooling

Several tree species, particularly from the more southern region of the Boreal Forest, avoid damage to their water-conducting tissues through the phenomenon of *supercooling* (Kasuga *et al.*, 2007). Supercooling (also known as undercooling see 6.3.2) is the cooling of a liquid below its freezing point without freezing. For this to take place the liquid has to be devoid of particles, which may cause ice-nucleation and thus induce a rapid phase change. Tree and shrub species that are capable of supercooling have absolutely pure water within their cells and avoid winter damage by supercooling of water. There is, however, a lower limit to which water can be supercooled. If the temperature should drop lower than −38°C, ice formation will occur within the plant cells and they will die, resulting in localized frost damage or even the death of the entire plant. Some trees, such as Red Oaks, keep their cellular liquids free of crystal-forming nuclei and so are able to survive at these low temperatures without tissue damage. Xylem parenchyma cells in many Boreal Forest tree species adapt to sub-freezing temperatures by just such deep supercooling. By this adaptation, cellular water in xylem parenchyma cells can remain in a liquid state without dehydration to nearly −40°C for long periods exceeding several weeks.

7.4.5 Antifreeze proteins

Throughout the seasons of the year the processes of acclimation in autumn and de-acclimation in spring are accompanied by a range of biochemical and physiological changes, including the synthesis of specific, low-molecular antifreeze proteins (AFPs). AFPs inhibit ice crystal growth by binding ice nuclei and thus decreasing the freezing point of the cellular contents, thereby enabling survival at sub-zero temperatures. It appears that the antifreeze proteins reduce the rate of ice propagation through the tissues. Antifreeze proteins were first discovered in the blood of antarctic fish (DeVries and Wohlschlag, 1969), where it was found that proteins were able to prevent freezing of the blood serum. Since then, antifreeze proteins have been identified in a range of animals as well as in bacteria, fungi, and flowering plants. In plants they were first discovered in winter rye (*Secale cereale*). These proteins are located mainly in the intercellular spaces (apoplast) of acclimated rye leaves and have been found to have molecular similarities with the antifreeze proteins discovered in antarctic fish. Apart from delaying the seeding of ice crystals, antifreezing proteins can also interact with ice crystals. The result of this hydrophilic action is to bind the protein to the ice crystal and inhibit its growth. At the same time, these proteins are able to modify the structure of ice crystals, forcing them to adopt a hexagonal shape. Altogether, these properties of AFPs can protect cells from the potential damage that could be caused by ice crystal growth.

7.4.6 Frost resistance in conifers

In the coldest forests the conifers are the predominant trees. As a group, they possess a powerful mechanism against winter desiccation in the form of their water-conducting tissues. Throughout all the conifers, narrow tracheids are present instead of the well-developed vessels that are found in broad-leaved trees. The tracheids overlap at their ends and water passes from one tracheid to another through bordered pits which comprise a valve-like centre— the *torus*, which moves under increased water movement against the bordered pit and restricts water movement (Figure 7.7).

Torus–margo pits are rounded or oval-shaped pits in narrow xylem tracheids. The *torus* is an impermeable, lignified thickening at the centre of the pit membrane, surrounded by a *margo*, which is

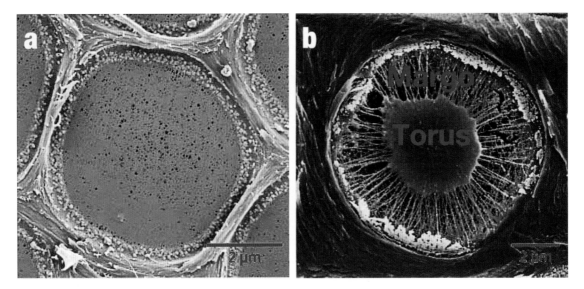

Figure 7.7 Variation in pit structure. (a) A homogeneous pit membrane of an angiosperm species, *Acer negundo*. (b) A margo–torus-type pit membrane of a conifer, *Calocedrus decurrens*. Homogeneous pit membranes, typical of angiosperm species, have a relatively uniform array of microfibrils, whereas in margo–torus pit membranes of tracheid-bearing conifers, the conductive and protective functions of the membrane are spatially distinct as a porous outer region (margo) that allows for movement of water between conduits and a central thickened plug (torus). (Reproduced with permission from Choat *et al.*, 2008).

a ring of radial slits or pores that can be much larger than pores in an angiosperm conduit. Tracheids therefore provide a uniquely effective means of controlling water loss throughout winter when soil water remains frozen. The homogeneous pit membrane of an angiosperm species can be compared with the tracheid bordered pit (Figure 7.7).

Apart from winter desiccation another danger for over-wintering trees in cold climates is embolism. This occurs when *cavitation* (breaking) of the water column causes air bubbles to coalesce and render the water-conducting tissues inoperable. This is generally considered to be brought about by the formation of gas bubbles on freezing and their subsequent expansion on thawing and as such is described as the *thaw-expansion hypothesis*. Conifers are generally resistant to freeze–thaw-induced embolisms, as bubbles formed in narrow tracheids are small and redissolve during thawing.

The unicellular conifer tracheid might be expected to have greater resistance to water flow than the multicellular angiosperm vessel, as its high-resistance end-walls are closer together. However, tracheids and vessels have similar resistivities for the same diameter, due to the unique torus–margo structure of the conifer pit membrane. The torus can be deflected to one side when desiccation threatens cavitation and embolism generation. The water, however, can continue to flow through the pores of the margo region which can impede the movement of bubbles through the tracheids. In addition the closing of the pit by the torus will make the embolism process slower.

7.5 Ice-encasement injury

A further risk for plants during arctic winters is *ice-encasement injury*. When warm, wet winter weather from the south intrudes into the arctic and boreal regions and soils are frozen, the precipitation rapidly turns to ice as it falls on frozen ground. Ice is a barrier to oxygen diffusion and ice-encased plants can be deprived in this way of access to air for months at a time, and sometimes for more than a year when large snow banks fail to melt in summer. Paradoxically, as a result of climatic warming, this risk is increasing with more incursions of winter rains onto frozen ground. Prolonged anaerobiosis, although dangerous in itself, creates a potentially more dangerous condition for tissues when they are

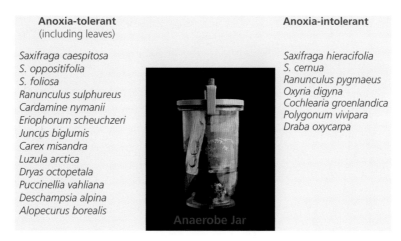

Figure 7.8 Ability of arctic species examined in Spitsbergen to survive 7 days of total anoxia in an anaerobe jar in the dark at ambient arctic temperatures. In the anoxia-tolerant species, the entire plant including leaves and flowers survived (see also Crawford et al., 1994).

once again in contact with oxygen. The Arctic is the region of the Earth's land surface where plants are most at risk of having to endure long periods without access to oxygen.

It is therefore not surprising that a uniquely high level of anoxia tolerance has been observed when testing plants *in situ* at high latitudes in Spitsbergen (79°N) (Figure 7.8). The cell damage that arises on re-entrance into air, i.e. the post-anaereobic state, is termed *post-anoxic injury*. This is similar to the *post-ischaemic injury* that can arise in animal and human organs when they are suddenly re-exposed to air after interruption of the blood supply, as after a heart attack or a transplant operation (Crawford et al., 1994).

The lack of oxygen, as a result of either ice-encasement or just flooding, causes the transition metals involved in aerobic metabolism (e.g. the iron in cytochromes) to become reduced to the ferrous state. On return to air this reduced iron reacts with oxygen to produce active oxygen species (superoxide and hydroxyl ions, etc.).

$$Fe^{2+} + O_2 \rightarrow Fe^{3+} + O_2^{-}$$

Super oxide dismutase

$$O_2^{-} + O_2^{-} + 2H^4 \rightarrow H_2O_2 + O_2$$

Ethanol that has accumulated under anoxia can then react with hydrogen peroxide and provide a rapid generation of acetaldehyde which is highly toxic to membranes.

Catalase
$$C_2H_5OH + H_2O_2 \rightarrow CH_3CHO + 2H_2O$$

7.5.1 Protection from post-anoxic injury

The activity of the enzymes needed for counteracting the dangers of the post-anoxic generation of reactive oxygen species (ROS), typically oxygen (O_2^{-}) and hydroxy radicals (*HO_2), can also be much diminished in anoxia-sensitive species. Consequently, the sudden re-emergence into air as ice melts and water tables subside presents an aerobic shock to unprepared tissues as membranes are attacked. Under such conditions many plants are liable to suffer *post-anoxic injury*. Tissues prone to this type of injury become soft and spongy on return to air, rapidly lose their cell constituents, and are irreversibly destroyed (Crawford, 2012).

Tolerance of anoxia and its post-anoxic consequences are energetically expensive, particularly in the use of carbohydrate reserves that have to be conserved to support over-wintering anaerobic metabolism and provide the anti-oxidants for defence in spring against post-anoxic injury. Many arctic species are rich in anti-oxidants such as vitamin C (ascorbic acid), glutathione, α-tocopherol, and various phenols which also act as anti-oxidants.

Due to this protection a number of arctic species and ecotypes are able to survive prolonged anoxia

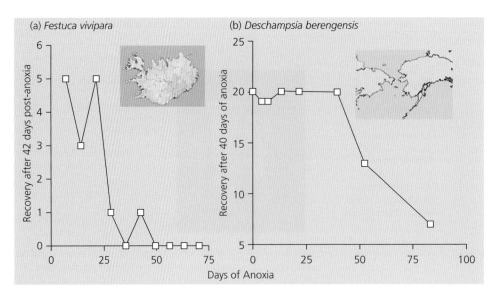

Figure 7.9 Tolerance of anoxia in high-latitude grasses using (a) a sub-arctic Icelandic population of pseudo-viviparous plantlets of *Festuca vivipara* and (b) seedlings from an arctic population of *Deschampsia beringensis*. The plants were kept under total anoxia for different lengths of time at 5°C and then allowed to recover in a cold room at 10°C. With *F. vivipira* six plants per treatment were used and with *D. beringensis* 25 plants per individual time treatment.

better than their more southerly congeners. A prolonged tolerance of anoxia has been found in grasses of northern origin (Figure 7.9). An outstanding example is Purple Saxifrage (*Saxifraga oppositifolia*), where arctic populations can survive more than a week without oxygen and do not even lose their flowers. By contrast, 4 days of anoxia is lethal to the entire plant in more southern populations of this species (Figure 7.8).

The species listed in Figure 7.8 had been placed, complete with root systems, on moist filter papers in anaerobe jars and kept in the dark at ambient air temperatures for 7 days in an environment that was totally devoid of oxygen. In this way it was possible to observe any deleterious effects either during anoxia or in the subsequent post-anoxic period. From 20 species tested in this manner 13 were found to be completely tolerant of anoxia and its after-effects (Crawford, 2008). It was remarkable that even in summer, at the height of the growing season, so many species did not show any damage to the growing leaves. Even the flowers were undamaged by the anaerobic incubation. Normally, green leaves lose turgor and wither rapidly when deprived of oxygen.

Tests on more southern populations of *Saxifraga oppositifolia* from Norway, Iceland, and Scotland failed to find any degree of anoxia tolerance to match that of the Spitsbergen populations (Crawford, 2012).

7.6 Plant form and climate

Climate has a direct effect on plant form and many attempts have been made to produce a generalized descriptive classification of plant life-forms. One of the most enduring systems is that devised over a century ago by the Danish botanist Christen Raunkiaer (Raunkiaer, 1904). The basis of the Raunkiaer classification system is the distance of the perennating bud above the ground in the adverse season.

By plotting the percentage of the flora that belongs to any of these classes it is possible to obtain a systematic comparison of how plant form responds to geographical changes in climate (Table 7.2).

7.6.1 Dwarf trees

One of the most remarkable examples of changes in plant form with climatic stress is the dwarfing of

Table 7.2 Comparison of classification of major growth forms of higher plants in the Swiss Alps and in Spitsbergen in relation to the height of the perennating bud in the adverse season. (Data taken from Raunkiaer, 1904).

Growth form	Position of perennating bud above ground	Percentage distribution in Greenland	Percentage distribution in Spitsbergen	Percentage distribution in Swiss Alps
Phanerophytes	Usually greater than 0.5 m	1	–	–
Chamaephytes	Low-growing plants—buds above ground level and below 0.5 m. Includes cushion plants	27.5	22	24.5
Hemicryptophytes	Aerial shoots degenerate to ground level during the adverse season	52.5	60	52.5
Cryptophytes	Buds and shoot apices survive the adverse season beneath the ground surface	18	15	4
Thereophytes	Plants that survive the adverse season as seeds	2	2	3.5

woody species in marginal habitats. It is a world-wide phenomenon and can be observed from the fringes of the Boreal Forest to the montane forests of the tropics.

Such distorted and prostrate or stunted tree forms are referred to as *Krummholz* (German *krumm*—crooked, bent, twisted; *holz*—wood). They are frequently found at or just beyond the treeline either on mountain slopes or at the Tundra-Taiga Interface. The dwarfed or prostrate tree forms often have the appearance of being severely pruned by wind and ice-blast. Sometimes this growth form is genetically determined, while in other cases it is due entirely to phenotypic plasticity. Dwarf trees that have the latent ability to resume a pole form should environmental conditions change are sometimes described as *Krüppelholz* (German—crippled wood) (see section 5.2.1).

In North America, Black Spruce (*Picea mariana*) is the main Krummholz-forming species and shows a greater plasticity in form than White Spruce (*P. glauca*). Black Spruce is also better adapted to poorly drained acidic soils and high permafrost tables. Due to its particular Krummholz-habit of prostrate growth it is able to survive the dangers of wind-throw in shallow soils and is more able to propagate vegetatively by layering than White Spruce (Gamache and Payette, 2005).

Scots Pine (*Pinus sylvestris*) and Norway Spruce (*Picea abies*) can also exist in the Krummholz form which they commonly adopt as a phenotypic response to exposure. In terms of latitudinal extent across the Eurasian Arctic, the Dwarf Siberian Pine (*Pinus pumilo*) is outstanding for the amount of terrain that it occupies and for being the most northerly pine species in the world, reaching 71°N in Yakutia.

7.6.2 The Arctic treeline

Many studies have sought to relate the position of both arctic and alpine treelines to the total carbon balance of the tree. Intuitively, it might be expected that woody plants, which devote a considerable part of their resources to the formation of non-productive trunks and stems, may be unable to support such a growth strategy when growing seasons are cool and short. However, an extensive world-wide study of the total carbon balance in trees at their upper altitudinal and latitudinal boundaries has shown the converse, namely that tree growth

Figure 7.10 Another example of a Creeping Willow which is commonly found in the Arctic reaching 79°N in Spitsbergen. Female Net-Leaved Willow (*Salix reticulata*) with flowering shoots emerging from a horizontal main stem.

Figure 7.11 Dwarf Birch (*Betula nana*) in late August at 73°N near Mesters Vig in north-east Greenland.

Figure 7.12 Ancient stand of Mountain Birch (*Betula pubescens* ssp. *czerpanovii*) on Tunga Hill in the upper reaches of the Hvita River, western Iceland. The light colour of the exposed rock is rhyolite, an igneous, volcanic (extrusive) rock of felsic (silica-rich) composition. Local tradition asserts that this is a very old undisturbed forest, reminiscent of the description given in the Íslendabók by Ari the Wise (1067–1148) who said at the time of settlement that Iceland was 'forested from mountain to sea shore'. (Photo courtesy of Már Johnsson).

near the timberline is not limited by carbon supply (Hoch and Körner, 2005).

The absence of trees from the Arctic cannot be related directly to low temperature. Rather, it would appear that it is the shortness of the growing season that does not allow trees to complete the annual development cycles of growth and bud and bark protection in time for the arrival of winter. In addition, the presence of a permafrost layer in the soil restricts root penetration, with the result that upright large trees of the Boreal Zone become unstable in high winds.

7.6.3 Prostrate trees

Botanically it has been customary to use somewhat arbitrary definitions in relation to defining trees, shrubs, and herbs and it cannot be denied that plant form and the presence or absence of erect trees is related to climate. Whether or not trees can be considered to be present in polar regions is merely the imposition of an arbitrary human judgement. Willows that grow upwards in the temperate zone survive better in the Arctic if their trunks lie flat along the ground (Figure 7.10, and this should be recognized as a natural development for woodland survival in this particular situation. The vast mats of willows that cover the Tundra and reach as far as 80°N in Spitsbergen can have horizontal trunks several metres long and be centuries old. For this achievement they have been called Spitsbergen's 'forest'. At ground level the colourful autumnal display of this prostrate woodland with carpets of Dwarf Birch and willow can be as striking as any North American Maple forest (Figure 7.11).

Birches are among the most cold tolerant of the broad-leaved deciduous trees of the Boreal Forest. In Scandinavia and across the Atlantic to Iceland and Greenland they constitute the bulk of the trees in the upper forest zone and at the treeline. In the northerly and upper regions the success of birch appears to be due to its ability to hybridize with the Dwarf Birch (*Betula nana*). This phenomenon can readily be seen in ancient birch woodlands throughout Scandinavia. Even in Iceland, where much of the native forest was removed shortly after the Norse settlement (*Landnám*) *ca.* AD 870, hillsides with intact stands of ancient undisturbed forest can still be found (Figure 7.12).

The birch trees in these forests, particularly in their upper regions, show marked hybridization between Common Birch (*Betula pubescens*) and Dwarf Birch (*Betula nana*). This cross is generally referred to as Mountain Birch (*B. pubescens* ssp. *czerpanovii*) (Figure 7.13). Taxonomically, this is no more than a name of convenience as the exact parentage of the many populations is subject to location. In different geographical regions the Mountain Birches have probably evolved individually and are composed of many varieties, hybrids, and subspecies of *B. pubescens* and sometimes also *B. pendula* as well as with different species of *B. nana*. In north-east Canada (Baffin Island), eastern Greenland, northern Europe, and north-western Asia, *B. nana* ssp. *nana* is the usual Dwarf Birch. In Iceland large specimens of the hybrid can reach a height of

Figure 7.13 Krummholz, tall form of Mountain Birch (*Betula pubescens* var. *czerpanovii*) photographed in western Iceland. (Photo courtesy of Már Johnsson).

Figure 7.14 Dwarf Birch photographed in Iceland; *left* with catkins, *right* typical orbicular foliage.

13 m as the limbs can often maintain the persistence of the leading shoots (Figure 7.13). In more exposed conditions these leading shoots can adopt prostrate growth forms.

In eastern Canada (Quebec and Labrador) *B. glandulosa* is the dwarf component, while in Asia and northern North America (Canada east to Nunavut), *Betula nana* ssp. *exilis* provides the Dwarf Birch genes (Wielgolaski and Sonesson, 2001). Downy and Mountain Birch extend farther north into the Arctic than any other broad-leaved trees. More commonly, populations of sub-arctic Mountain Birch are usually small and contorted due partly to hybridization with Dwarf Birch (*B. nana*) and partly to growing in exposed situations (Krüppelholz, see section 7.6.1).

Within the Krummholz form, a variety of typical growth types can be found (Väre, 2001; Holtmeier, 2003). These can be either one stemmed (monocormic) or many stemmed (polycormic). In addition there is frequently a twisted form that extends itself close to the ground often developing semi-upright knees and is sometimes considered as the variety *appressa* (Figure 7.15).

This type is probably genetically fixed. Polycormic forms of Mountain Birch are also found typi-

Figure 7.15 Creeping Krummholz form with knees of the Mountain Birch growing at the treeline in central Norway. This form is sometimes referred to as var. *appressa*.

cally in areas with nutrient deficiencies, or where there has been disturbance or attack from the autumn moths *Epirrata autumnalis* and *Operophtera brumata*, which feed on the leaves of the Mountain

Birch   Aspen   Hazel

Figure 7.16 Scratched barks of Common Birch (*Betula pubescens*), Aspen (*Populus tremula*), and Hazel (*Corylus avellana*) showing the chlorophyll-containing tissues that are found under the outer layer of bark.

Birch and have population peaks approximately every 10 years. In many cases, however, the polycormic forms will show improved growth with nutrient application and it has therefore to be assumed that here the Krummholz form is phenotypic and strictly should be called Krüppelholz.

Irrespective of how the Krummholz or Krüppelholz arise they have two principal ecological advantages for tree survival at the timberline. First, there is the benefit in having foliage near the ground, as this decouples low-stature plants from air temperatures above the trees which are cooler than temperatures near the ground (Grace *et al.*, 2002). Second, the ratio of photosynthetically productive tissues to non-productive tissues is increased especially in the polycormic form (many stemmed). In addition, the individual plant is not dependent for survival on one upright tree trunk.

7.6.4 Stem photosynthesis

The increase in photosynthetic tissues in polycormic Krummholz does not just come from the foliage. A further advantage of the polycormic form is the increase in the number of stems that are present per plant, as opposed to trees that have only a single trunk. Birches, in common with a small number of other trees, have chlorophyll-containing tissues under their bark (Figures 7.16–7.18).

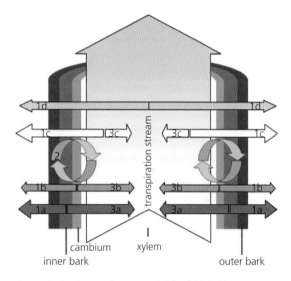

Figure 7.17 Diagram of sources and sinks of CO_2 inside a tree stem. 1, Diffusion of CO_2 out of the stem from inner bark, brown arrow (a); cambium, light brown arrow (b); xylem ray cells, white arrow (c); or imported in xylem sap, blue arrow (d). 2, Fixation of CO_2 by corticular photosynthesis (green arrows), which can utilize CO_2 from all four sources (a, b, c, d) above. 3, CO_2 diffusing into the transpiration stream. Sources of this CO_2 can be inner bark, brown arrow (a); cambium, light brown arrow (b); or xylem ray cells, white arrow (c). (Reproduced with permission from Teskey *et al.*, 2008).

Photosynthesis that takes place under bark that is sufficiently translucent is termed *corticular photosynthesis*. A particular advantage of this property in trees is that it not only permits photosynthesis to

Figure 7.18 Transverse section of a sector of a 1-year-old alder stem: red auto-fluorescence under blue light is due to chloroplasts in secondary cortex, secondary phloem, secondary xylem, and medulla. Scale bar = 200 mm. (Reproduced with permission from Armstrong and Armstrong, 2005).

Figure 7.19 Mountain Birch in snow at the end of winter at Åbisko, Sweden (68°N) in a situation where corticular photosynthesis would have advantages. Note snow melting around base of trunk.

continue after the leaves are shed but also allows the trees to profit from the high levels of internal CO_2 that can be found in the internal tissues of stems and dissolved in the xylem sap (Teskey et al., 2008). Birches, poplars, and aspens, which are all northern trees, appear to be particularly well endowed with chlorophyll in their corticular tissues (Figure 7.16).

Compared on a unit area basis, the bark chlorenchyma in *Betula pendula* can contain up to 55% of the chlorophyll of the concomitant leaves when grown under 100% sunlight and even 66% when trees are grown under low light (20% of full sunlight). Light penetration through the periderm of birch twigs and branches is age-dependent and ranges from roughly 24% of the incident sunlight in young twigs to 1–3% in 5-year-old main stems. In winter the refixation of respiratory carbon dioxide (as opposed to the atmospheric source) amounts to a very effective carbon capture process, recycling from 60–80% of the potential respiratory over-winter carbon loss (Pfanz, 1999). This adaptation may account for the ecological success of birches as well as aspens and alders in the short growing seasons of the northern Taiga.

Although there is little experimental evidence as yet, it is highly probable that a further benefit from corticular photosynthesis will be to provide a source of oxygen for reducing internal stem anaerobiosis.

Winter corticular photosynthesis for trees in northern habitats may therefore have a number of advantages, in particular when the ground is saturated or flooded or when the bases of the trees are trapped with snow, ice, and water and root aeration is impeded (Figure 7.19). Birch, alder, and aspen are therefore the species most likely to benefit from any additional supplies of oxygen that may be made available by sub-cortical photosynthetic activity.

7.7 Conclusions

Examination of plants that survive the winter in the frozen lands of the Tundra and Taiga reveals a very wide range of adaptations, which are not just for surviving the vicissitudes of frost and snow and long periods with limited access to water but include many other aspects of life in the cold as it affects survival and reproduction. Low temperature survival calls for specialized adaptations not just in winter but also in summer. Plants, ever since they came on land, have had to adapt to the threat of water deprivation. As this is one of the main constraints for plant survival in cold climates it could be considered that land plants have had a long evolutionary history involving many adaptations in relation to their water economy, and this has pre-adapted them to the adversities and fluctuations of arctic and boreal environments.

CHAPTER 8

Demography and reproduction

8.1 Reproduction at high latitudes

Bringing forth the next generation is a hazardous process for all living things. Nowhere is success in this process more affected by environmental fluctuations than in polar regions. For most species, the brevity of the climatic window for reproduction is short and this, coupled with a series of adverse seasons, can result in years in which raising another generation fails. The terrain in which local populations exist is often topographically and climatically limited. This is not always obvious at ground level. The limits to open ground in summer become apparent only when a more extensive view of the total landscape is obtained from the air (Figure 8.1).

Demographically, reproduction and demography are inextricably linked. This association is particularly evident in relation to species persistence at high latitudes. Where annual reproductive activity is limited or environmental disasters frequent, longevity provides extended opportunities for population recovery. In addition to longevity, migration has to be considered. For those species that have the capacity to move, a strategic retreat with a timely abandonment of the current year's breeding attempt can have a significant survival value.

Given the ever-present risks of starvation at high latitudes, it is not surprising that there is a tendency for arctic populations to fluctuate from over-abundance to near extinction. The population fluctuations of lemmings and other microtines are legendary. Uncertainties abound in arctic population studies. The term biodemography has now come into use to encompass the insight that is brought into population studies when ecology and genetics are included in demographic studies.

8.2 Human demography in the Arctic

Our recently increased knowledge of the genetic history and demography of the human species (*Homo sapiens*) has increased awareness of the speed with which the human populations extended their settlements into the Arctic even before the final retreat of the Pleistocene glaciations (see Chapter 3).

Currently, there is considerable debate concerning the attributes and circumstances that made this possible. It is unlikely that any one attribute can be identified, as conditions vary between the diverse regions where the human race now lives. In relation to reproduction, it is remarkable that our potential life expectancy exceeds by a considerable margin beyond the age of active female reproductive activity. Such seemingly lengthy longevity may have been preferentially selected, not because it increased the birth rate, but for the ability it gave to a population for educating and providing family assistance through more than one generation.

Similarly, the slower rate of maturation of children in *Homo sapiens*, as compared with the Neanderthals (*Homo neander-thalensis*), may have contributed to the greater time we spend in youth in the learning process (Smith *et al.*, 2010). The success of *Homo sapiens* in colonizing even the more inhospitable parts of our planet might be related therefore to the possession of a number of such pre-adaptive traits which prepared us in advance for living in challenging environments.

There are also certain acquired skills, such as the ability to sew and make weatherproof clothing, and to anticipate the provisions necessary for surviving winter at high latitudes. Such skills would be enhanced in a highly integrated family or tribal life

Tundra-Taiga Biology. First Edition. R.M.M. Crawford © R.M.M. Crawford 2013.
© Published 2013 by Oxford University Press.

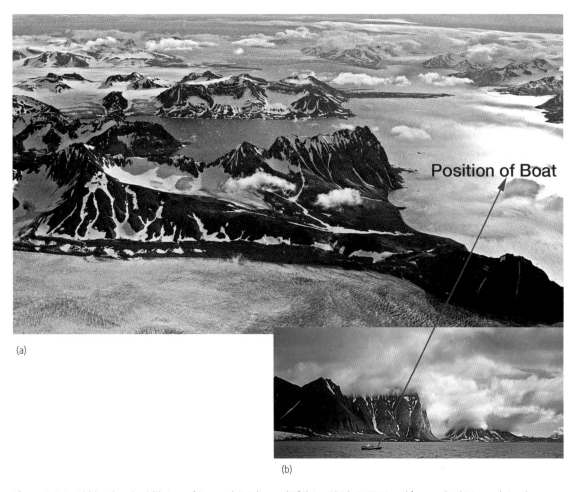

Figure 8.1 Aerial (a) and sea level (b) views of Hornsund, Spitsbergen (77°N), in mid-July 2005. Viewed from sea level it is not obvious how much of the land remains snow- and ice-bound in summer.

by the presence of older members, with accumulated knowledge and experience together with their role in looking after and educating the young.

Human survival at high latitudes has always had to contend with periods of adversity. The causes can be varied and range from climate fluctuations which either deplete resources or make them inaccessible or other disturbances such as disease and pollution. The native peoples of the Arctic have been described as living in a 'land of feast and famine' (Ingstad, 1992). Famine can set in when either the caribou do not arrive in accordance with their normal expected migrations, or when bad weather has prevented hunting for seal and walrus. Accustomed to these vicissitudes, arctic peoples have always been adept in conserving resources. However, when adverse conditions persist for more than one season starvation is almost inevitable.

Such a case arose on St Lawrence Island (Figure 8.2) in the Bering Sea between 1878 and 1880 when passing sea captains reported finding the bodies of the inhabitants lying where they had died in their houses. Bad weather, starvation, and disease had brought about the mortality of over 90% of the population (Mudar and Speaker, 2003).

Life in northern regions is very susceptible to sudden disturbances, such as those caused by climatic or other environmental conditions. In particular, sensitivity to volcanic eruptions has often been noted irrespective of where they occur. The relatively low

Figure 8.2 Location of St Lawrence Island in the Bering Sea, which had one of the worst recorded cases of famine death among northern peoples.

intensity of solar radiation in the Arctic can be significantly diminished by volcanic ash, even when eruptions take place thousands of miles away. As might be expected they are even more detrimental when the eruptions are local.

Figure 8.3 Population of Iceland 1750–1850. Note the 20% reduction in population after the Laki eruption in 1783 due to famine and disease, and a recovery that took place in only 30 years. (Data from the Icelandic Statistical Bureau).

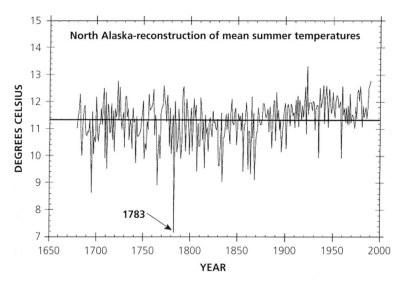

Figure 8.4 Reconstruction of mean summer (May–August) temperatures for northern Alaska. The temperature data are taken from a number of sources including tree-ring density series. The 7.23°C estimated summer temperature for 1783 is over 5 SE below the mean of 11.33°C. Given that May and June in the oral history are not described as being unusual, July and August could have been mostly below freezing, as suggested by local oral history. (Reproduced with permission from Jacoby et al., 1999).

Such a situation arose when the great Laki eruption In Iceland (1783–84) lead to the eventual death of one-fifth of the Icelandic nation (Figure 8.3). The eruption so poisoned the fragile and shallow volcanic soils that a large proportion of the domestic animals were killed. This, together with a virulent outbreak of smallpox, a lack of fresh vegetables, and a subsequent famine, caused the death of some 10,000 people, reducing the population to probably less than 40,000 (Mackenzie, 1811).

The Icelanders, however, proved themselves capable of surviving this unprecedented disaster and recovered their population numbers in just over two generations (Figure 8.3).

Disastrous as the Laki eruption was for Iceland, it was an even greater calamity for the Inuit in the extreme north-west of America. A detailed study of volcanology, climatology, history, anthropology, and tree rings, as well as accounts of travellers, have only recently been found and made possible a documentation of the disaster that befell the Alaskan Inuit. The disaster that took place in Alaska was synchronous with the Icelandic Laki eruption of 1783 (Jacoby et al., 1999). Tree-ring data (Figure 8.4) indicate that 1783 was the coldest year in Alaska for at least 400 or possibly even 900 years. It appears that this extraordinary cold period was the cause of the death that befell all but a few people in the scattered Inuit communities along the Alaskan north coast.

8.3 Longevity at high latitudes

8.3.1 Insect longevity

The Arctic is not a welcoming habitat to species with short lifespans. Insects are mostly short-lived, although the queens of some ant species are notable exceptions. The Arctic is, however, a home to some insects whose survival is due to the fact that they have a long life cycle in which their larval development is spread over a number of years. Outstanding among these long-lived species is the Arctic Woolly Bear Moth (*Gynaephora groenlandica*), a species that is restricted to northern Canada and Greenland. For the greater part of its life, the larvae remain frozen, emerging only after snow-melt for a feeding season that lasts only about 3 weeks, before they once again spin a chrysalis and become dormant (Kukal and Kevan, 1987). The total lifespan of this insect has been variously estimated, with some exceptional records of 14–17 years. However, a careful analysis of morphological characters, including larval head capsule, suggests that a more typical length of life cycle is 6–7 years. (Morewood and Ring, 1998).

8.3.2 Avian longevity

Numerous examples of longevity are found among birds with northern distributions. The Arctic Tern (*Sterna paradisaea*) has been recorded as living for 34 years (Figure 8.5). Even greater longevity occurs in the Northern Fulmar (*Fulmaris glacialis*) of which there are three sub-species: *F. glacialis glacialis* (Figure 8.6), which breeds in the High Arctic regions of the North Atlantic, *F. glacialis auduboni*, which breeds in the Low Arctic and boreal regions of the North Atlantic, and *F. glacialis rodgersii*, which breeds on the coast of eastern Siberia and Alaska.

Studies carried out on Northern Fulmars in the Orkney Islands to the north of Scotland indicate a mean adult lifespan of about 32 years. In North America an age of 31 years is not uncommon. Several Northern Fulmars banded as adults in 1951 in Orkney were still breeding in 1990, at ages likely to have been greater than 50 years (Dunnett, 1992). In Spitsbergen, the oldest age yet recorded for a Northern Fulmar is 13 years.

Figure 8.5 An Arctic Tern hovering over its nest. Arctic Terns have been recorded as living for up to 34 years. (Photo courtesy of E.R. Meek).

Figure 8.6 Northern Fulmar (*Fulmaris glacialis glacialis*) in the north Canadian Arctic. (Photo courtesy of Clare Kines).

This predominantly northern bird lays but one egg each year. Despite this minimal clutch size fulmars have shown a remarkable ability to increase their number in recent years, presumably due to longevity. They have a low mortality rate of 5% per annum. Fulmars only begin breeding at an exceptionally old age. Most do not breed until they are at least 8–10 years old, and sometimes even later.

8.3.3 Mammal longevity

Longevity combined with bringing forth a low number of offspring at any one pregnancy is a feature in bears, horses, muskoxen, and seals. Frequently, only one offspring is reared each year, a characteristic that is accompanied by having a relatively long reproductive life. Among the larger arctic mammals the Polar Bear (*Ursus maritimus*) can achieve an age of 32 years. Muskoxen (*Ovibos moschatus*) can live for 20 years, and the Arctic Fox (*Vulpes lagopus*) 14 years when kept in captivity, but under natural conditions life expectancy is less.

Individual longevity is clearly a major factor influencing fitness for species with repeated reproductive cycles (iteroparity). Theories of life-history evolution suggest that increased longevity allows individuals to have many attempts at breeding, and in the case of the more intelligent animals (including human beings) this allows the possibility of gaining experience in rearing offspring.

Attempts at quantifying this concept of learning in relation to reproductive success have been made with female reindeer (Weladji *et al.*, 2006). All females were found to show increased reproductive success with age up to their penultimate year of life. Long-lived females had more successful breeding attempts during their life and had higher reproductive success at all ages, especially during the last year of life (individual quality component) than short-lived females.

8.3.4 Plant longevity

Longevity, from a reproductive point of view, is beneficial for all arctic life, whether it be animal or vegetable. The Tundra and also the Taiga both have habitats where there are considerable plant populations that are not actively reproducing sexually. This is especially evident in mature forests and the matt vegetation of many Tundra heaths and bogs. The ecological stability of many of these habitats, and the fact that they provide shelter for more actively reproducing species, means that the longevity of the ageing or moribund component has a demographic relevance for the ecosystem as a whole in maintaining ecological stability.

In terms of longevity, plants, even under arctic conditions, have a survival capacity that verges almost on immortality. Given the hazards that exist for reproduction at high latitudes it might be expected that longevity (even without a learning process) is beneficial for plants as well as for birds and mammals. This must not be taken to mean that populations of species that are found in the Arctic are longer-lived than their southern counterparts. It is merely the fact that species with rapid generation turnover can be expected to become rarer with increasing latitude. This is particularly obvious in arctic floras where annual species are few.

Two features of plant life histories contribute to longevity in the Arctic. The first is *clonality*, by which plants spread vegetatively and perpetuate the existence of the original individual. Rejuvenation is achieved by subsequent detachment from the parental region of the clone. The second facilitator of longevity is the seed bank, which not only perpetuates species survival but also ensures a preservation of genetic diversity.

A study of the tussocks of Cotton Grass (*Eriophorum vaginatum*) in central Alaska estimated the minimal age of mature tussocks over a number of sites to range from 122–187 years, confirming a general impression that plants of this species are relatively long lived (Mark et al., 1985).

Aspens (*Populus tremuloides*) and various arctic Sedges (*Carex* spp.) are but two examples where clonal age has been estimated in millennia. In the Siberian Arctic estimates of over 3,000 years have been made for the longevity of clones of *Carex ensifolia* ssp. *arctisiberica* (Jonsdottir et al., 2000). More examples are continually coming to light as molecular markers present ever-greater information.

8.3.5 Trade-offs between plant longevity and reproduction

The longevity of many clones makes it difficult to deduce from direct observation whether or not sexual reproduction takes place at spaced out or irregular intervals. Nevertheless, various attempts have been made to assess the degree of trade-off that might exist between survival (or longevity) and reproduction in arctic plants (Philipp et al., 1990). In two relatively long-lived species, *Dryas integrifolia* and *Silene acaulis*, biomass (dry weight) allocation to reproduction as measured in Greenland was on average 0.8 and 3.2%, respectively, while the shorter-lived *Ranunculus nivalis* allocated up to 17% of its total dry weight to reproduction. Dry weight allocation to reproduction in *Dryas integrifolia* and *S. acaulis* was found to be even lower on the High-Arctic Ellesmere Island than on Greenland, i.e. 0.2 and 1.6%, respectively (Maessen et al., 1983). The few studies that have explicitly addressed the allocation to sexual reproduction in arctic clonal plants suggest that it is generally lower on a genet basis in clonal plants than in non-clonal plants.

In a study of long-established *Carex* spp. clones in the north-west Russian Tundra it has been possible to detect a distinct trade-off between sexual reproduction and clonal growth (Jonsdottir et al., 1999). In estimating the relationship between flowering frequency and the frequency of young ramets it was found there was a negative correlation between flowering frequency and the growth of young ramets (Figure 8.7). This was particularly evident in the cases with high frequency of flowering where there was a marked reduction in vegetative reproduction.

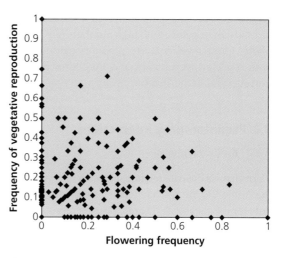

Figure 8.7 Frequency of young ramets' vegetative reproduction and flowering in *Carex* spp. in an extensive survey over 17 sites on the north-western Russian Tundra. (For detail see Jonsdottir et al., 1999).

Such findings prompted the suggestion that in areas where flowering frequency is low, a high degree of clonal growth would result in an increased persistence of the clones and therefore lead to an increased probability that the vegetative genet will eventually flower and reproduce sexually.

Using molecular methods it has been possible to estimate clonal diversity and compare this with genet longevity in populations of four arctic-alpine plant species with contrasting life histories. Investigations on amplified fragment length polymorphism (AFLP) in plants from northern Norway and Sweden, as well as on *Carex curvula* from the European Alps, found that it is possible to estimate genet size in plants of *Dryas octopetala*, *Salix herbacea*, and *Vaccinium uliginosum* (de Witte et al., 2012). The sizes varied from just a few centimetres to 18 m. From measurements of mean growth rates this would imply that some clones had attained considerable ages, which it was thought for the largest genets could lie between 500 and 4,900 years.

Such longevity would imply that some extant populations would contain individuals that would have lived through significant episodes of climate change. Most surprisingly this degree of immortality was not found to be accompanied by any significant loss in biodiversity. Thus, despite the longevity of some individuals, clonal diversity remained high, with most individuals existing as small, relatively young genets. The persistence of the long-lived individuals, together with high numbers of younger plants found in this study provides evidence for both repeated recruitment and long-term population survival in the Tundra.

8.4 Predator–prey interactions

8.4.1 Arctic herbivory

Plants have a delicate relationship with herbivores in the Arctic. On the one hand, being eaten, even if it is only for the sugar-rich flowers, can severely suppress reproduction. On the other hand, not being eaten can result in a deleterious accumulation of plant cover. Whether this cover is living or dead, it will interfere with the nutrient cycle as in the cold northern soils the absence of an active microflora and fauna makes nutrient cycling largely dependent on warm-blooded herbivores. Increased soil shading from too much vegetative growth also causes the permafrost level to rise, with adverse effects on plant growth and survival.

In recent times grazing by reindeer in Spitsbergen has had a severe effect on the Mountain Avens (*Dryas octopetala*) in removing the flowers before they came to fruition (Cooper and Wookey, 2003). As a result, the recruitment of *Dryas octopetala* on the Brøgger Peninsula in Spitsbergen has been severely hindered.

This situation is not unique to Spitsbergen and also extends to other species of flowering plants and has existed for millennia. It has recently been found (Brochmann, *pers. comm.*) from examination of remains of the Woolly Mammoth (*Mammuthus primigenius*) preserved in the now retreating ice, that they grazed on the flowering shoots of the Globe Flower (*Trollius europaea*) (Figure 8.8).

The deleterious effects of over-grazing by geese have also been noted on vegetation in the Arctic both in North America and Europe. For geese, successful reproduction is dependent on rearing the goslings on a nutritionally rich diet. This can only be obtained from certain salt marsh sedges and grasses which maintain high sugar concentrations as a protection against salt and frost injury.

In arctic salt marshes along the western shores of the Hudson Bay the geese have removed 90% of the annual above-ground primary production as well as grubbing up the buried stolons (Figure 8.9). In

Figure 8.8 Globe Flower (*Trollius europea*), the flowers of which were found to have featured in the diet of mammoths.

areas where the vegetation has not been fully destroyed the germinable seed density (Figure 8.10) in ungrazed plants was six times greater than in grazed plots (Kuijper *et al.*, 2006).

Moderate goose grazing on Creeping Salt Grass (*Puccinellia phryganodes*) enhances plant production but if it is intensified beyond a certain threshold can completely destroy the plant cover (Jefferies, 1997).

The preferred species are the stoloniferous graminoids of the salt marshes which are rich in soluble sugars and amino acids, of which *Carex subspathacea*, *C. aquatalis*, and *Puccinellia phryganodes* are the principal species.

For mosses, the damage is less severe. One defence response that some plants have to over-grazing is the generation of metabolic grazing deterrents in their foliage (see section 6.4.3).

Resources in the Arctic are limiting and the length of food chain between the different trophic levels is

DEMOGRAPHY AND REPRODUCTION 171

Figure 8.9 Damage to coastal stands of willow by over-grazing, with grubbing of inter-tidal marsh in early spring by Lesser Snow Geese on the shores of the Hudson Bay. (Photo by the late Professor R.L. Jefferies).

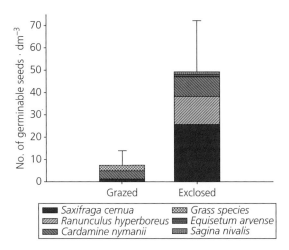

Figure 8.10 Density of germinable seeds and other propagules found in the topsoil layer in and outside exclosures ($n = 4$ with SE) in Ny Ålsund, Spitsbergen (79°N). (Adapted from Kuijper et al., 2006).

short. The limited choice of prey and vegetative fodder creates a tendency for certain components in the food chain to be more exposed to predation than in the more complex and richer food chains found in warmer climates. The damaging effects of goose grazing on particular plant communities has just been discussed. The same effect can also exist in relation to predator–prey relationships in animals. Such is the pressure of predation that it can frequently negate successful reproduction in both plants and animals.

The Arctic by being a land of 'feast and famine' can provide in turn ideal conditions for reproduction with periods free from disturbance. Once predators have for necessity moved elsewhere in search of more plentiful prey, there is in turn a period of respite for the successful re-establishment for the lower order members of the food chain.

Small mammals in particular have to face the hazards of predation. It has been estimated that a single Snowy Owl can consume 1,600 lemmings in one year. Owls and skuas can therefore drastically deplete the lemming population, just as high densities of reindeer with their tendency to seek out flowers with their high sugar contents can drastically reduce the reproductive ability of some keystone vascular plants.

The ancient indigenous human populations in the Arctic are sometimes portrayed as living in a harmonious equilibrium with their environment. However, they too appear to have seriously depleted local populations of animals on which they were dependent. Archaeological evidence from the frequency with which dwellings were abandoned as people moved on to fresh hunting grounds suggests that arctic hunters exhausted the prey around their settlements (Krupnik, 1993).

It is difficult for us today to have any realistic estimation of just how easily satisfied any group of prehistoric hunters might have been with their kill. Some insight is, however, given from early European travellers who observed the hunting habits of the northern Indian tribes in the area between the Hudson Bay and the Mackenzie River. Such a traveller was Samuel Hearne who was one of the earliest explorers in the Canadian Arctic between 1769 and 1772. He travelled for many months, living entirely with the northern Indians and keeping a careful and detailed diary (Hearne, 1795). He noted that reindeer tongue was a highly valued delicacy and that they would kill far more animals than they otherwise needed in order to enjoy this particular morsel. If this type of hunting was widely practised, then even a small population of hunters might have a significant effect on local populations of their prey.

8.4.2 The Arctic Fox and its prey

Numerous studies relate the interactions between lemming population cycles and the abundance of their common predators, e.g. Arctic Foxes (*Vulpes lagopus*), Snowy Owls (*Bubo scandiacus*), and Long-Tailed Skuas (*Stercorarius longicaudus*). Where there are no lemmings there are no Snowy Owls or Long-Tailed Skuas. However, this does not hold for the presence or absence of the Arctic Fox. In areas where there are lemmings the populations of Arctic Foxes (Figure 8.11) vary with the fluctuations of the lemming population usually with a lag of 1 year (Angerbjorn et al., 1999).

Such fox populations that track their rodent prey dynamics are termed 'lemming foxes' or 'inland foxes' as the ecotype is typically reported from inland Tundra and alpine areas. However, Arctic Foxes can live successfully in areas in which there are no lemmings.

In areas such as Iceland and Spitsbergen, cyclic major fluctuations in Arctic Fox numbers are largely lacking. Nevertheless there can be found a degree of variation in the density at a local level. On the Brøgger Peninsula and Kongsfjorden region in Spitsbergen, Norway, a study carried out from 1990 to 2001 found that fox numbers increased from 1990 to 1995 whereupon they decreased sharply, before increasing again and leveling off in 2001. Increasing numbers of foxes during the first part of the study paralleled increasing numbers of Svalbard Reindeer

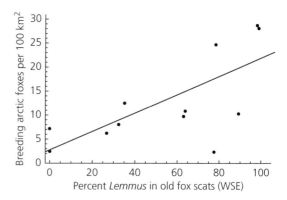

Figure 8.11 Relationship as found in northern Siberia between the density of breeding Arctic Foxes and the percentage of lemming (*L. sibiricus*) in 1-year-old fox scats. The fit of the line indicated that the numerical response of Arctic Foxes to changes in lemming population density had a time lag of approximately 1 year. WSE, Whole scat equivalents. (Adapted from Angerbjorn et al., 1999).

(*Rangifer tarandus platyrhynchus*), as their carcasses in winter provided improved scavenging. In addition there were ever increasing numbers of nesting Barnacle Geese (*Branta leucopsis*) in summer.

Despite the the transitory nature of their summer prey, the Spitsbergen coastal fox populations have been reported to be relatively stable, with little variation in den occupancy, although litter sizes did vary. Birds and eggs account for a substantial part of the diet and eggs are cached for winter. It has been reported that cached eggs have been eaten even when they were a year old. In winter the Spitsbergen Arctic Foxes

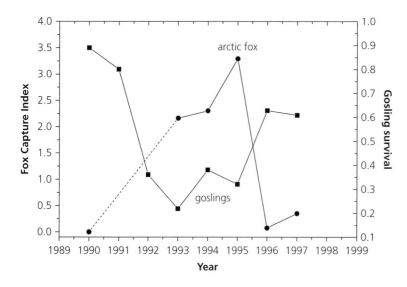

Figure 8.12 Relationship between the 'Fox Capture Index' (number of Arctic Foxes captured per 100 trap-days) on Brøgger Peninsula, Kongsfjorden (solid circles), and gosling survival in Kongsfjorden (solid squares). Gosling survival was negatively correlated with number of foxes captured per 100 trap-days. (Reproduced with permission from Fuglei et al., 2003).

rely on catching Ptarmigan and scavenging reindeer carcasses. The presence of reindeer carcasses in winter and increasing numbers of nesting Barnacle Geese (*Branta leucopsis*) in summer have done much to enable the foxes to maintain their population. Nevertheless increasing fox numbers can reduce the number of goslings reared (Figure 8.12).

When reindeer carcasses are few this can cause a reduction in the fox population. Temporal variability in the availability of reindeer carcasses has been suggested as one of the main factors when the Spitsbergen fox population declines. There can therefore be fluctuations in Arctic Fox population even when they are not feeding on lemmings (Fuglei *et al.*, 2003).

Within Spitsbergen a distinction has to be made nevertheless between coastal and inland fox populations. Foxes using the coastal dens have higher occupancy rates than inland dens, as might be expected from seasonal availability of prey resources. There is also a higher density of fox dens along the coast than in the inland sites. It has been estimated that sea birds contribute to the feeding for about half of the annual pup production. Juvenile Arctic Foxes in Svalbard have lower survival rates and a high age of first reproduction compared with other populations and it has been suggested that this may be caused by a lack of unoccupied dens and a saturated population (Eide *et al.*, 2012).

8.5 Phenology and reproduction

Matching the various aspects of biological activity with changes in the seasons is a necessity for most plants and animals. Given the brevity of the arctic summer, accurate timing for arriving at the nest site is essential for species that breed at high latitudes. The most noticeable events are probably the spring and autumn migrations of birds and the spring flowering and onset of autumn colours in plants. The mechanisms that trigger these responses have to be in some measure independent of vagaries in weather.

For both animals and plants, timing involves more than just sensing an improvement or decline in the weather. There has to be a capacity to anticipate the changes in the seasons and for this the most common method is a photoperiodic response to changes in day length. For plants at least this is more accurately described as a change in the length of the night, as interrupting the period of darkness with a short period of illumination can negate the photoperiodic effects of long nights.

8.5.1 Timing of goose migrations

The arrival of migrant birds in the Arctic, such as geese, at their arctic nesting sites is timed to coincide with the beginning of the growing season. As such long-distance migrations involve a considerable change in latitude, this in itself imposes an alteration in the perception of day-length signals in relation to the progression of the seasons. There is therefore a need for a compensating mechanisms if optimal timing is to be achieved. The nutritional needs of large birds such as geese imposes a further complicating factor beyond just arriving at the nesting site on time. Arriving in the Arctic the birds need to be ready to breed and for this they make strategic stops in staging areas. The amount of time spent in these stop-over sites also demands a timing element as to when is the appropriate time to move on.

8.5.2 Role of staging areas

Numerous studies have demonstrated a reproductive advantage amongst northern nesting geese in arriving early in breeding areas in good condition. Many observations have shown that pre-nesting feeding in Spitsbergen by females can contribute substantially to their nutrient stores and their subsequent reproductive success.

Barnacle Geese (*Branta leucopsis*) in common with many other species of arctic-nesting geese use staging areas close to their nesting sites which are referred to more specifically as as *pre-breeding areas*. Studies of such an area on the south-facing slopes of Ingeborgfjellet, Varsolbukta (Figure 8.13), on the west coast of Spitsbergen (77°N) have shown that the timing of migration to these sites has a great significance for replenishing body stores during this last period of the migration (Hübner, 2006).

Barnacle Geese can afford to arrive early in Svalbard without suffering loss of body condition, provided that they stage in favourable pre-breeding areas before moving on to the breeding sites.

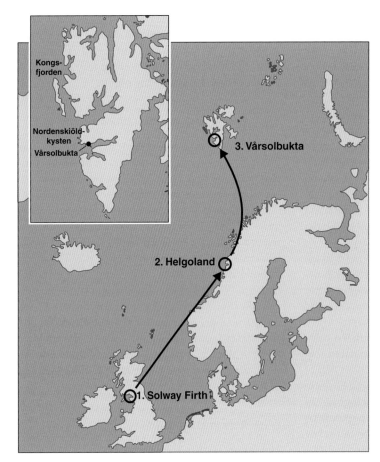

Figure 8.13 Migration route of Barnacle Geese from Scotland to Spitsbergen via Helgoland, their main staging ground in Norway. (Reproduced with permission from Hübner, 2006).

Not only does such a staging period in Varsolbukta allow the birds to replenish their reserves, it also has the advantage of increasing the accuracy of timing an appropriate arrival in their arctic breeding ground and a greater degree of certainty of finding adequate snow-free ground for grazing.

It is self-evident that the timing mechanism that governs the migration movements in birds is complex. Nevertheless, the basic stimulus for seasonal responses to the environment is one that is shared by most animals, namely a hormonal response to alterations in day length.

8.5.3 Sensing the seasons

The circadian pacemaking system of birds in common with the daily rhythms of all animals is generated by stimuli of endogenous origin which are

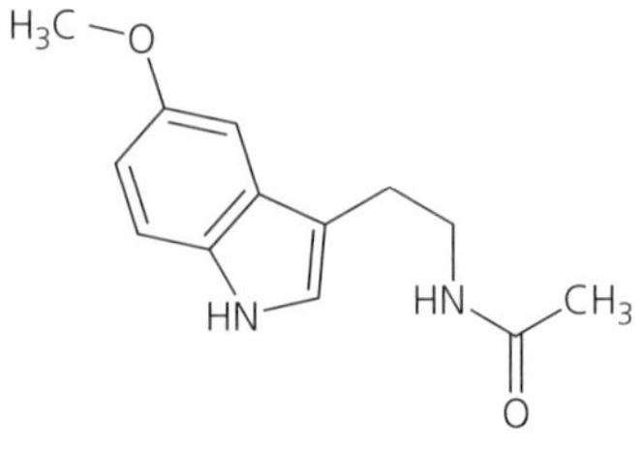

Figure 8.14 Melatonin—*N*-acetyl-5-methoxytryptamine—sometimes known as the 'sleep hormone'. The same substance when found in plants is referred to as phyto-melatonin.

triggered by day length which affects both behaviour and physiology. In mammals the light input is perceived by the eye retina, with periods of darkness leading to the production of the sleep-inducing hormone *melatonin* (Figure 8.14).

The pineal gland is the source of melatonin in all animals, including birds (Figure 8.15). The secretion

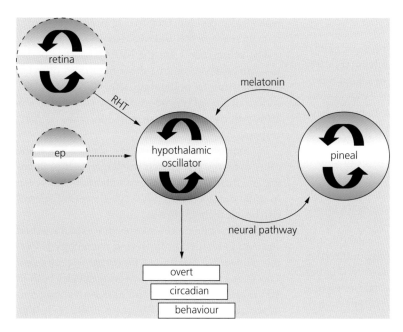

Figure 8.15 Oscillatory components of the avian circadian pacemaker and some of their interactions, as well as sites of photoreceptors influencing the pacemaking system (green background). ep, Encephalic photoreceptors; RHT, retinohypothalamic tract. (Reproduced with permission from Gwinner and Brandstätter, 2001).

of the hormone is suppressed by light and synthesized during periods of darkness with the effect of inducing sleep, hence its sobriquet 'the sleep hormone'.

In addition to sleep, many other biological effects of melatonin are produced through activation of melatonin receptors. It also acts as a powerful antioxidant with a particular role in the protection of nuclear and mitochondrial DNA.

In birds the situation is more complex, with additional photoreceptors within the brain (encephalic photoreceptors—ep). A third element also exists in birds in the hypothalamic oscillator. The hypothalamus is a portion of the brain that has the function of linking the nervous system to the endocrine system via the pituitary gland (Gwinner and Brandstätter, 2001). These components interact jointly to stabilize and amplify each other to produce a highly self-sustained circadian output (Figure 8.15). Their relative contribution to observable rhythmicity appears to differ between species.

It has been suggested that this variability is functionally important for enabling High-Arctic birds to retain synchronized circadian rhythms. The necessary *Zeitgeber* stimulus (i.e. the environmental cue for setting a biological clock) will be weak for migrating birds and therefore they need to adjust their circadian system quickly to altered environmental conditions during migration. The pineal melatonin rhythm, apart from being involved in generating the avian pacemaking oscillation, is also capable of retaining day-length information after isolation of the animal. Hence, it appears to participate in photoperiodic after-effects. These responses suggest that intricate circadian clocks have evolved to help birds adjust to complex environments (Gwinner and Brandstätter, 2001).

It appears that this adaptability of the circadian pacemaking system in birds comes from the interaction of three major components: (1) the pineal gland, which rhythmically synthesizes and secretes melatonin; (2) a hypothalamic region, the hypothalamic oscillator; and (3) the retinae of the eyes plus the encephalic photoreceptors (Figure 8.15). It is suggested that these components jointly interact, stabilize, and amplify each other to produce a highly self-sustained circadian output (Gwinner and Brandstätter, 2001).

The relative contribution to overt rhythmicity appears to differ between species and the system may change its properties even within an individual depending, for example, on its stage in the annual cycle or its photic environment.

8.5.4 Plant phenology

The short and unpredictable growing seasons that characterize most arctic and alpine habitats carry specific risks in terms of the timing of sexual reproduction for plants as well as animals. Several studies of flowering in cold northern habitats have sought to define reproductive strategies in relation to the varying phenological patterns of plants. A study of variation between three Nordic Butterwort (*Pinguicula*) species growing in Swedish Lapland has detected phenological differences that can be related to optimizing opposing strategies for ensuring reproductive success (Molau, 1993).

The Alpine Butterwort (*Pinguicula alpina*) (Figures 8.16) is an example of an early-flowering outbreeder, while the Common Butterwort (*P. vulgaris*) (Figures 8.17 and 8.18) is an opportunistic late-flowering inbreeder, and the Hairy Butterwort (*P. villosa*) represents an intermediate behaviour. The species were found to be sympatric and reproductively isolated due to different ploidy levels. In Molau's view, the length of the growing season combined with the different breeding patterns is critical for reproductive success and cannot be simply explained by the classical survival strategy concepts such as *r and K selection* (see Glossary). Instead, Molau hypothesized that arctic and alpine flowering plants are more realistically divided into two opposing categories namely *pollen riskers* and *seed riskers*.

The *pollen riskers* are the early flowering plants that are outbreeding but show high rates of ovule abortion due to pollination failure. Nevertheless, some viable seed is produced every year and as it is cross-pollinated will also provide some genetic dispersion.

By contrast the late-flowering species are at the other extreme. The flowers are largely self-pollinating and there is little or no ovule abortion. However, in unfavourable years these late-flowering species risk losing their entire seed crop.

Another arctic-alpine species that exhibits a dual strategy in relation to flowering phenology is Purple Flowering Saxifrage (*Saxifraga oppositifolia*). This is one of the earliest flowering species both in alpine and arctic habitats and has also a strong tendency to out-crossing (Figures 8.19–8.21). In this case the actual time of flowering depends on the

Figure 8.16 Alpine Butterwort (*Pinguicula alpina*), an early-flowering outbreeding species of this insectivorous genus which is common in the Arctic as well as in the European Alps.

Figure 8.17 Common Butterwort (*Pinguicula vulgaris*), an example of an opportunistic late-flowering inbreeder, which is a common insectivorous species in the more northern and western parts of Europe. Note the insects adhering to the leaves.

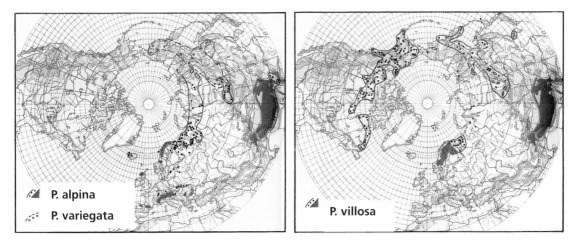

Figure 8.18 Distributions of Butterwort species (*Pinguicula alpina*, *P. variegata*, and *P. villosa*). These species are notable examples of true arctic-alpine species with significant presence in the Arctic as well as the European Alps. (Reproduced with permission from Hultén and Fries, 1986).

Figure 8.19 Variation in form and flowering in arctic forms of Purple Saxifrage (*Saxifraga oppositifolia*) as found in Ny Ålesund, Spitsbergen. *Left* Tufted form that grows on dry ridges; *right* prostrate form that grows on late wet sites.

thawing of snow banks and those populations that live on cold coastal areas where snow lies late may have their flowering dangerously delayed, in that although they may flower, it can be too late for seed production.

High-Arctic populations of *Saxifraga oppositifolia* have an early-flowering, tufted ecotype that inhabits dry ridges, sometimes only a few metres distant from prostrate late-flowering populations growing in low-lying, cold, wet areas with late-lying snow cover. Intermediate types are also common (Figure 8.19).

In the cold shore sites the prostrate plants do not give an impression of being particularly robust.

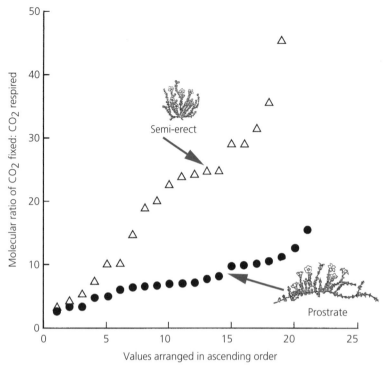

Figure 8.20 Molecular ratio of carbon dioxide fixed (net photosynthesis) to respiration in shoots of *Saxifraga oppositifolia* of two distinct ecotypes. Solid symbol: measurements made on semi-erect plants taken from a dry beach ridge; hollow symbols: measurements at the same time from an adjacent low beach late wet site with prostrate creeping plants. Note the semi-erect plants from the early dry site are carbon savers and the plants from the late wet site are carbon spenders.

However, appearances are misleading and this frail-looking prostrate saxifrage has a capacity adaptation (Figures 8.20 and 8.21) which allows it to out-perform the more robust type on the beach ridge in terms of gross photosynthetic capacity, respiration, and shoot growth (Crawford *et al.*, 1995).

Physiologically the plant on the beach ridge is more drought tolerant and can be described as a *saver* conserving carbohydrate, usually with the presence of a well-formed tap root.

The plants of the shore habitat by contrast are *spenders* and have no tap root. They use a much greater portion of their energy gains immediately for rapid growth (see 5.3.1 and Figures 8.20 and 8.21). The two forms have therefore developed opposing strategies which aid their survival in their particular microhabitats. Plants of the tufted and creeping ecotypes of Purple Saxifrage maintain their distinctive aspects as seen in their floral morphology (flower and scape arrangement) when grown in cultivation, and although inter-fertile have been recognized as distinct types (Lid and Lid, 1994).

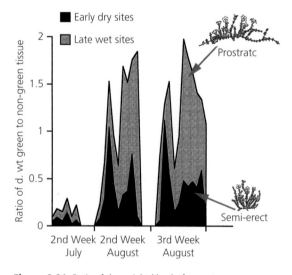

Figure 8.21 Ratio of dry weight (d.wt) of green to non-green tissues in plants of the semi-erect and prostrate forms of *Saxifraga oppositifolia* inhabiting, respectively, dry warm ridge sites and late wet shore sites at Kongsfjord, Spitsbergen. Note the greater proportion of green tissue in the prostrate form that developed during the peak of the growing season which will compensate in part for the shortness of the growing season on the wet shore site.

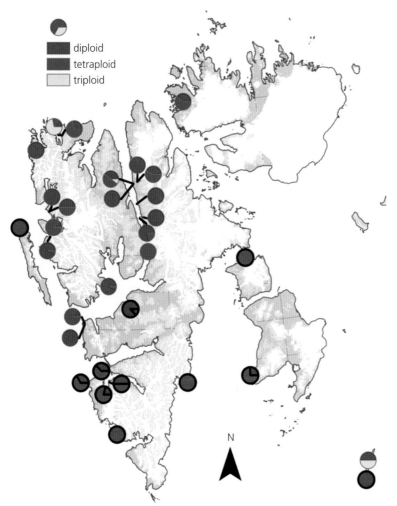

Figure 8.22 Distribution of diploid, triploid, and tetraploid individuals of *Saxifraga oppositifolia* in Svalbard based on 193 samples viewed within the 20 sampling sites. (Reproduced with permission from Eidesen *et al.*, 2013).

It has long been thought that Purple Saxifrage throughout most of its range occurred most commonly as a diploid (2n = 26), although the existence of tetraploids (2n = 52) were known. Recent research, however, has shown that the autotetraploid occurs more commonly than was previously realized (Figure 8.22). In addition it has been possible to demonstrate a significant correlation between ploidy level, ecology, and growth form in *S. oppositifolia*. Diploids have a preference for ridge habitats but show an overall much wider ecological niche and phenotypic plasticity than tetraploids.

The tetraploids display a narrower niche, which is slightly shifted towards more alkaline habitats with less bare ground and higher vascular vegetation cover (Figures 8.22 and 8.23). Nearly all tetraploids have the prostrate form and have a preference for snow cover during winter and are found mostly in low coastal areas (Eidesen *et al.*, 2013).

When grown under controlled temperatures the prostrate form had a significantly higher mortality than the cushion form (Crawford *et al.*, 1995).

It is now evident that autoploidization within *S. oppositifolia* has led to an expansion of the species niche. The annual distribution and duration of snow is strongly dependent on topography and is of great importance for small-scale vegetation structure in the Arctic (Eidesen *et al.*, 2013). Surviving under snow and having only a short growing season can also be accomplished by a different

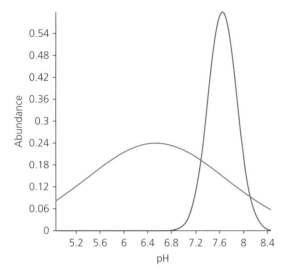

Figure 8.23 A Gaussian response model fitted to the abundance of diploids (red line) and tetraploids (blue line) of *Saxifraga oppositifolia* in 163 vegetation plots along a gradient of soil pH (six plots with triploids discarded). (Reproduced with permission from Eidesen *et al.*, 2013).

Figure 8.25 *Ranunculus pygmaeus*, a Dwarf Buttercup that can survive more than one growing season buried under snow.

Figure 8.24 Blue Heath (*Phyllodoce caerulea*), a species in danger of disappearing in Scotland due to a lack of winter snow. This same species survives fluctuations in winter snow level in northern Japan apparently facilitated by hybridization with *P. aleutica*.

strategy, namely having a much reduced size of plant. There are a number of arctic plant species that can survive under snow banks for one or more growing season (Pielou, 1994). This long-term endurance of snow cover may have enabled some plants in the past to withstand prolonged periods of climatic deterioration. Examples of these diminutive species with extreme nival endurance are found in the Blue Heath (*Phyllodoce purpurea*) (Figure 8.24), the Polar Willow (*Salix polaris*), *Potentilla hyparctica*, *Sibbaldia procumbens*, *Saxifraga oppositifolia*, and the Pygmy Buttercup (*Ranunculus pygmaeus*-Figure 8.25). All the above species are capable of missing out on an entire growing season and survive for two or more years covered by snow.

8.6 Plant reproduction

Longevity is clearly aided by various forms of asexual reproduction. One form of asexual reproduction is *Apomixis*. In its most narrow sense this is more accurately described as *agamospermy*, i.e. the production of seeds without fertilization.

This means of asexual reproduction represents a rejuvenation of the individual plant and is most common in polyploid species. The Arctic is particularly noted for a high incidence of recently evolved polyploids (Brochmann *et al.*, 2004). Clonal growth forms can vegetatively reproduce plants that do not age physiologically as they are not encumbered by a senescent growth form after detachment from the parent plant.

8.6.1 Clonal genetic diversity

Alpine Bistort (*Polygonum vivipara*) (Figure 8.26) has often been considered as a species in which asexual reproduction by bulbils is so predominant that sexual reproduction by seed can be almost ignored. However, recent research on High-Arctic, sub-arctic, and alpine populations of *P. vivipara* has detected intermediate to high levels of genetic diversity within these various populations as well as showing physiological differentiation between locations.

Plants from Scandinavia require considerably longer photoperiods for floral induction than plants from lower latitudes (the Alps). Although *Polygonum vivipara* appears to reproduce almost exclusively asexually by bulbils, seed development can occasionally be observed even in arctic and alpine populations and may be sufficient to account for the differentiation of ecotypes and medium to high levels of genetic diversity found with the arctic and alpine populations (Bauert, 1996).

Intermittent or rare production of fertile seeds has been found in two circumpolar arctic saxifrage species: Drooping Saxifrage (*Saxifraga cernua*) (Figure 8.27), and Leafy Stem Saxifrage (*S. foliosa*). Previously

Figure 8.26 Alpine Bistort (*Polygonum vivipara*), a common and widespread species of the High Arctic extending south in high mountain ranges. The flowers rarely produce viable seed and reproduction is asexual by means of bulbils, which are readily eaten by Pink-Footed Geese.

Figure 8.27 Drooping Saxifrage (*Saxifraga cernua*), an arctic species that flowers regularly but seldom produces seed and relies mainly on asexually produced bulbils for reproduction.

both species had been assumed to have abandoned sexual reproduction in favour of vegetative bulbils. Field studies in northern Sweden, however, have found that both species have an androdioecious mating system (a species that has both male and hermaphrodite individuals), a phenomenon which is extremely rare in natural plant populations (Molau, 1992). The two gender classes within each species are maintained by self-incompatibility in hermaphrodites and a higher rate of vegetative propagation in female-sterile plants.

Hermaphrodites are rare, especially in *S. cernua*, and most populations consist exclusively of female-sterile plants. The scarcity of fruiting material in herbaria may be a result of the low frequency of collecting due to the inconspicuous nature of the plants at this stage. The tendency for asexual plants to have a more northern range than closely related sexual plants has been shown to be true for 76% of examined species (Bierzychudek, 1985). This aspect of the distribution of asexual reproduction has been described as *geographic parthenogenesis*.

A computer model designed to examine the causes of asexual plants being favoured in the north and at higher altitudes as well as in marginal resource-poor environments has suggested that as individuals move from areas where they are well adapted to where they are poorly adapted, sexual reproduction may reduce fitness (Peck *et al.*, 1998). The mal-adaptation of sexually reproducing migrants due to lower population densities when they move to marginal areas can lead to a loss of fitness. Asexual populations will be protected from this tendency and this may account for their success in marginal habitats in the northern hemisphere and elsewhere and may be the ultimate cause for *geographic parthenogenesis*.

8.6.2 Flowering in the Arctic

Like deserts after rain, arctic and alpine habitats are noted for their remarkable floral displays. The diminutive nature of plant vegetative organs in these marginal areas is not accompanied by a proportional reduction in the size of the flowers. Consequently, summer flowering displays are highly visible. Despite the abundance of flowers that is often found in montane and arctic habitats, it is necessary to determine the effectiveness of sexual reproduction, first in producing viable seed and second in establishing sustainable populations under the potentially inhibiting combination of cold and short-growing seasons that are found at high latitudes and altitudes.

In a detailed study in Greenland of flowering success made before the more recent onset of climatic warming, it was found that, despite prolific flowering in *Dryas integrifolia*, *Silene acaulis*, and *Ranunculus nivalis*, in all three species seedling production was very uncertain due to the unpredictability and quality of the growing season. Few seedlings survived and the population structure of the established species indicated that the input of new individual plants was episodic (Philipp *et al.*, 1990).

Since this study was carried out thermal conditions in the Arctic have changed with earlier snow-melt. Already differences are visible to those who have been studying these regions over the past few years. Figure 8.28 shows the extent of flowering of Purple Saxifrage near the Polish Research Station in Horn Sound in 2002. Such was the unexpected extent of this prolific flowering that it attracted much comment.

The rapid flowering that takes place in the Tundra after the melting of snow and ice is mostly achieved by long-lived perennial plants. This prompt flowering, often only days after the plants have become free of snow and ice, is a result of two partially related adaptations. The most immediately observable aspect is the short pre-flowering period at the beginning of the growing season which takes place in the long days of mid-summer. However, this ability to produce flowers quickly is dependent on a second adaptation, namely the possession of pre-formed flowering buds.

When the flowering requirements of Ice Grass (*Phippsia algida*), a High-Arctic and High-Alpine snow-bed grass species (Figures 8.29 and 8.30), were studied in controlled environments (Heide, 1992), seedlings flowered rapidly in continuous long days at temperatures ranging from 9–21°C, provided they had previously had a short-day period for the initiation of flower primordia at the same temperatures.

In common with many other arctic-alpine species, Ice Grass has the characteristics of a regular long-day plant for the actual act of flowering but the initiation of flowering requires short days. This dual response to

Figure 8.28 A vegetation response to climatic warming at high latitudes. Vigorous early summer flowering display of coastal populations of Purple Saxifrage (*Saxifraga oppositifolia*) in 2002 at Hornsund, Spitsbergen (77°N) appears to be related to an early retreat of sea-ice.

Figure 8.29 Ice Grass in Spitsbergen. (Photo courtesy of Norman Hagan, Norwegian Botanical Association).

day length is found in other species similarly adapted to short growing seasons that occur in the vicinity of snow banks. *Carex bigelowii* from the Rondane Mountains in southern Norway has also been shown to require short days and moderate temperatures for at least 10 weeks followed by long days for optimum flowering (Heide, 1992). Similarly, a number of arctic-alpine *Carex* species—*Carex nigra, C. brunnescens, C. atrata, C. norvegica*, and *C. serotina*—all had a dual induction requirement for flowering (Heide, 1997).

Another more extreme situation exists where the initiation and development of flowering buds and flowers can take several years in order to bring flower buds to maturation. An example of this type of reproductive behaviour has been found in the Alpine Bistort (*Polygonum vivipara*) (Figure 8.31). This species can take as long as 4 years for each leaf and inflorescence to progress from initiation to functional and structural maturity.

Due to the protracted duration of leaf and inflorescence development, five cohorts of primordia, initiated in successive years, are borne simultaneously by an individual plant which can limit the number of flowers and asexual bulbils that are produced in any one year (Diggle, 1997).

Another arctic species that takes several years to produce a mature flower from the initial flower

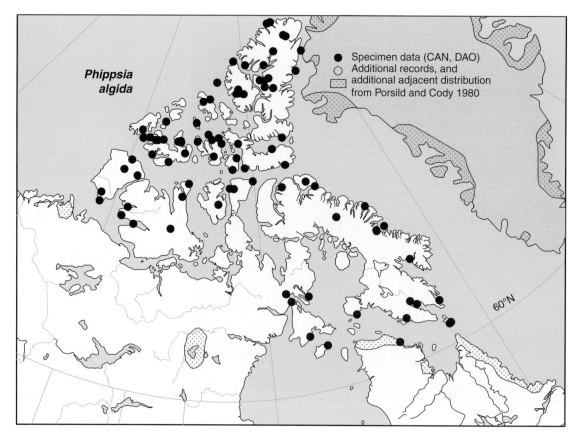

Figure 8.30 Canadian distribution of the circumpolar arctic-alpine Ice Grass (*Phippsia algida*). DAO, Department of Agriculture Canada Herbarium. (Reproduced with permission from Flora of the Canadian Archipelago, National Research Council of Canada, 2007).

primordium is the the Arctic Wintergreen (*Pyrola grandiflora*) (Figure 9.1).

8.6.3 Annual plants

In the Arctic most species are perennials and annual species are few. It is therefore of interest to examine the annual species that do manage to survive in the Arctic. Probably the largest annual plant in the Arctic is the Mastodon Plant (*Tephroseris palustris* ssp. *congesta* syn. *Senecio paludosus*) (Figure 8.32). In the early stages of growth, the leaves, stem, and flower heads are covered with translucent hairs, producing a 'greenhouse effect' close to the surface of the plant, essentially extending the growing season by a few vital days by allowing the sun to warm the tissues and preventing the heat from escaping. It is claimed that the species is '*semelparous*' (flowering once) and dies completely after having flowered and produced seed. However, at high latitudes, specimens in which the flowers of the first year are killed by early frost may flower and fruit the following year (Aiken et al., 1999 (onwards)).

The species is widespread, occurring throughout the northern hemisphere from Canada across Beringia to Eurasia. The most northerly occurrence is at 77°N on Prince Patrick Island, the farthest west of Canada's Arctic Queen Elizabeth Islands in the Canadian Arctic, which is surrounded by sea ice all year round and consequently is one of the most inaccessible islands of the Canadian Arctic (Figure 8.32).

Another annual species found in the Arctic which deserves special discussion is Iceland Purslane (*Koenigia islandica*) (Figure 8.33). The genus as a whole comprises six species derived from montane

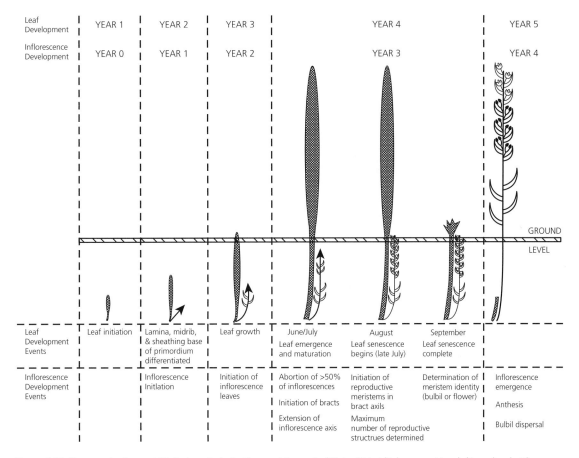

Figure 8.31 Five-year developmental trajectory of a leaf-axillary-meristem-unit of Alpine Bistort (*Polygonum vivipara*). (Reproduced with permission from Diggle, 1997).

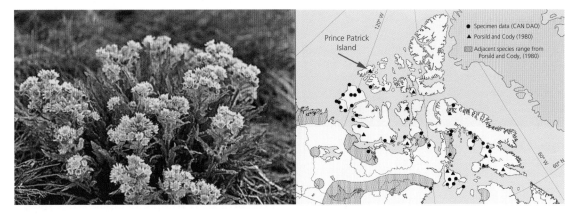

Figure 8.32 *Left* The Mastodon Plant (*Tephroseris palustris* var. *congesta*), the largest annual plant to be found in the Canadian Arctic. *Right* Distribution of the Mastodon Plant in the Canadian Arctic. (Reproduced with permission from Aiken *et al.*, 1999).

Figure 8.33 Iceland Purslane (*Koenigia islandica*), a species with a capacity for completing its life cycle in the short growing season of the north. This image is of a plant growing in a patchy moss community in Iceland. The plant is approximately 3 cm high (see 1 cm scale). It is the only annual species to survive in the Arctic and for the size of the plant succeeds in producing relatively large seeds in a very short growing season.

Figure 8.34 Upland Eyebright (*Euphrasia frigida*), an annual species that achieves rapid growth and profuse flowering in northern regions due in part to being a semi-parasitic species.

ancestors of a small group of high-mountain-dwelling species displaying adaptive radiation into diverse areas with short growing seasons.

Five out of the six species are now confined to high mountain areas in south-eastern Asia, primarily in the Himalayas, whereas the sixth, *Koenigia islandica*, has spread to the arctic and alpine areas in the northern hemisphere and has even penetrated to the fringes of Antarctica in the del Fuego.

The genus *Euphrasia* (Eyebrights) is also common in the Arctic and comprises 450 species and many hybrids, most of which are annuals (Figure 8.34). In northern habitats such as Spitsbergen they can behave as biennials. Experiments with *Euphrasia frigida* in sub-artic alpine meadows have shown that plants attached to a host were at least an order of magnitude taller than unattached plants, and individuals produced a total above-ground biomass up to three orders of magnitude greater (Seel and Press, 1993).

8.6.4 Two outstanding grass species

The Arctic never ceases to provide astonishing examples of species with a capacity to survive in areas that would appear to be marginal for a continued occupancy. Accidental arrivals come and go, but long-term survival over a wide geographical range is worthy of note. A series of case histories for outstanding plants that live in marginal areas can be found elsewhere (Crawford, 2008). The discussion here is therefore limited to just two outstanding examples.

Creeping Salt Grass (*Puccinellia phryganodes*)

The triploid sterile Creeping Salt Grass (*Puccinellia phryganodes*) is outstanding for the extent of its distribution throughout the Arctic as a major component of salt marshes (Figures 8.35 and 8.36). An account even exists of stolons of this plant being found at sea, frozen into ice floes, which were rescued, brought ashore, and when thawed out and planted proved to be still viable (the late R.L. Jefferies, *pers. comm.*). Dispersal throughout the Arctic of fragments of this species frozen into the pack ice must therefore be considered as a possible method of long-distance colonization. In addition to the sterile form, sexually active populations of this species are found. In Beringia it exists as the fertile diploid ($2n = 14$), while across Canada, Greenland,

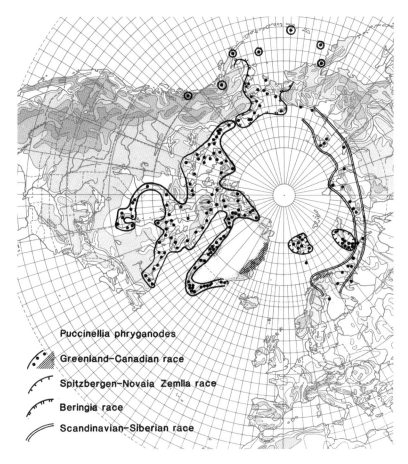

Figure 8.35 Circumpolar distribution of Creeping Salt Grass (*Puccinellia phryganodes*). (Reproduced with permission from Hultén and Fries, 1986).

northern Norway, Spitsbergen, and Fennoscandinavia, it is the sterile triploid 2n = 21 that occurs.

Alpine Meadow Grass (Poa alpina)

A very successful circumpolar species is *Poa alpina*, commonly known as Alpine Meadow Grass or Alpine Blue Grass (North America). The species occurs in two forms, *P. alpina* var. *alpina* which reproduces sexually and *P. alpina* var. *vivipara* which produces pseudo-viviparous vegetative spikelets (Figure 8.37) that are then shed at the end of the growing and root themselves in the ground.

This ability to survive for prolonged periods without oxygen (Figure 8.38), even while they are small plantlets, and then emerge again into air without showing any injury makes then well adapted to withstand the arctic winter. A capacity to endure prolonged ice-encasement, which deprives their tissues of access to oxygen, is coupled with a high degree of anoxia tolerance.

8.7 Conclusions

The outstanding feature of the Arctic biota is the frequency of catastrophic reductions in populations. The ability to repair decimated populations by sexual reproduction or, as in the case with plants, also

Figure 8.36 View of an arctic salt marsh at Longyearbyen, Spitsbergen. The brown colour of the marsh is typical of much arctic vegetation even in summer. The dominant species is Creeping Salt Grass (*Puccinellia phryganodes*) (see inset).

asexually is all the more astonishing given the climatic and resource limitations of high-latitude habitats. This ability of the arctic biota to recover from near extinction may be due to the reduced inter-specific competition that can be enjoyed in the sparseness of the arctic landscapes. Once a prey population is reduced to a low level, the predators diminish in number or else move elsewhere. When the lemming population has been decimated in a region by Snowy Owls and Skuas the predators have to depart. This then allows the lemmings a period of relative tranquillity for the recovery of their populations. Similarly, when over-grazed habitats cease to provide either the quantity or the quality of forage that is required, the herbivore population has to move.

Despite the well-demonstrated resilience to climatic fluctuations and disturbance of the arctic biota as a whole, including its human populations (see

Figure 8.37 The pseudoviviparous Alpine Meadow Grass (*Poa alpina* var. *vivipara*) bearing vegetatively produced plantlets.

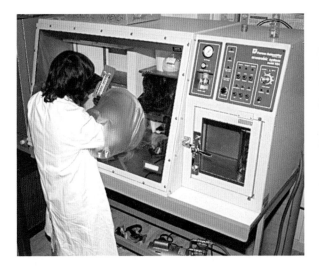 

Figure 8.38 Use of an anaerobic incubator (a) where plants are held in the dark under an atmosphere of 90% oxygen-free nitrogen and 10% hydrogen continually circulated over a platinum catalyst. (b) The plantlets of *Poa alpina* var. *viviparm* in the glass vessel have just been taken from the incubator and found to have suffered no injury even though they had been kept under total anoxia for 3 weeks at 20°C in total darkness.

Chapter 3), it would be wrong to be complacent about the continued ability of the cold-adapted biota to maintain their populations in a changing world.

Attention has to be given to the indisputable fact that the Arctic is currently warming to a greater extent than elsewhere. What may eventuallly be more damaging and disturbing is not rising temperatures but the possible continued exposure to ever-increasing levels of pollution and disturbance emanating from lower latitudes.

CHAPTER 9

Evolution in the Arctic

9.1 Pre-adaptation for arctic survival

The Earth with a mean temperature of 15°C is a cold planet for metabolic systems that depend on carbon metabolism. The making and breaking of covalent carbon bonds does not take place spontaneously at the temperatures that sustain life on this planet. One of the most remarkable aspects of evolution has been the creation of enzymes which can metabolize carbon compounds at temperatures well below those at which carbon chemistry can function in an abiotic system. Our world is therefore metabolically cold and the polar regions represent merely the lower end of what can be regarded as an unpromising thermal environment.

Put more positively, life that can exist in a world with a mean temperature of only 15°C is already pre-adapted to the challenges of surviving at high latitudes. This is amply demonstrated in plants which are able to survive in polar regions provided that there is some access to water, even if only for a few weeks in the year.

In terms of biodiversity, the *Cenozoic Climatic Decline* which might have been expected to diminish the diversity of the flowering plants had just the reverse effect. By increasing the thermal gradient from the poles to the tropics the Cenozoic Climatic Decline created a greater variety of thermal habitats and increased the diversity of flowering plants which in turn created new opportunities for animal evolution.

In addition to the evolution of metabolic and morphological adaptations to low temperatures, there has been an evolution of behaviour in the polar biota as part of adaptation to a lower temperature regime. Plants have readily adapted to an enforced dormant season—a behaviour that is partially matched in animals by hibernation. Even before the Arctic cooled, the polar night would have enforced on green plants, and many animals, at least a warm dormant season. In the recent post-glacial period various animals, and most notably birds, have evolved seasonal migration patterns, some of which can only be described as trans-global, in which they exchange the arctic polar night for the summer light of the southern hemisphere. Whether this took place to some extent in pre-glacial times to escape the dark season remains unknown. Selection for changes in behaviour can arise relatively rapidly. If the seasonal movement of a population leads to improved breeding success then the more mobile population will quickly replace the more sedentary.

Alterations in behaviour, when added to other selected traits such as morphological adaptation, feeding habits, and migrations patterns, introduce a dynamic dimension to the ability to respond with considerable rapidity to fluctuations in habitat and climate.

The human species (*Homo sapiens*) is a striking example of a species that evolved in the warm climate of Africa, yet within a period of 10,000 years after leaving the tropics was capable of establishing hunting settlements on the shores of the Arctic Ocean (see Chapter 3).

9.2 Arctic evolutionary influences

The Arctic stands apart from temperate regions in the advantages it presents in terms of summer day length for growth and reproduction. Although the growth rate of plants at low temperatures is slow, the vegetation—particularly in coastal regions and marshes—is rich in sugars and therefore highly nutritious for herbivores. In particular, the sedges

Tundra-Taiga Biology. First Edition. R.M.M. Crawford © R.M.M. Crawford 2013.
© Published 2013 by Oxford University Press.

and grasses of the salt marshes provide rich feeding for geese.

For plants the general imposition of slow growth rates has the advantage that both intra- and interspecific competition is reduced. Slow growth rates are only a disadvantage if there is a danger of being shaded out of existence by neighbours. However, in the Arctic where all plants grow slowly, shading is not usually a danger. Consequently, the Arctic is exceptional for plants as it is one of the few areas in the world where light is not limiting and there is usually sufficient water for this to be exploited throughout most of the short growing season. There is nevertheless the possibility that if climatic warming hastens snow melt in spring, then adverse summer droughts may follow.

Life in a cold climate raises the question as to whether thermal energy and time should be considered as resources. Heat is electro-magnetic radiation, and like light can be absorbed by living tissues. The extent to which organisms can absorb or dissipate thermal energy can be maximized or minimized through phenotypic plasticity in relation to morphology, pigmentation, and orientation. Cold-climate plants and animals are certainly not lacking in this respect.

There are warmer habitats within cold landscapes where thermophilous elements occur that are not found in surrounding areas. These arise usually from topographic features combined with strong continentality. Greenland (see section 2.2) has probably the highest concentration of such arctic hotspots (Elvebakk, 2005). Nevertheless, within any one habitat, heat is not a resource for which plants normally compete. It is, however, one of the most decisive environmental factors influencing resource acquisition.

In relation to evolution it is worthwhile considering whether or not time should be regarded as a resource. Time is neither matter nor energy and as it exists into infinity is never depleted and therefore cannot be considered a consumable resource. Life proceeds at a slower pace in the Arctic and therefore this may justify viewing the availability of time to be a potential resource in low-temperature biology. Whether or not this is justified will depend on the life history of the species under discussion. There are many examples of both plant and animal species in the Arctic that take their time in reproducing,

Figure 9.1 Arctic Wintergreen (*Pyrola grandiflora*), a species that can take several years from the initiation of a flower bud to the eventual production of flowers.

either in reaching puberty in the case of animals or in the case of some arctic plants taking several years to produce a functioning inflorescence and producing seed (see also Chapter 8).

For plants, time also affects competition. When time is available, metabolic rates can be slower and growth, development, and phenology can be adapted to proceed at a gentler pace when competition for resources is reduced. The Arctic Wintergreen (*Pyrola grandiflora*) can take several years to complete its reproductive cycle from the initial formation of a flowering shoot to the dispersal of seed (Figure 9.1). It is therefore indifferent to the brevity of any one growing season in relation to its eventual capacity to produce flowers and seeds.

9.3 Evolution of seasonal bird migrations

The origins of seasonal migration have long been the subject of scientific speculation. Those who consider the problem in the broadest sense often begin by asking whether migration evolves towards new breeding areas or towards survival areas in the non-breeding season. Sometimes this can be phrased as the 'southern-home-theory' or 'northern-home-theory'

(Salewski and Bruderer, 2007). In relation to the Holocene history of the arctic avifauna, it would seem logical to assume that the home territory is in the south and the routine of a migratory advance into Arctic summer nesting grounds began only as the ice retreated. It is difficult to explain the fidelity of many birds to their winter territories in any other way. An outstanding example is found in the Northern Wheatears and their African wintering territories.

Similarly, the summer migration of Barnacle Geese to Greenland and Spitsbergen from two distinct and separate over-wintering grounds in Scotland—in Islay and the Solway Firth—would suggest a long-established ancient post-Pleistocene migration habit (Figure 9.2).

It has long been suspected from bird-ring recoveries and observations at sea that the Arctic Tern (*Sterna paradisaea*) despite its small size (< 125 g) is the record holder in terms of long-distance migration, with its annual flight from the boreal and arctic regions of the North Atlantic to the South Atlantic. Tracking of 11 Arctic Terns fitted with miniature (1.4 g) geolocators has revealed that these birds do indeed travel huge distances (Figure 9.3). Some

Figure 9.2 Barnacle Geese (*Branta leucopsis*) overwintering in Islay, Scotland. This population migrates to Greenland to nest in summer. (Photo courtesy of E.R. Meek).

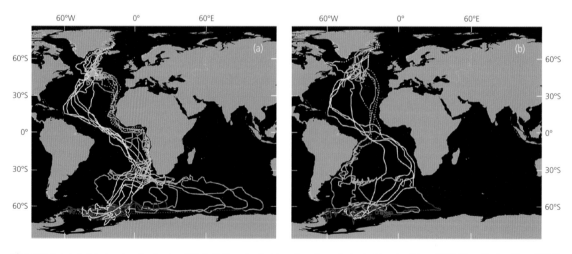

Figure 9.3 Interpolated geolocation tracks of 11 Arctic Terns tracked from breeding colonies in Greenland (*n* = 10 birds) and Iceland (*n* = 1 bird). Green, autumn (post-breeding) migration (August–November); red, winter range (December–March); yellow, spring (return) migration (April–May). Two southbound migration routes were adopted in the South Atlantic, either (a) West African coast (*n* = 7 birds) or (b) Brazilian coast. Dotted lines link locations during the equinoxes. (Reproduced with permission from Egevang *et al*., 2010).

individuals annually fly more than 80,000 km (49,600 miles) (Egevang et al., 2010).

Surprisingly, size does not appear to be a factor in the evolution of the migratory habit, as it is found in diminutive passerines such as the Northern Wheatear (*Oenanthe oenanthe*) as well as in the more massive-bodied swans and geese.

9.3.1 The Northern Wheatear

The Northern Wheatear is a small (approx. 25 g), insectivorous migrant breeding from the eastern Canadian Arctic across Greenland, Eurasia, and into Alaska (Figure 9.4). The recent discovery of the astonishingly long migration journeys undertaken by this minute bird has brought to light one of the longest migratory flights known—30,000 km (18,640 miles), from sub-Saharan Africa to their Arctic breeding grounds (Bairlein et al., 2012).

There are two populations of this species in the New World, one centred on Alaska in the western Arctic and another in eastern Canada. Using ultralight geolocators, it has been shown that individuals from these New World regions over-winter in northern sub-Saharan Africa. During the autumn, the Alaskan birds travel through northern Russia and Kazakhstan before crossing the Arabian Desert, to over-winter in northern sub-Saharan Africa, a one-way journey that averages 14,500 km each way.

Meanwhile, the birds from eastern Canada cross an approximately 3,500 km stretch of the North Atlantic before continuing south to sub-Saharan Africa. On this journey they travel on average 290 km per day. There is no evidence that breeding populations in the New World have established over-wintering sites in the western hemisphere (Figure 9.4).

Such findings have provided the first evidence of a migratory songbird capable of linking African ecosystems of the Old World with Arctic regions of the New World (Figure 9.4). These remarkable journeys, particularly for birds of this size, last between 1 and 3 months depending on breeding location and season (autumn/spring) and result in mean overall migration speeds of up to 290 km per day. Stable-hydrogen isotope analysis of winter-grown feathers sampled from breeding birds generally support the view that Alaskan birds over-winter primarily in eastern Africa, and eastern Canadian Arctic birds over-winter mainly in western Africa.

In terms of the evolution of migration, it has been suggested that this extraordinary fidelity to the African winter territory and a failure to winter in America may have arisen as a consequence of a gradual Pleistocene expansion of the breeding range of the species, progressively extending around the Arctic as a result of post-glacial climatic warming.

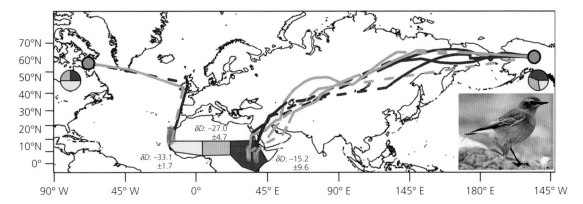

Figure 9.4 Migration routes and wintering grounds of three Northern Wheatears breeding in Alaska (AK) and one in the eastern Canadian Arctic (Canada (CN): grey dot, breeding area; blue, autumn migration; orange, spring migration; dashed lines, uncertainty in migration routes close to equinoxes). Pie charts indicate the proportion of individuals (Alaska: n 1/4 9, Canada: n 1/4 4) originating from one of the three predefined wintering regions (red, western; orange, central; yellow, eastern—based on stable-hydrogen isotope (∂D) values in winter-grown feathers and ∂D values within each wintering region (mean + SD shown). (Adapted from Bairlein et al., 2012).

9.3.2 Swan migrations

At the other end of the weight scale in avian migration are geese and swans which are remarkable for the size of the body that they can transport across continents and oceans. Bewick's Swan (*Cygnus columbianis bewickii*) and the Whistling (or Tundra) Swan (*C. columbianis columbianis*) are small-bodied swans that make long migrations.

The longest migration of any swan is that of the Whistling or Tundra swans that fly from breeding grounds in northern Alaska to wintering regions in eastern USA, a distance of 6,000 km (3,730 miles). In the Old Word the closely related Bewick's Swan is a mid-weight swan at around 6 kg. The Bewick Swan is the smallest of the northern swans and among swans is the species with the longest migratory path in Europe. Each spring they fly from their wintering grounds in temperate Europe and Asia to breed in the Arctic Tundra, a distance of around 4,000 km (2,400 miles). A relative heavy-weight is the Whooper Swan (*C. cygnus*). Despite their relatively great weight they are capable under favourable circumstances of accomplishing the journey from Iceland to Scotland in a little as 13 hours. In general, however, this species rests on the sea during its migration (Figure 9.5).

The largest of all swans is the Trumpeter Swan (*C. buccinator*) at around 11 kg, with large males having a wing span of 2.97 m (10 ft). They nearly became extinct in the first half of the 20th century as a result of hunting, but have now partially recovered, with a significant population in Alaska. Compared with the diminutive Northern Wheatears, most northern swans have a more leisurely migration. Tundra Swans (*C. columbianis columbianis*) caught in Pennsylvania and Virginia and marked with satellite transmitters were found to spend approximately 3.5 months on breeding areas, 3.5 months on wintering areas, and 5 months feeding on migratory staging areas, which emphasizes the importance of migratory habitats for the successful migration of these birds (Wilkins *et al.*, 2010).

The variation in body mass and the size and form of the wings found in migrating swans prompts the question as to what is the optimal combination of the anatomical features in relation to the lifestyle of the various swan species. It has been argued (Powell and Engelhardt, 2000) that optimal flight theory exerts an evolutionary influence on wing span, body cross section, and mass in relation to aerodynamic variables such as air density and drag, in achieving the most energy-

Figure 9.5 Whooper Swans (*Cygnus cygnus*) are among the heaviest of migrating birds. They require ready access to large areas of water, especially when they are still growing, as their legs cannot support their body weight for extended periods of time. The Whooper Swan spends much of its time swimming, straining the water for food, and eating plants that grow on the bottom. This group in north-east Greenland will probably return to Scotland for the winter. (Photo courtesy of Dr. H. Körner).

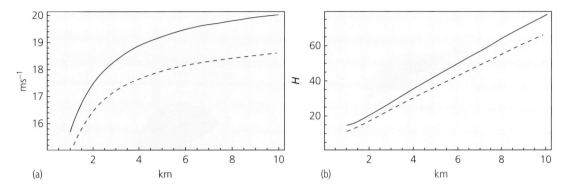

Figure 9.6 Comparison between Tundra Swans (solid line) and Trumpeter Swans (dotted line) for (a) flight velocities and (b) altitudes for the same expected travel distances. In general, Trumpeter Swans are predicted to be 1–2 m s^{-1} slower than Tundra Swans. (Reproduced with permission from Powell and Engelhardt, 2000).

efficient velocity for long-distance migration. For short-range (2–10 km) foraging flights the theory has to be expanded to include the costs for take-off and energy savings or losses in connection with climbing to altitude.

The theory predicts clear differences between the relatively light Tundra (or Whistling) Swan and the heavier Trumpeter Swan. Consequently, it can be expected that for flights between 2 and 10 km Trumpeter Swans can be expected to fly approximately 5–10 m lower in altitude and 1–2 m s^{-1} more slowly than Tundra Swans (Figure 9.6).

Moreover, the total energy required for these foraging flights is approximately 150% larger for a Trumpeter than a Tundra Swan at 80 vs 120 kJ of direct mechanical energy for a 5 km flight), suggesting that Trumpeter Swans may be less inclined to take-off than Tundra Swans. These factors have been taken to suggest that Trumpeters would be more vulnerable to hunting than native Tundra Swans (Powell and Engelhardt, 2000).

9.3.3 Semi-Palmated Sandpiper and Godwits

Remarkable as the above journeys are, they appear mere parochial sorties when compared with the migration of the Semi-Palmated Sandpiper (*Calidris pusilla*) and some populations of Bar-Tailed Godwits (*Limosa lapponica*).

The Semi-Palmated Sandpiper breeds in the southern Tundra in Alaska and Canada (Weber, 2009) and migrates long distances in very large flocks to winter in coastal South America as well as in the southern USA (Figure 9.7).

The *baueri* race of Bar-Tailed Godwits breeds in western Alaska and spends the non-breeding season a hemisphere away in New Zealand and eastern Australia. The *menzbieri* race breeds in Siberia and migrates to western and northern Australia. Although the Siberian birds are known to follow the coast of Asia during both migrations, the southern pathway followed by the Alaska breeders has long remained unknown.

It would appear, however, that the timing of the southward migration indicates the absence of a coastal Asia route by the *baueri* race. Instead the birds appear to take a direct route across the Pacific (Figure 9.8). Evidence for this is based on the absence of land sightings en route. This, together assessments based on flight simulation models, taking into account a highly developed capacity to take on extreme fat loads, as well as the use of a wind-selected migration route from Alaska adds support for a non-stop trans-oceanic flight (Battley *et al.*, 2012).

9.3.4 Costs and benefits of bird migration

Quantifying the costs and benefits of migration distance is critical to understanding the evolution of long-distance migration. In migratory birds, life history theory predicts that the potential survival costs of migrating longer distances should be balanced by benefits to lifetime reproductive success.

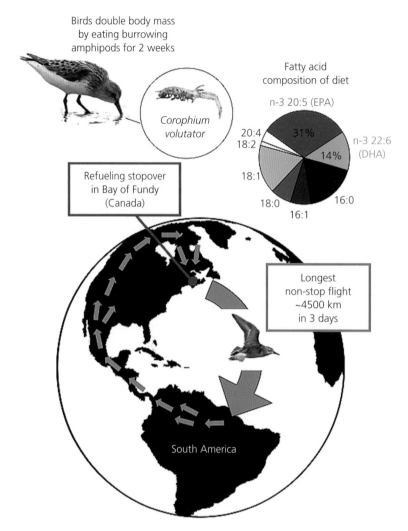

Figure 9.7 The Semi-Palmated Sandpiper (*Calidris pusilla*) stops in the Bay of Fundy (New Brunswick, Canada) during its autumn migration from breeding areas in the Arctic. This crucial stopover allows the migrant to store large lipid reserves by eating seasonally abundant amphipods (*Corophium volutator*) buried in the mudflats. (Reproduced with permission from Weber, 2009).

Given that birds that migrate to the Arctic in summer have a cosmopolitan experience of the world, some assessment has to be made as to what attracts them to the Arctic as their breeding ground. Less predation could lead to greater fecundity of the more northern nesting birds so that they become the more numerous section of the over-wintering populations. Adequate feeding and freedom from disturbance is clearly an advantage that can be found in the Arctic.

Merely to view the breeding success of arctic migrants as the ultimate explanation of the evolution of this type of behaviour has a certain demographic logic. However, in examining the costs of making the journey to the Arctic it is necessary to assess the physiological evolution of a metabolism that is capable of sustaining the birds for such strenuous journeys. Birds are well placed to profit from flying as it offers a more rapid form of transport than swimming and the energy required for a journey of equivalent length is less than that which would be used in running.

Relevant adaptations for long-distance flying which decrease the energy costs include:

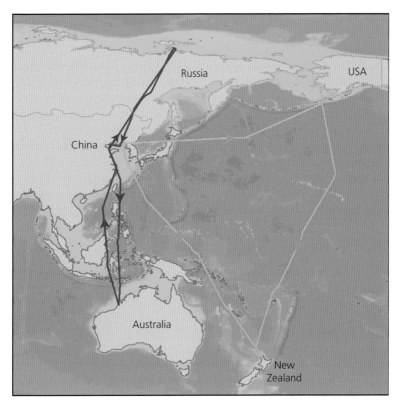

Figure 9.8 Round-trip migrations of individual Bar-Tailed Godwits of the *L. l. baueri* and *L. l. menzbieri* sub-species. The *baueri* individual (yellow track) was tracked for 29,280 km, including three main migratory flights of 10,270 km (in 7.0 days) between New Zealand and the Yellow Sea (China), 6,510 km (4.9 days) between the Yellow Sea and Alaska (USA), and 11,690 km (8.1 days) between Alaska and New Zealand. The *menzbieri* (blue track) was tracked for 21,210 km, including four main migratory flights of 5,620 km (4.2 days) between Australia and the Yellow Sea, 3,990 km (2.1 days) between the Yellow Sea and Siberia (Russia), 4,090 km (2.6 days) between Siberia and the Yellow Sea, and 6,270 km (5.3 days) between the Yellow Sea and Australia. (Reproduced with permission from Battley *et al.*, 2012).

(1) high efficiency for muscle contraction;
(2) pointed wings and low wing-loading;
(3) travelling in V-formations—as each bird flies slightly above the bird in front there is a reduction in wind resistance;
(4) storing fuel as energy-dense lipids;
(5) atrophy of non-essential organs.

On long trans-oceanic flights birds have no access to water or nutrition and have therefore to rely on their body fat and protein stores to provide both fuel and life support. Although fat is the most energy-dense metabolic fuel, the insolubility of its component fatty acids could create problems for the transport to working muscles at a rate sufficient to support the highly aerobic exercise required in order to fly. Recent evidence indicates that migratory birds express large amounts of fatty acid transport proteins onto the muscle membranes as well as into the cytosol (the liquid component of the cytoplasm surrounding the cellular organelles), either through endogenous mechanisms and/or diet. Feeding on a performance-improving diet is of the utmost importance before setting off on a long fight.

The Semi-Palmated Sandpiper (*Calidris pusilla*) stops in the Bay of Fundy (New Brunswick, Canada) during its autumn migration from breeding areas in the Arctic (Figure 9.7). This crucial stopover allows the migrant to store large lipid reserves by eating seasonally abundant amphipods (*Corophium volutator*) buried in the mudflats just before the sandpiper crosses the Atlantic Ocean to South America. The *Corophium* mud shrimp contains unusually large amounts of n-3 polyunsaturated fatty acids (45% of total lipids) and is found only in the Bay of Fundy and along the coast of Maine. Dietary n-3 fatty acids not only are used as an energy source but also act to increase the capacity for endurance exercise. Fatty acid chain length, degree of unsaturation, and placement of double bonds can all affect the rate of mobilization of fatty acids from adipose tissue and the muscle utilization of fatty acids (Guglielmo, 2010).

The ability to rapidly process lipids is also vital. Muscle performance is improved by restructuring membrane phospholipids and by activating key genes of lipid metabolism that control the peroxisome proliferator-activated receptors (Weber, 2009).

9.4 Plant diversity and evolution at high latitudes

Traditionally, the Arctic has been considered as an evolutionary freezer where a depauperate flora of post-Pleistocene origin with minimal genetic diversity survived in relative isolation. However, the advent of molecular methods has now accumulated a wealth of recent evolutionary, phylogenetic, and phylogeographic information (Brochmann and Brysting, 2008) which has shown that there is no noticeable lack of genetic variation in flowering plants at high latitudes (Table 9.1). Although individual plants do not

Table 9.1 Examples of species that maintain a high biodiversity at high latitudes.

Species	Observations	References
Arctagrostis latifolia	Three chromosome races within Alaska	(Mitchell, 1992)
Armeria maritima	Several sub-species in Greenland and Arctic America but only one species on the Pacific coast from Vancouver to California	(Lefebvre and Vekemans, 1995)
Cassiope tetragona	Molecular evidence for variation in intra-population diversity (see Figure 9.11). A diploid species (2n = 26)	(Brochmann and Brysting, 2008)
Carex bigelowii	High levels of clonal diversity (genetic diversity) within populations	(Jonsson *et al.*, 1996)
Cerastium arcticum	High levels of genetic diversity at high latitudes with allopolyploidization	(Brochmann *et al.*, 1996)
Draba spp.	Allopolyploidization prevalent in closely related species with minor variations in habitat preferences	(Brochmann *et al.*, 1992)
Dryas octopetala	Exposed ridge and snow bank populations adapted to different lengths of growing season	(McGraw and Antonovics, 1983)
Pedicularis lanata	Breeding system with the highest capacity for outcrossing in the study area and also the greatest morphological variation within its geographic distribution	(Philipp, 1998)
Pinus sylvestris (arctic populations)	In populations within the Arctic Circle there is strong differentiation between populations with respect to adaptation to the very short growing season	(Savolainen, 1996)
Polygonum vivipara	High levels of genetic diversity despite asexual reproduction	(Diggle *et al.*, 1998)
Saxifraga cernua	High levels of genetic diversity despite asexual reproduction	(Gabrielsen and Brochmann, 1998)
Saxifraga oppositifolia	Intra-specific variation at high-latitude sites	(Abbott *et al.*, 1995; Crawford, 1995)
Saxifraga cesopitosa	High levels of genetic diversity at high latitudes with allopolyploidization	(Tollefsrud *et al.*, 1998)
Silene acaulis	Patterns of spatial variation in genetic structure within populations reflecting the workings of micro-evolutionary processes	(Abbott *et al.*, 1995)
Polygonella spp.	Species within Pleistocene refugia maintain higher levels of biodiversity	(Lewis and Crawford, 1995)
Puccinellia phryganodes	High levels of clonal variation within and among populations	(Jefferies and Gottlieb, 1983)

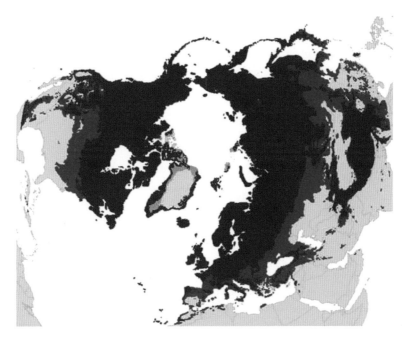

Figure 9.9 Estimation of loss of genetic diversity and range reduction as exemplified by data for present and future (2080) potential distribution of Bog Bilberry (*Vaccinium uliginosum*). Potential distribution lost is shown in red, stable in purple, and future new habitats in blue. In this example, a 26% range reduction in two-dimensional mapped area was estimated for the A2 emission scenario and CCM3 global circulation model. This representation, however, does not take into account changes in altitudinal representation of the capacity of the species to adapt. (Reproduced with permission from Alsos *et al.*, 2012).

in general migrate, they have the great advantage as compared with most animals that they can achieve genetic dispersal over great distances by gene dispersal through seed and pollen transfer.

There is also an ever-growing corpus of recent genetic evidence for the existence of refugia in certain favourable locations within frozen landscapes (see section 2.2).

Despite the well-documented evidence for the resilience of arctic plants to past climatic oscillations, doubts have been expressed suggesting that if current models for climatic change are correct then this might lead by 2080 to a loss of distributional range for many northern species (Figure 9.9). Such a loss of range could then be expected to lead to a loss in species diversity and thus to loss of genetic diversity, crucial for their long-term persistence. Loss of genetic diversity would probably vary considerably among species, depending on dispersal adaptation and genetic differentiation among populations (Alsos *et al.*, 2012). However, these somewhat pessimistic prognostications are based on models that use only two-dimensional maps and take no account of changes in altitudinal distribution.

In considering the impact of climatic change in relation to the fate of the European alpine flora, similar fears have been dismissed. Here it has been argued that the extent of the retreat of ice from the nival zone will leave ample room for plant migration to higher altitudes (Körner, 2003). It should also be remembered that many different populations exist with mapped species areas, and for plants and also birds those in marginal areas have a long history of enduring climatic fluctuations and as a result have inherited a degree of genetic variation that facilitates adaptation to changing environments (Crawford, 2008).

9.4.1 Plant polyploidy in the Arctic

Over 80% of the flora of Svalbard is polyploid. From such observations it is frequently concluded that polyploid species are better adapted physiologically to cold climates than diploid species. The Arctic is an excellent model system for the study of polyploidy as it is one of the Earth's most polyploid-rich areas. It is also noted for the fact that the level of polyploidy increases northwards within the Arctic. However, a detailed examination of distribution has shown that there is no association between polyploidy and the degree of glaciation for the arctic flora as a whole (Brochmann et al., 2004). The frequency of diploids is much higher in the Beringian area, which remained largely unglaciated during the last ice age, in strong contrast to the heavily glaciated Atlantic area. This result supports the hypothesis that polyploids are more successful than diploids in colonizing after deglaciation. All Svalbard polyploids examined so far are genetic allopolyploids with fixed heterozygosity at isozyme loci.

As has been pointed out (Crawford, 2008), rapid evolution at the diploid level provides a facility for crossing and avoiding inter-breeding barriers and can therefore promote further diversification after expansion from different refugia. This in itself can then provide new raw materials for allopolyploid formation. The evolutionary success of polyploids in the Arctic may thus be based on their fixed-heterozygous genomes, which buffer against inbreeding and genetic drift through periods of dramatic climate change. It has therefore been claimed that, from a long-term evolutionary point of view, polyploidy is a dead end (Mayrose et al., 2011). It is also

Table 9.2 Examples of flowering plants that have been considered to be ancient, indigenous (autochthonous) species widespread in the Arctic (Tolmatchev, 1960). Note the low ploidy level of the chromosome numbers.

Species	Ploidy
Saxifraga oppositifolia	(2n = 26)
Saxifraga hyperborea	(2n = 26)
Saxifraga tenuis	(2n = 20)
Ranunculus pygmaeus	(2n = 16)
Dryas octopetala	(2n = 18)
Kalmia procumbens	(2n = 24)
Cassiope tetragona	(2n = 26)
Diapensia lapponica	(2n = 12)

noteworthy that the autochthonous (ancient and indigenous) species of the Arctic are usually diploid or occasionally tetraploid (Table 9.2).

Given the long-term survival and diversity of many diploid species in the Arctic (Table 9.2; Figures 9.10 and 9.11) it can be argued that a high level of

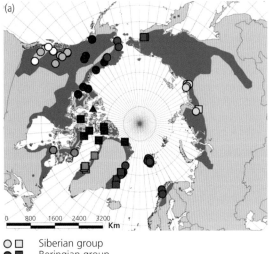

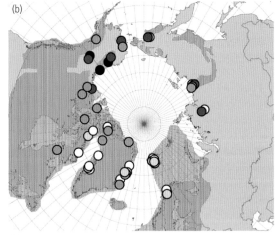

○ □   Siberian group
● ■   Beringian group
● ■ ▲ Canadian group
○ □   E Canadian/W Greenlandic group
● ■   E Greenlandic/Scandinavian group
○   C. tetragona ssp. saximontana
○   C. mertensiana
■   Geographic distribution of C. tetragona

○ 0.105 – 0.124
○ 0.125 – 0.142
◐ 0.143 – 0.160
● 0.161 – 0.178
● 0.179 – 0.197
▨ Last Glacial Maimum extension

Figure 9.10 Genetic variation in the diploid Arctic Heather (*Cassiope tetragonoa*) (2n = 26). (a) Geographic origin and genetic grouping of 68 *Cassiope* populations (see Eidesen et al., 2007). Different colours identify different main structure groups based on 265 AFLP markers. Different symbols with the same colour identify sub-groups within each main group of *C. tetragona* ssp. *tetragona* based on 171 AFLP markers. The geographic distribution of *C. tetragona* is shown in grey. (b) Intra-population variation in *C. tetragona* ssp. *tetragona* measured as Nei's unbiased diversity. (Reproduced with permission from Brochmann and Brysting, 2008).

Figure 9.11 The diploid Arctic Bell-Heath (*Cassiope tetragona*) growing at Ny Ålesund, Spitsbergen (79°N).

polyploidy, although common at high latitudes, is not necessarily an obligatory adaptation for survival in the arctic environment.

It has often been suggested (Stebbins, 1971) that the presence of so many polyploid species in the Arctic is merely the consequence of plant migrations caused by periods of cooling and warming, with the climatic disturbance bringing species into proximity in unusual combinations.

In any argument on the role of polyploidy in relation to plant fitness for life at high latitudes it should be noted that it has never been demonstrated that polyploid species are more cold-hardy or better adapted to short growing seasons than diploid species.

It has to be admitted, however, that the species with the high-altitude record at high latitudes (see Table 9.1) is *Cerastium arcticum*—which is normally either an octoploid (2n = 72) or even a dodecaploid (2n = 108). Equally, it should be noted that the runner-up at only 20 m behind for this High Arctic location is *Saxifraga oppositifolia*, which is commonly a diploid (2n = 26) although autotetraploids (2n (52) are known at low altitudes (see Figures 8.22 and 8.23).

The genetic advantages of polyploidy include restoring fertility to hybrids and the ability to form new species within the home range of their parents. Polyploidy also has obvious advantages in masking the effects of harmful alleles. More questionable, however, is the assertion that polyploids benefit from having a greater store of genetic variability merely because there are more alleles at any one locus. If harmful alleles are masked by polyploidy, then the same must also be true for useful alleles.

It is, however, now realized that hybridization by whole-genome duplication (allopolyploidy) can generate a wide range of novel and epigenetic variation. Many plant species traditionally considered to be diploid have recently been demonstrated to have undergone rounds of genome duplication in the past. Such species are now referred to as *paleopolyploids*. Polyploidy and inter-specific hybridization (with which it is often associated) have long been thought to be important mechanisms of rapid species formation. The widespread occurrence of polyploids, which are frequently found in habitats different from that of their diploid progenitors, would seem to indicate that polyploidy is associated with evolutionary success in terms of the ability to colonize new environmental niches.

Recent genomic studies have provided fresh insights into the potential basis of the phenotypic novelty of polyploid species. It has been shown that in the polyploid arctic-alpine *Cerastium alpinum* group and close relatives, the dynamic nature of polyploid genomes paves the way for evolutionary novelty and are an aid to the ecological success of many polyploid species. The situation is complex as immediately after polyploidization, modification of the polyploid genome can begin through *gene silencing* (Brysting et al., 2011). The term *gene silencing* is given to the process of the 'switching off' of a gene by a mechanism other than genetic modification. A silenced gene is one that would be normally expressed but is switched off, often by epigenetic changes such as DNA methylation. There are therefore a number of genetic, epigenetic, and transcriptional changes associated with polyploidy in plants which might contribute to the evolutionary success of polyploid plants (Hegarty and Hiscock, 2008).

Polyploidy can, however, be an impediment to gene flow in meta-populations.

It is possible to speculate that neighbouring diploid populations with different environmental preferences will have the advantage of being able to profit more readily than polyploids from gene flow, as interchange of alleles from one population to another is less likely to be masked. It is therefore perhaps no coincidence that the ancient (autochthonous) arctic species (see Table 9.2) are mainly diploid (or else have only a low level of polyploidy), and they can therefore adapt rapidly to climatic change by exchanging readily expressed genes between different ecotypes of the same species. This prompts the suggestion that mutualism, rather than competition between ecotypes, could be a significant feature in species survival only if genetic exchange leads to readily expressed adaptations. This would be more likely to take place between adjacent diploid than between polyploid populations but requires experimental evidence from further study.

Conflicts between adaptations at the physiological, biochemical, and morphological level inevitably limit the choice of adaptations that can be expressed by any individual plant. No matter how much genetic information is present within the genome of any individual it is of little use if it cannot be expressed. It would appear logical therefore to suggest that polyploid populations could under certain circumstances, because of their genetic stability, be less likely to adapt promptly to changing circumstances than diploids through a reduced capacity to exchange genetic material. In short, the adaptive significance of polyploidy, although a powerful influence in contributing to speciation, may nevertheless impede rapid physiological adaptation and can be described again as a dead-end in relation to changing climatic conditions, which have long been a feature of the arctic environment.

9.4.2 Genetic history and plant migration in the Arctic

Not only can DNA studies show that there are plants that survived at several locations north of the major ice sheets (see section 2.2) but also they provide the means to trace their evolutionary history and specific migration routes. A 10-year-long study of the widespread, polymorphic, and highly successful arctic colonizer the Opposite-Leaved Purple Saxifrage (*Saxifraga oppositifolia*) investigated the chloroplast DNA variation at a number of sites throughout the circumpolar distribution of this species (Figures 9.12–9.14). Fifteen different haplotypes of chloroplast DNA (cpDNA) were identified which provided a geographical phylogenetic history of the species on a circumpolar basis.

There appear to be two major lineages. One is a 'Eurasian lineage' distributed westward from the Taymyr Peninsula in north-central Siberia, through Europe and Greenland to Newfoundland and Baffin Island.

The second possible route is a Beringian lineage also emanating from Siberia but distributed eastward from the Taymyr Peninsula, through northeast Siberia and North America to north Greenland. Haplotypes basal to each lineage co-occur only in north-central Siberia (Taymyr). A possible explanation of this is that the Opposite-Leaved Purple Saxifrage originated in the southern Altai Mountains which is the centre of distribution for Opposite-Leaved Saxifrage species (B.A. Yurtsev, *pers. comm.*).

Figure 9.12 Geographical distributions of two widespread sub-species of *Saxifraga oppositifolia* ssp. *oppositifolia* (solid line) and ssp. *glandulisepala* (broken line). (Reproduced with permission from Abbott and Comes, 2004).

EVOLUTION IN THE ARCTIC   203

Two major cpDNA clades in *Saxifraga oppositifolia*

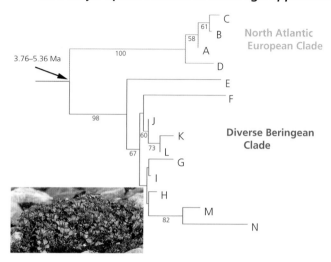

Figure 9.13 Phylogenetic reconstruction of *Saxifraga oppositifolia* evolution showing percentage boot-strap support of particular clades and indicating the approximate chronology of the division between the shallow North Atlantic-European clade and the more diverse Beringean clade between 3.76 and 5.36 million BP at the period of Pliocence cooling. (Image constructed from data of Abbott *et al.*, 2000 and Abbott and Comes, 2004).

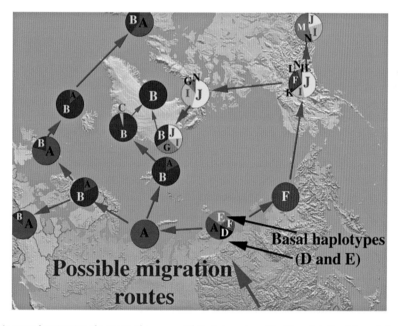

Figure 9.14 Possible routes for migration of major *Saxifraga oppositifolia* clades as traced in differentiated haplotypes (letters A–M), showing the postulated location of where the species migrating from the southern Altai Mountains probably reached the Arctic Ocean between 3.76 and 5.36 million BP as the Pliocene climate cooled. (Data from Abbott *et al.*, 2000 and Abbott and Comes, 2004).

With the onset of *Cenozoic climatic cooling* the purple saxifrage would have migrated northwards into the territory vacated by the retreating boreal forest, eventually reaching the shores of the Arctic Ocean. In this way *S. oppositifolia* would have entered the Arctic in western Beringia between 3.76 and 5.36 million years ago (Figures. 9.12–9.14) during the late Tertiary before migrating east and west, to eventually obtain a complete circumpolar distribution (Abbott *et al.*, 2000).

The two cpDNA lineages were estimated to have diverged from their most recent common ancestor 5.37–3.76 million years ago, during the early to mid-Pliocene. A nested clade analysis was conducted in an attempt to determine how past episodes of range fragmentation, range expansion, and long-distance dispersal may have influenced the geographical distribution of cpDNA haplotypes. In Alaska, a known refugium during the last ice age, for the species with high levels of cpDNA diversity this may be partly explained by divergence between populations that were isolated in different ice-free regions (Abbott and Comes, 2004).

Another genetic biogeographical history of *Saxifraga oppositifolia* sampled areas including central Asia and the south-eastern European mountain ranges has detected two major plastid DNA lineages likewise suggesting a long and independent vicariant history. However, no specific site of origin was identified, but it was suggested that possibly the Alps and central Asia, respectively, were the likely ancestral areas of the two main lineages, and that contact areas between the two clades in the Carpathians, northern Siberia, and western Greenland were secondary (Winkler *et al.*, 2012). Further sampling in the mountains of south-eastern Asia will be required before a definitive history of this ancient Opposite-Leaved Saxifrage species is completely elucidated.

9.4.3 Disjunct plant distributions

A long-standing problem in the study of plant geography in relation to the Arctic is the cause of disjunct distributions. There are species found in the Arctic that also occur at a considerable distance, in the European Alps and elsewhere, with no intermediate locations to suggest a migrational trail. One such example is Arctic Cotton Grass (*Eriophorum scheuchzeri*), a sedge that is found in Greenland, Iceland, and Scandinavia, with the nearest southern population being found in Switzerland. In general such species are often referred to as *arctic-alpines*.

Recent advances in molecular biology have now made it possible to trace species migratory pathways including those of some of the so-called *arctic-alpines*. These molecular biology advances have also dispelled some ancient myths.

Two classical arctic-alpine species with a widespread circumarctic distribution but only a few isolated occurrences in the European Alps are Mountain Sandwort (*Minuartia biflora*) and Pygmy Buttercup (*Ranunculus pygmaeus*) (Figure 9.15). An analysis of amplified fragment length polymorphism (AFLP) and chloroplast DNA sequence data was used to unravel the history of their immigration into the Alps and to provide data on their circumpolar phylogeography. Surprisingly, in spite of similar ecological requirements they were found to have strikingly different immigration histories into the Alps. In *M. biflora*, the alpine populations are most probably derived from source populations located between the alpine and Scandinavian ice sheets, in accordance with the traditional biogeographic hypothesis. In contrast, the alpine populations of *R. pygmaeus* (Figure 9.16) cluster with those from the Tatra Mountains and the Taymyr region in northern Siberia, indicating that the distant Taymyr area served as the source for the alpine populations. Both these species showed different levels of genetic diversity in formerly glaciated areas. In contrast to the considerable AFLP diversity observed in *M. biflora*, *R. pygmaeus* was virtually non-variable over vast areas, with a single phenotype dominating throughout the Alps and another distantly related one dominating the North Atlantic area from Greenland across Spitsbergen to Scandinavia. The same pattern was observed in the chloroplast DNA sequence data. Thus postglacial colonization of *R. pygmaeus* was accompanied by extreme founder events (Schönswetter *et al.*, 2006). For this species the most probable course of events was that during cold stages of the Pleistocene, the Arctic and alpine vegetation zones came into close contact and could have mutually exchanged taxa. This latter theory is well supported for several other arctic-alpine species (Schönswetter *et al.*, 2003).

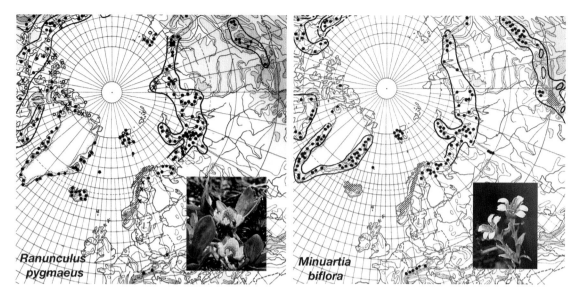

Figure 9.15 Arctic-alpine disjunct distribution in Pygmy Buttercup (*Ranunculus alpinus*) and Mountain Sandwort (*Minuartia biflora*). (Map adapted from Hulten and Fries, 1986).

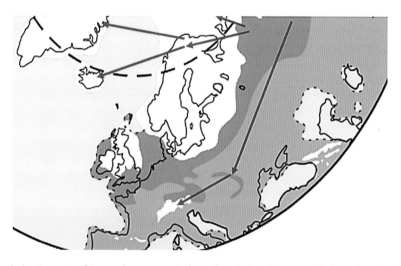

Figure 9.16 Deduced migration routes of *Ranunculus pygmaeus* to the northern Arctic and European Alps from where the distant Taymyr region served as the source for both regions. (Map based on data from Schönswetter et al., 2006)

In recent years it has been noted that a number of species that are found in Mongolia and other areas in Asia can also be seen in Switzerland (Kadereit *et al.*, 2008). The initial botanical reaction in Switzerland was how and when did they migrate from Switzerland to Mongolia? It has sometimes been said in Switzerland that Purple Saxifrage 'is Switzerland's gift to the Arctic'. However, those who have made a systematic study of the flora of the Alps have long been aware of the links of the flora of the European Alps with the flora of Asia. An early phytogeographical study found that 27% of the Swiss flora also grew in Asia and constituted an Altai or Arctic-altai element (Brockmann-Jerosch, 1908).

In this migration Purple Saxifrage would not have been alone. In a study of the relationships between the flora of the European Alps and Asia (Kadereit and Comes, 2005) it was found that the flora of the Alps (and other European high mountains) contains a 'northern element' which through a northern connection is linked to Asia, and it appears that in most such cases Asia can be identified as the centre of origin of many alpine and arctic genera.

There are also examples of species that have taken other routes. Such is the case with the Globe Flower (*Trollius europaeus*). This species would appear to have migrated from a refugium in Switzerland northwards to reach the arctic regions of Fennoscandinavia (Despres *et al.*, 2002). In this species the alpine populations retain most of the ancestral genetic variability, in a moderately fragmented habitat, and therefore exhibit the maximum within-population diversity.

9.5 Arctic and boreal mammal lineages

If plants survived in glacial refugia at the Last Glacial Maximum it is probable that some animal members of the arctic communities would also have been part of the refugial biota. A wide-ranging phylogeographic study of the Tundra Shrew (*Sorex tundrensis*) has demonstrated that this species is divided into five main mitochondrial DNA phylogenetic lineages with largely parapatric distributions. In addition to a single Nearctic clade (Alaska), four Palearctic clades are identified from the Western Urals, through Kazakhstan, and east to Chukotka. Date estimates obtained by use of a molecular clock, corrected for potential rate decay, suggest a common ancestor for all contemporary Tundra Shrew populations. Relatively high genetic divergence between phylogroups indicates that the observed phylogeographic structure was initiated by historical events that pre-dated the Last Glacial Maximum. It is therefore assumed that, being more cold- and arid-tolerant, the Tundra Shrew underwent expansion during an early cold phase of the Last Glacial Period and spread through its recent range earlier than most other Siberian Red-Toothed Shrews (*Sorex* spp.) Comparative phylogeographic analysis of Siberian shrews and rodents suggests that evolutionary histories of these species is associated with azonal or open habitats and is therefore distinct from what is observed in the forest species (Bannikova *et al.*, 2010).

9.5.1 Polar Bear evolution

The Polar Bear (*Ursus maritimus*) (Figure 9.17) is essentially a member of the marine ecosystem and therefore is in a similar nutritional position to the ocean-feeding birds that nest in the Arctic. Although it is only marginally nutritionally dependent on terrestrial ecosystems, it nevertheless has to be considered as a land mammal for its reproductive needs.

A reliance on perennial sea ice and a strongly pagophilic (ice-dependent) prey base, mainly seals, suggests that Polar Bears as they now exist with entirely carnivorous dentition could not have evolved in a world in which the Arctic Ocean remained unfrozen for extensive periods of the year.

There is nevertheless a debate as to when the Polar Bear (*Urus maritimus*) diverged completely from its ancestor the Siberian Brown Bear (U. arctos). Fertile hybridization between Grizzly Bears (*Ursus arctos horribilis*) and Polar Bears (Figure 9.18) can take place in captivity but is extremely rare in the wild as both species are well adapted to their different habitat requirements and consequently classified as separate species. They have, however, interbred in the past (Edwards *et al.*, 2011) and may do so again in the future.

Those who have examined divergence between Brown Bears and Polar Bears from an anatomical point of view place the divergence from coastal

Figure 9.17 Polar Bear (*Ursus maritimus*), a species about whose evolution there is some debate. (Photo courtesy of Clare Kines).

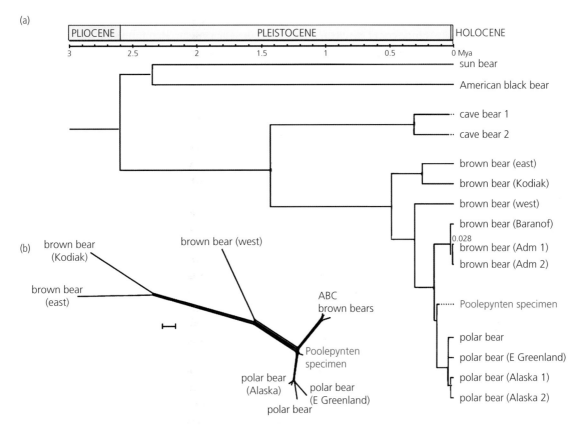

Figure 9.18 Phylogenetic and chronographic reconstruction of Polar Bear evolution. (a) Maximum clade probability tree inferred from mitochondrial genome sequences. Numbers at selected nodes indicate mean ages in millions of years. The geological time scale with major relevant epochs is shown above the tree. (b) Phylogenetic network of complete mitochondrial genomes of 11 Polar and Brown Bears. (Adapted from Lindqvist et al., 2010).

Brown Bears in Siberia to the middle to late Pleistocene (Talbot and Shields, 1996). From studies of fossil bear skulls it has even been suggested that the present entirely carnivorous dentition of the modern Polar Bear was only attained in the last 100,000 years (Kurten, 1995).

Recently, a lower jawbone (left mandible) estimated to be 130,000–110,000 years old was excavated *in situ* at Poolepynten on Prins Karls Forland, a narrow strip of land on the far western edge of the Spitsbergen archipelago (Figure 9.19), named after an English whaler Jonas Poole who repeatedly visited the region in the early 17th century. Comparisons of this well-preserved mandible with Brown Bear and other available sub-fossil Polar Bear remains, as well as a large collection of extant Polar Bears from Spotsbergen, has shown that it falls within

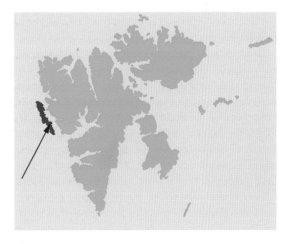

Figure 9.19 Location of Poolepynten, the site of the discovery of an early Polar Bear skull, on Prins Karls Forland in the Spitsbergen archipelago.

ie range of modern Polar Bears and probably belonged to an adult male (Lindqvist *et al.*, 2010).

Investigations of the trophic relationships of this ancient Poolepynten specimen with present-day Polar Bears from Svalbard were carried out using the stable isotope content in the canine tooth. This showed that the jaw was from an individual that had a feeding ecology similar to present-day Polar Bears, at the top of the Arctic marine food chain. Furthermore, analyses of the stratum containing the sub-fossil jawbone uncovered a bivalve and foraminifera fauna, reflecting an arctic, open marine environment influenced by glacier input and advection of warm North Atlantic water as today.

Molecular reconstructions of the matrilineal history of Brown and Polar Bears have shown two striking features of their genetic relationships (Figure 9.18). The first is punctuated evolution by dramatic and discrete climate-driven dispersal events. The second is opportunistic mating between these two species as their ranges overlapped, which has left a strong genetic imprint. In particular, a likely genetic exchange with extinct Irish Brown Bears appears to have left a significant imprint on the origin of the modern Polar Bear matriline. This suggests that inter-specific hybridization may be more common than previously considered and has been a mechanism by which the bear species adapted extremely rapidly, both morphologically and physiologically. It also appears probable that this was the means by which the species dealt with marginal habitats during periods of environmental deterioration (Edwards *et al.*, 2011).

In addition to the hypothesis of a very recent split, a divergent view has been proposed. Using an extended dataset of mitochondrial genomes, there has even been suggested a split between Brown and Polar Bears in the middle Pleistocene, about 600,000 (338,000–934,000) years ago. This provides more time for Polar Bear evolution and supports previous suggestions (Hailer *et al.*, 2012). It is doubtful, however, that Polar Bears, as we know them now, could have evolved when the Arctic Ocean remained unfrozen for large portions of the year as it did during the warm period of the middle Pleistocene.

It is more probable that Polar Bears branched off from Brown Bears and became isolated on Siberian coastal enclaves some time during the mid to late Pleistocene, gradually becoming an increasingly specialized carnivore that hunted solely on sea ice. Using next-generation sequencing technology, it has been possible to generate a complete, high-quality mitochondrial genome from the stratigraphically validated 130,000- to 110,000-year-old Polar Bear jawbone mentioned previously (Lindqvist *et al.*, 2010). This shows that the phylogenetic position of this ancient Polar Bear lies almost directly at the branching point between Polar Bears and Brown Bears (Figure 9.18).

9.5.2 Brown Bears

The Brown Bear (*Ursus arctos*) is geographically widespread, with many sub-species across both Eurasia (Figure 9.20) and North America. It is a polymorphic species as diverse in its habitats as it is varied in its diet. Eating habits vary from being largely vegetarian to exploiting local animal sources when available.

Depending on location and sub-species, the Brown Bear diet can vary from moths to catching fish and small mammals. Large mammals such as deer and elk may also be taken, particularly by the larger sub-species such as the Grizzly Bear (*U. arctos horribilis*) (Figure 9.21). In most cases the bears prey on mainly young, infirm, or injured animals.

The classical taxonomic recognition of many sub-species of Brown Bear has prompted molecular studies of the genetic diversity of present-day Brown Bears. Despite the numerous purely taxonomic sub-species, DNA studies have shown (Figure 9.20) that the European Brown Bear populations can be geographically placed into four main clades. Evidence of extinct populations has indicated that during the Holocene the diversity of *U. arctos* was greater in the past than it is now.

Four closely related haplogroups have been identified in northern continental Eurasia. Several haplotypes have been found throughout the whole Eurasian study area, while one haplogroup is restricted to Kamchatka. The haplotype network estimated divergence times and various statistical tests indicate that bears in northern continental Eurasia recently underwent a sudden expansion, preceded by a severe bottleneck. This Brown Bear population was therefore most likely founded by a

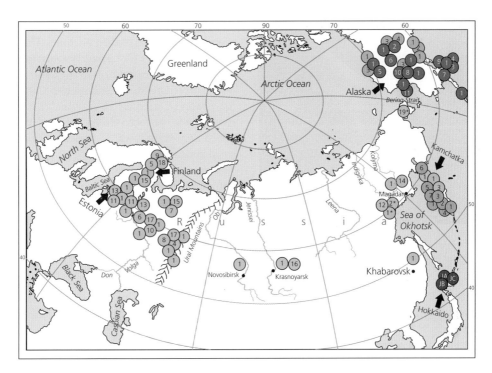

Figure 9.20 Geographical distribution of major Brown Bear (*Ursus arctos*) mitochondrial cytochrome *b* haplotypes in northern continental Eurasia, Japan, and Alaska. Numbers with the major clades denote sub-types. (Adapted from Korsten et al., 2009).

Figure 9.21 Grizzly Bear (*Ursus arctos horribilis*), Kenai Peninsula, Alaska. (Photo courtesy of Charles Woodley).

small number of bears that were restricted to a single refuge area during the Last Glacial Maximum. This pattern has been described previously for other mammal species and as such may represent one general model for the phylogeography of several Eurasian mammals (Korsten et al., 2009).

9.5.3 Reindeer evolution

Reindeer are thought to have originated probably 2 million years ago in Beringia and moved from there further into both North America and western Eurasia. By the time of the Last Glacial Maximum they were widespread south of the ice sheets both in Eurasia and North America as well as in Beringia, subsequently leading to their current circumpolar Holarctic distribution (Flagstad and Roed, 2003).

The highly developed mobility of reindeer has enabled this species to not only spread throughout the Taiga and boreal Forests of Eurasia and North America but also reach even the remotest parts of the Arctic. Throughout this vast area various forms have been described and recognized as sub-species based on differences in their morphology.

These differences in their morphology include the Eurasian Tundra Reindeer (*R. t. tarandus*) with their typical long antlers, Alaskan Caribou (*R. t. grantii*), and Canadian Barren Ground Caribou (*R. t. groenlandicus*). There are also the woodland forms, with

Figure 9.22 Woodland Caribou from the Nahanni herd in the Canadian North West Territories. (Photo courtesy of M. Pealow).

Eurasian Forest Reindeer (*R. t. fennicus*) and the North American Woodland Caribou (*R. t. caribou*) (Figure 9.22), which have a larger body size, long legs, and short and heavy antlers, although a significant number of the females lack antlers. In the northern isolated arctic islands there are sub-species in Spitsbergen and Novaya Zemlya, as well as in some North Canadian Islands. The arctic ecotypes are smaller and have been recognized also as sub-species on the basis of their anatomy. These include Svalbard Reindeer (*R. t. platyrhynchus*) (see Chapter 4, Figure 4.20), Peary Caribou (*R. t. pearyi*), and the now extinct Greenland sub-species Canadian Barren-Ground Caribou (*R. t. groenlandicus*). Small size reduces the surface:volume ratio of the animal, which may point towards a selective pressure for a small body at high latitudes.

A taxonomy based on morphological features does not necessarily reflect the evolutionary history of a species, as features such as body size may be imposed either phenotypically or by local selection pressure. The true phylogenetic relationships as inferred from studies of mitochondrial DNA (mtDNA) are more likely to reflect historical patterns of fragmentation and colonization. Such a study (Flagstad and Roed, 2003) detected three major haplogroups presumably representing three separate populations during the Last Glacial Maximum (Figure 9.23).

The most influential one has contributed to the gene pool of all extant sub-species and seems to represent a large and continuous glacial population extending from Beringia and far into Eurasia.

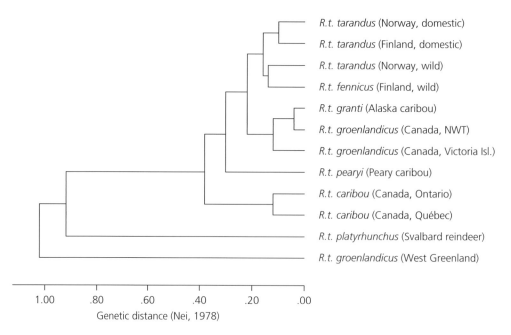

Figure 9.23 Cluster analysis of the genetic distance (Nei, 1978) between sub-species of reindeer and caribou based on variability in microsite loci. (Reproduced with permission from Roed, 2007).

A smaller, more localized refugium probably became isolated in connection with ice expansion in western Eurasia. A third glacial refugium was presumably located south of the ice sheet in North America, possibly comprising several separate refugial populations. Significant demographic population expansion was detected for the two haplogroups representing the western Eurasian and Beringian glacial populations. These data demonstrate that the current sub-species designations based on anatomy do not reflect the mtDNA phylogeography of the species. The morphological differences among extant sub-species probably evolved therefore as local adaptive responses to post-glacial environmental changes. A possible exception to this is the North American Woodland Caribou (Figure 9.22), which seems to have had its origin in a sub-species-specific refugium to the south of the continental ice sheet.

9.5.4 Genetic isolation in arctic island reindeer

In comparison with the mainland, caribou/reindeer populations on arctic islands tend to be isolated and sedentary which always carries a risk of reducing genetic diversity (Cote et al., 2002).

To characterize the possible genetic consequences of this sedentary disposition on levels of genetic diversity, comparisons have been made between populations from two adjacent areas on Spitsbergen (Figure 9.24) based on data from up to 14 microsatellites. The mean number of alleles per locus in Svalbard reindeer was 2.4 and mean expected heterozygosity per locus was 0.36. The latter value was significantly lower than in Canadian caribou and Norwegian reindeer. Large samples of females ($n = 743$) and small samples of males ($n = 38$) from two sites approximately 45 km apart showed significant genetic sub-division, which could be due to local

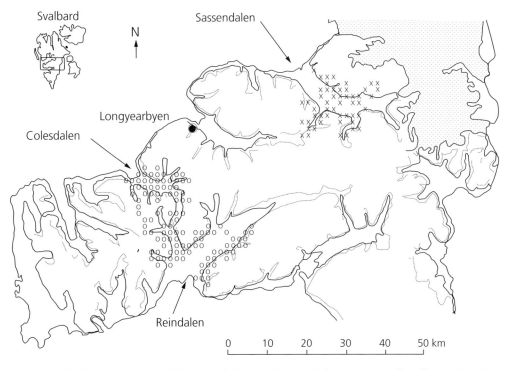

Figure 9.24 Nordenskjöldland, Spitsbergen, (78° N), showing the location of sites for the live capture and culling of female reindeer in neighbouring Colesdalen and Reindalen (open circles) and the more distant site at Sassendalen (crosses). Note the selection of site Colesdalen for sampling and compare with its 17th-century historical selection for reindeer hunting by whalers stranded over winter in 1630–1 (see Figure 4.9). (Reproduced with permission from Cote et al., 2002).

population fluctuations or limited gene flow. This study therefore concluded that Svalbard reindeer appeared to conform to the general conception of island populations being genetically depauperate due to genetic drift. Female Svalbard reindeer tend to be sedentary, so the home ranges of daughters could adjoin those of their mothers. This territorial attachment is described as *female philopatry* and could contribute to the level of genetic sub-division observed in these isolated populations even if some males undergo breeding dispersal. It has also to be noted that in Spitsbergen the predominance of glaciers and high mountain ranges can further isolate local populations.

9.6 Human evolution at high latitudes

The human race (*Homo sapiens*), in the short time of its existence, has adapted with amazing facility to the thermal gradient from the poles to the tropics. The Palaeolithic hunter settlements in northern Siberia (see Chapter 3) dating from 27,000 BP represent not only a cultural but also a behavioural ability suited to aquiring and storing resources in these habitats. In addition, survival would have demanded the evolution of a physiological capacity to survive in an environment where cold and a severely testing winter season would have imposed major diet limitations for early hunter gatherers.

9.6.1 Skin colour

It has been argued that selection for a light-coloured skin in northern peoples would have been necessary in order to facilitate the photogenic source for vitamin D. During exposure to sunlight, ultraviolet B photons enter the skin and photolyse 7-dehydrocholesterol to pre-vitamin D_3 which in turn is isomerized by the body's temperature to vitamin D_3. Many modern peoples depend on sun for their vitamin D (Figure 9.25). In northern regions the location of sites for the live capture and culling of reindeer might therefore have been expected to select for a darker skin colour which would have been maladaptive for the absorption of UV to such an extent that a vitamin D deficiency would ensue.

Global studies of skin colour have established a tendency for lighter coloured skin with increasing

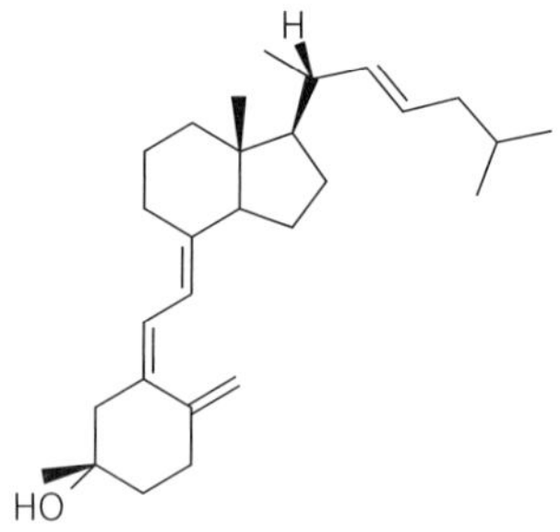

Figure 9.25 Structure of vitamin D_2.

latitude. It has also been shown (Jablonski, 2004) that whereas there is sufficient UV radiation for adequate provision of pre-vitamin D_3 in the tropics for lightly coloured skins, at higher latitudes there are zones where skin pigmentation can give rise to vitamin D deficiency. This study also showed that the highest correlation between skin reflectance and UV levels was observed at 545 nm, near the absorption maximum for oxyhaemoglobin, suggesting that the main role of melanin pigmentation in human skin is the regulation of the effects of UV radiation on the contents of cutaneous blood vessels located in the dermis. The respective levels of pigmentation in different locations represent a compromise solution to the conflicting physiological requirements of photo-protection and vitamin D synthesis.

It must, however, be emphasized that skin coloration in the human species is adaptive and labile and has changed more than once in human evolution. Because of this, skin coloration is of little value in determining accurate phylogenetic relationships among modern human groups, although it can be concluded that over a large area of the Earth there is a clinal gradation of skin coloration among indigenous peoples and that is correlated with UV radiation levels. Despite the general tendency for light-coloured skins at higher latitudes, there are some notable exceptions in the Arctic. Polar peoples, such as the Inuit, have access to rich sources of vitamin D in fish oils and the liver and fat of marine and terrestrial mammals. Polar people store frozen carcases over winter and therefore have an all-year-round supply of vitamin D. Consequently, a darker skin colour or a body wrapped in protective clothing for the greater part of the year is not a disadvantage.

There is also an advantage for the human skin to be pigmented in regions where stressful climates can be a dermatological hazard. Pigmented skin can have an enhanced permeability barrier function. There is also a greater degree of cohesion in the *stratum corneum*, the outermost layer of the epidermis. Here, large, flat, polyhedral, plate-like envelopes filled with the protein keratin help to keep the skin hydrated by preventing water evaporation. Pigmented skin also has a reduced susceptibility to infection. These enhanced properties of pigmented skin can be attributed to the lower pH of the outer epidermis, possibly due to the persistence of acidic melanosomes in the outer epidermis. It has even be suggested that interfollicular pigmentation in early hominids may possibly have evolved in response to stress to the permeability barrier and that protection from UV radiation was a secondary feature of this adaptation (Elias *et al.*, 2010). Once the use of the needle for sewing developed and it was possible to have clothing which provided efficient thermal protection, human migration and permanent settlement in regions with cold climates became possible and took place rapidly as the hunter-gatherer peoples followed the Pleistocene megafauna. Under such conditions, clothing, especially when it is necessary for the protection of the whole body for the greater part of the year, will negate the effect of sunlight in promoting vitamin D. Consequently there would have been a reduction in the potential for natural selection on human skin colour to provide protection against UV (Jablonski, 2004).

9.6.2 Vitamin C availability in the Arctic

A potential hazard for human survival at high latitudes is a lack of vitamin C (ascorbic acid) giving rise to the condition described as *scurvy*. The lack of this vitamin, which is essential for the synthesis of collagen in human beings, brings about a severe debilitating condition which is frequently fatal. It was a condition that was greatly feared by arctic travellers, sailors, and anyone undertaking long voyages. It is now known that scurvy arises under these conditions due to a deficiency of mainly vitamins C and B, sometimes compounded in arctic travellers by an overdose of vitamin A from eating seal liver. As the condition develops there is a progressive breakdown in the cellular structure of the body. In 1520 Magellan lost 80% of his crew while crossing the Pacific. Richard Walter, the chaplain on a British squadron in the Pacific in 1748, described its ghastly symptoms: 'skin black as ink, ulcers, difficult respiration, rictus of the limbs, teeth falling out and, perhaps most revolting of all, a strange plethora of gum tissue sprouting out of the mouth, which immediately rotted and lent the victim's breath an abominable odour.' Fridthof Nansen, in his account of the voyage of the *Fram* in his book *Furthest North*, described the fear of the crew as the ship's doctor once a month made his examination which consisted in the weighing of each man, the counting of blood corpuscles, and estimating the amount of blood pigment. This work was watched with anxious interest, as every man thought he could tell from the result how long it might be before scurvy overtook him. Despite the human inability to synthesize ascorbic acid, the human race has had resident populations that have survived in the Arctic for thousands of years. The explanation for their survival was summed up by the Norwegian arctic explorer Helge Ingstad (see Figure 3.11) in his account of life with the Nunamiut in inland Alaska. In describing the consumption of caribou he recorded that 'every single part of the animal is eaten except the bones and the hooves. The coarse meat, which in civilization is used for joints and steaks, is the least popular.... The heart, liver, kidneys, stomach and its contents, small intestines with contents (if they are fat), the fat around the bowels, marrow fat, etc.' (Ingstad, 1951). This description represents the diet of an inland Nunamiut population, which differed from that in the settlements within reach of the sea where people would also eat raw fish, whale, and seal. The consumption of the entire animal has been the key to obtaining an adequate supply of vitamin C. The traditional staple diet of polar peoples included the organs and brains of sea mammals. The vitamin C-rich thin inner skin (*muktuk*) of the Beluga Whale (*Delphinapterus leucas*) was traditionally considered a great delicacy.

Other sources, especially for inland populations, included raw caribou and in particular the liver and stomach contents. To be able to use such a diet can be considered as behavioural evolution which, if the problems of digestion and fat metabolism are ignored, might be argued not to be a genetic character but an acquired adaptation There are, however, other aspects of metabolic and diet adaptations that are clearly genetically evolved.

9.6.3 Metabolic adaptations to diet

The adoption of agriculture as the mainstay of human culture created an evolutionary distinction between the peoples of the far north where farming was not possible and those to the south where cereals became the basis of the economy. The boundary originally would have approximated to the southern limits of the Taiga and is one of the most notable ecological frontiers in relation to human settlement and adaptation in northern regions.

The southern limit of the Taiga (or boreal forest) is marked in the soil by being the geographical limit to the process of podzolization. The generation of podzol soils takes place under coniferous forests where the acid humic substances in the forest litter dissolve organic matter, together with aluminium and iron, and leach them downwards in the soil profile (Figures 9.26 and 9.27). Russian peasants thought that this bleached zone marked the remains of former forest fires and gave the name podzol which means 'ashes below' (Russian *pod*—below; *zola*—ashes). The leached minerals and clay particles become deposited at a lower horizon where they can accumulate and impede drainage by forming an iron and clay pan and lead eventually to the formation of *gley soils*. The ultimate fate is then bog development (paludification; see Chapter 5). Before the advent of mechanized agriculture and the ability to carry out deep ploughing to break up the impermeable iron and clay plan at the base of the soil, such soils were almost impossible to cultivate and limited the northern spread of arable farming.

In Russia, the boreal forest (Taiga) has had an important historical role. The forest and its infertile podzol soils were never cleared in the past for agriculture. They did, however, provide a refuge from mounted invaders, first from the Mongol Hordes and then the Teutonic Knights when they swept across steppe and the fertile lands of Ukraine, causing the Medieval Russian Kingdom to move north-east to the Vladimir region at the interface between the podzol soils of the north and east and the more fertile lands

Figure 9.26 A podzol soil formed under Scots Pine (*Pinus sylvestris*). The southern limit of nutrient-deficient podzol soils approximated in the days before deep mechanized ploughing to the northern limits of productive crop growing for cereals. Before trade became extensive this imposed a diet difference for human settlers north and south of the podzol line.

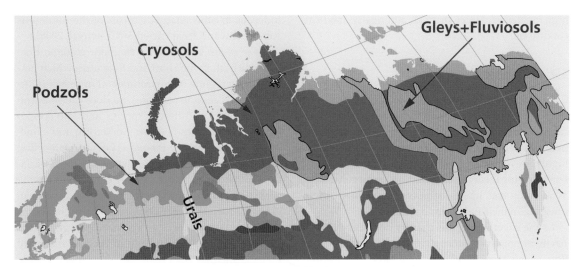

Figure 9.27 Location of major soil type zones in northern Europe. (Source: Soil Atlas of Europe, European Commission).

to the south. It is in this region, rich in timber which disadvantaged mounted soldiers, yet still with a limited capacity to grow some grain, that a Slav refuge and the fortresses and churches of Medieval Russia were built in the early 12th century. The northern boreal forests and their podzolic soils (Figure 9.26) had created a cultural barrier that persisted for over most of the 6,000 years that have passed since the development of farming. For this reason there has been a very distinct difference in the diet of the human populations north and south of the podzol line.

The cultural differences between these zones are most clearly seen in Russia, with its long history of national divisions between the peoples that settled to the south of the podzol line in the southern fertile cereal-producing steppe and the hunter-gatherers of the north who until recently pursued a lifestyle that in many ways differed little from their Palaeolithic ancestors.

A wide-ranging study of the influence of diet, subsistence, and eco-regions on single nucleotide polymorphisms in diverse human populations has revealed strong signals in polar regions in relation to a number of pathways involved in energy metabolism (e.g. pyruvate metabolism, glycolysis, and gluconeogenesis) (Hancock *et al.*, 2010). Among the genes in the pyruvate pathway, particularly strong signals were observed in the gene coding for mitochondrial malic enzyme, which catalyzes the oxidative decarboxylation of malate to pyruvate.

Overall there is a very noticeable genetic distinction between peoples who rely on the grain, tubers, and vegetable products that grow in warmer climates and those that live in areas too cold for agriculture and instead have a high meat content in their diet.

These dietary divisions remain today in the genetic inheritance of these distinct groups. A diet that relies on use of grain as a principal carbohydrate source is very different from one that is dependent on fat-rich animals. The development of farming quickly led not just to the baking of bread but also to fermentation and the production of alcohol. The long history of alcohol consumption by even the earliest civilizations appears to have triggered a change in the presence of the various polymorphic forms of the alcohol dehydrogenase enzymes between populations that lie in the agricultural regions of the world. By contrast, the modern descendants of the northern peoples who have survived in a land where agricultural activity is minimal show a low tolerance of alcohol largely due to a limited ability to metabolize ethanol as a result of differing properties in their alcohol dehydrogenase inheritance. The advent of easy access to alcohol in recent times for the northern peoples of America and Eurasia has unfortunately led to a high incidence of liver cirrhosis and related medical conditions.

9.7 Conclusions

Current interest in biological diversity places great emphasis on the tropics and in particular the tropical rain forest where large numbers of species compete and at the same time manage to co-exist due to highly developed niche specialization. The pressures driving this type of evolution are largely biotic as animals and plants compete to maintain their presence in their particular habitats.

As the examples in this chapter demonstrate, the Arctic is not an *evolutionary freezer*. The fauna and flora of the Arctic show much variation both within and between species. The driving force that creates biological diversity, however, is predominantly physical and due to variations in the environment. There is still competition for optimal sites for nesting and feeding in birds and mammals, just as there is between plants for sheltered and stable sites. Nevertheless, it is the thermal conditions that eventually determine survival.

When conditions for survival are at their limit, as they are in the Arctic, even a small drop in temperature or delay in the arrival of spring can have drastic effects on population viability and will prove to be a rigorous selective force. It is not too much to speculate that this is related to the existence at high latitudes, as in other marginal areas, of the numerous sub-species and local varieties (Crawford, 2008).

Polar regions experience frequent climatic fluctuations both in the short and long term which perpetuates a regime of environmental uncertainty. As the thermal regime oscillates, so does the selection pressure on the different species and sub-species operate, sometimes favouring one form and sometimes another. It can therefore be speculated that under such conditions the existence of these variations, although possibly creating competition in the short term, in the long term can be viewed as a form of *mutualism* as the variation between populations, and even species, acts as a genetic refugium that benefits the long-term future of related groups of animals and plants. The survival of the Polar Bear in an ice-free Arctic is currently viewed with great anxiety. Arctic bears in general have had a long and varied evolutionary history from before, during, and after the Pleistocene ice ages. Irrespective of what may happen to the Polar Bear in its present form, the Arctic, even in a warmer world, will continue to provide a diverse and possibly more favourable range of habitats for the terrestrial biota. Bears as a whole will not disappear. The Polar Bear is still capable of hybridizing with both the Siberian Brown Bear and the Grizzly Bear and there may even evolve a wider range of bear species cable of utilizing the changing and possibly more productive high-altitude ecosystems.

CHAPTER 10

Disturbance, pollution, conservation, and the future

10.1 Disturbance at high latitudes

The quest for ever more natural resources from the Arctic raises the probability of increasing disturbance and pollution. Disturbance is not a new feature in the Arctic environment. The nature of arctic landscapes makes periodic disturbances inevitable through avalanches, rock falls, and mud slips (Figure 10.1). Northern soils are also being constantly affected by frost-heave, cryoturbation, solifluction, and more recently *thermokarsting* as ice-rich soil melts and erodes.

The *arctic biota* has been described as intrinsically delicate in terms of its biodiversity. Low diversity can be an underlying feature of small fragmented populations, making them susceptible to disturbance (Elton, 1927). This is more true of animals than it is of plants. Reindeer populations on isolated islands in the Arctic Ocean (see section 9.5.3) can be deficient in diversity compared with those in continental areas (Cote *et al.*, 2002). Plants, however, have an ability to track their ecological niche after climate change. From DNA finger-printing (amplified fragment-length polymorphism) of nine plant species growing in Spitsbergen it has been shown that long-distance colonization of this remote arctic archipelago has occurred repeatedly from several source regions (Alsos *et al.*, 2007). In addition to pollen dispersal, plant propagules are easily carried by wind and drifting sea ice.

The combination of environmental stress and disturbance imposes a particularly testing blend of adversity for both plant and animal survival. Undoubtedly, a long history of living in such a situation has created a biota that is adapted to uncertainty and disturbance. The arctic biota would not exist today if it were not for the capacity of depleted populations to restore their numbers during recovery periods. Whether or not this built-in resilience which has preserved life at high latitudes in the past will prove adequate to restore degraded ecosystems in the future is as yet an unanswered question.

10.1.1 Physical and biological fragility

In examining survival in any habitat it is essential to distinguish between physical and biological fragility. A salt marsh or a tundra heath may be physically fragile and suffer loss of terrain through storm erosion or excessive grazing and trampling, yet despite this material damage, there may be no loss in biological diversity throughout the region as a whole. The plants that grow in these physically disturbed sites are usually adequately adapted to their surroundings and even require a certain degree of disturbance to provide opportunities for regeneration and habitat renewal.

Removal of herbivores can restore an initial lushness to vegetation with improved flowering. However, as evidence for ecological amelioration this can be deceptive as the improvement is frequently not maintained (Crawford, 2008). Without herbivores, nutrient cycling in the Arctic is much reduced as the soil microflora are not sufficiently active for this purpose at high latitudes. The plants and animals that grow in these physically stressed sites are usually adequately adapted to their surroundings and even require a certain degree of disturbance to protect them from invasion by other species which could out-compete them if a stable and constant environment were to be imposed as a means of conservation.

Figure 10.1 Landscape fragility in the Arctic as seen in Bell Sound, Spitsbergen (77°N).

10.1.2 Species under threat in the Arctic

Hunting of arctic terrestrial mammals, such as Polar Bear, Reindeer (Caribou), and Muskoxen, and in times past the now extinct species of the arctic megafauna, was the inducement that first brought human settlers to the Arctic (see Chapter 3). Examination of the vast numbers of mammoth skeletons that were used in the construction of late Palaeolithic dwellings in the Tundra has provided evidence for the suggestion that the ancient megafauna was brought to extinction not just by changing climatic conditions, but also by human activity (see Chapter 2).

The Arctic has long been an area that was explored for its assets. The hunters of the Palaeolithic followed the remnants of the Pleistocene megafauna relentlessly to their last refugia in the north (see Chapter 3). Equally, in medieval times the northern forests and the Tundra provided sources for satisfying the desire for furs. Before the ivory trade with India was developed, the Arctic was the principal source of walrus ivory. These unfortunate animals were subjected to wholesale slaughter just for their tusks. The quest for much-sought-after pelts, such as beaver skins, drastically reduced their populations, as did the pursuit of Polar Bears for their hides, which were prized objects for Vikings and other northern traders from Greenland to Siberia. The demand continued in modern times, with a particularly rapid rise in recorded numbers of Polar

Bears killed in the 1960s before legal protection was provided.

Aerial photography has contributed greatly to population monitoring of the larger arctic animals. Hunting restrictions are being increasingly imposed but vary from region to region. For some species more pro-active measures such as reintroduction to new or former habitats are possible and this has been successfully implemented both with Reindeer and Muskoxen (see section 4.6.5).

10.2 Conservation case histories

10.2.1 Polar Bears

Considerable concern has been expressed about the future of the Polar Bear given the marked decrease that has been reported in recent years in the extent of Arctic sea ice (Figure 1.8). Given the dependence of present-day Polar Bears on ice for catching seals, such concern appears well justified and the Polar Bear is on the red data list of the International Union for Conservation of Nature (IUCN).

Accepting the prediction that the maritime hunting capability of Polar Bears is likely to be impaired, it is of interest to know to what extent the bears can supplement their stored fat reserves from terrestrial forage such as berries. The ideal place to investigate this question is on the western side of the Hudson Bay during the ice-free periods when bears are fasting on land. However, it has been difficult to quantify their relative dependence on stored adipose tissue to meet their nutritional needs as opposed to berries and other forage.

An enterprising study has, however, been made of the the stable isotope ratios for carbon (∂C^{-13}) in the carbon dioxide content of the breath of 300 fasting bears (Hobson *et al.*, 2009). Some of the bears were known to have recently fed on berries and others had not. A simple isotopic distribution model would have predicted that bears metabolizing adipose tissue derived from seals would have breath ∂C^{-13} values close to −24.7‰ and those metabolizing berries exclusively would have values close to −32.6‰, as this would be the expected value for a terrestrial plant carbon source. However, the distribution of the ∂C^{-13} values were remarkably similar, ranging from −24.2 to −24.8‰. It would appear therefore that the bears that had fed on berries while fasting on land received an insignificant amount of energetic benefit from the vegetation diet.

Perhaps this result might have been expected, as Polar Bears are capable of fasting for several months in a state of *shallow hibernation* when they can exist on their body fat and even dispense with drinking. They are even capable of surviving for up to 8 months without eating and can easily walk over 5,000 km in this fasting state.

The distances that Polar Bears can cover contributes to the problems of making accurate estimates of populations in different parts of the Arctic. A study being carried out by the IUCN has divided Polar Bear territories into 19 different areas (Figure 10.2) where the sub-populations display seasonal fidelity to particular regions. However, DNA studies show that these geographic classifications do not necessarily represent reproductive isolation (Paetkau *et al.*, 1999).

The 13 North American sub-populations range from the Beaufort Sea south to the Hudson Bay and east to Baffin Bay in western Greenland and account for about 70% of the global population. The Eurasian population is broken up into the eastern Greenland, Barents Sea, Kara Sea, Laptev Sea, and Chukchi Sea sub-populations, though there is considerable uncertainty about the structure of these populations due to limited mark and recapture data.

The effects of global warming are most profound in the southern part of the polar bear's range, and this is indeed where significant diminution of local populations has been observed. The western Hudson Bay sub-population, in the southern part of the range, also happens to be one of the best-studied Polar Bear sub-populations. This sub-population feeds heavily on Ringed Seals (*Phoca hispida*) in late spring, when newly weaned and easily hunted seal pups are abundant. The late spring hunting season ends for Polar Bears when the ice begins to melt and break up, and they then fast or eat little during the summer until the sea freezes again.

Due to warming air temperatures, ice-floe breakup in western Hudson Bay is currently occurring 3 weeks earlier than 30 years ago, reducing the duration of the Polar Bear feeding season. The body condition of Polar Bears has declined during this period; the average weight of lone (and likely pregnant)

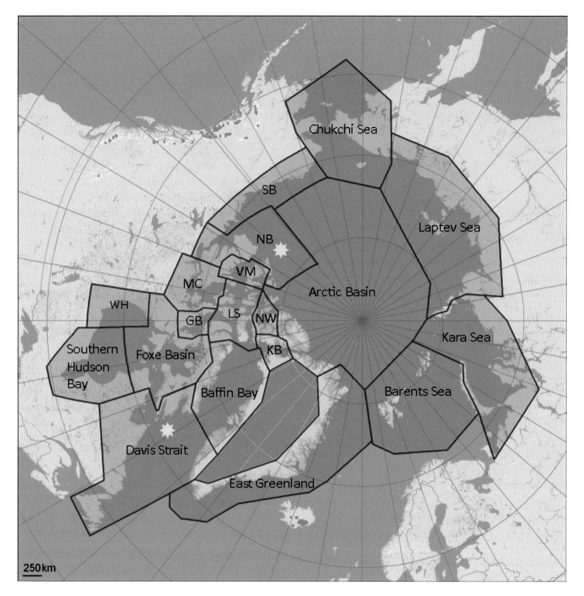

Figure 10.2 Polar Bear population map showing location of the 19 regions where Polar Bear counts are currently being undertaken. Yellow stars indicate a rising population. (Reproduced with permission from IUCN Report, 2010, http://pbsg.npolar.no/en/status/population-map.html).

female Polar Bears was approximately 290 kg (640 lb) in 1980 and 230 kg (510 lb) in 2004. Between 1987 and 2004, the western Hudson Bay population declined by 22% (IUCN, 2010).

In Alaska, the effects of sea ice shrinkage have contributed to higher mortality rates in cubs, and have led to changes in the denning locations of pregnant females. Recently, Polar Bears in the Arctic have undertaken longer than usual swims to find prey, resulting in a number of recorded cases of drowning (Monnett and Gleason, 2006).

10.2.2 Muskoxen conservation

Muskoxen (*Ovibos moschatus*), a remnant of the Pleistocene megafauna, probably evolved about

1 million years ago in the Tundra of north-central Asia, reaching North America over the Bering Land Bridge about 90,000 years ago (Blix, 2005). Post-Pleistocene climate and vegetation changes have now reduced the area over which this animal once roamed with the result that the modern wild herds of Muskoxen are now native only to the New World, being found solely in northern North America and Greenland (Figure 10.3).

These remaining wild herds show little genetic diversity at both the nuclear and mitochondrial levels (Groves, 1997; Holm *et al.*, 1999). Examination of ancient DNA sequences as found in the skeletons of Holocene Muskoxen (Figures 10.4 and 10.5) has shown that Muskox genetic diversity was previously much higher during the Pleistocene and has undergone several expansions and contractions over the past 60,000 years. North-east Siberia was of key importance in this study as it held a large diverse population until there was a local extinction at 45,000 BP. Subsequently, Muskox genetic diversity recovered at *ca.* 30,000 BP, then contracted again at *ca.* 18,000 BP, finally recovering in the mid-Holocene (Campos *et al.*, 2010). It was concluded from this study that the arrival of human populations in the Muskoxen ranges did not affect their mitochondrial diversity. It was also notable that both Muskoxen and human populations expanded concomitantly into Greenland.

These findings have supported the argument that the population dynamics of Muskoxen are more likely to have been associated with non-anthropogenic causes such as environmental change rather than the earlier view that it was due to an 'over-kill' by early hunters. This is also the argument advanced

Figure 10.3 Present native distribution and introductions of Muskoxen in arctic areas of Canada, Greenland, and Norway. The species has also been successfully introduced to Alaska, Quebec, and the Taimyr Peninsula of Russia (green areas and dots). It was introduced to Dovrefjell (red arrow) from Myggbukta on Greenland (red dot). (Reproduced with permission from Ytrehus *et al.*, 2008).

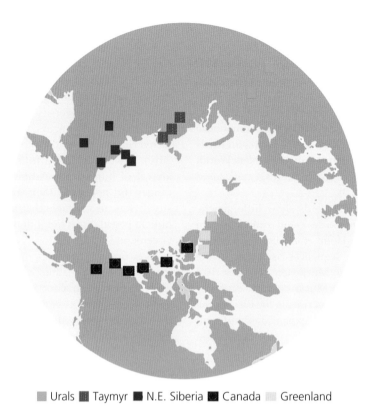

Figure 10.4 Geographic origin of Muskox samples that yielded ancient DNA sequences. (Adapted from Campos et al., 2010).

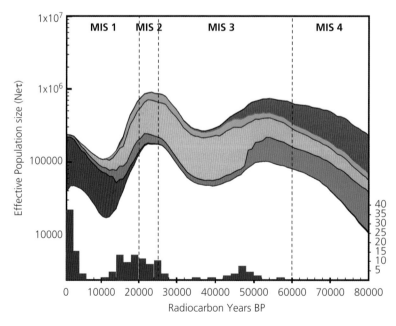

Figure 10.5 Statistical representation from the geospatial analysis of 135 ancient and 14 modern Muskox mtDNA control region sequences. The x-axis is in units of radiocarbon years in the past, and the y-axis is a function derived from the product of the effective population size and the generation length in radiocarbon years. Colours indicate geographical origins of the samples: blue, Greenland; red, north-east Siberia; orange, Taymyr; green, Urals; light bluegrey, Canada. The bar graph shows the number of radiocarbon-dated samples in bins of 2,000 radiocarbon years. (Reproduced with permission from Campos et al., 2010).

in relation to the history of mammoth survival (see Chapter 2).

Although many species went extinct at the end of the Pleistocene, others, such as Caribou and Bison, survived to the present. The Muskox appears to be in a borderline situation, being relatively abundant during the Pleistocene but only surviving naturally to modern times in Greenland and the North American Arctic archipelago.

Species survival north of the main ice sheets during the Pleistocene has often been claimed as evidence for the existence of ice-free refugia. This appears to account for the distribution of chloroplast DNA haplotypes in northern Greenland for Purple Saxifrage (*Saxifraga oppositifolia*) (see Chapter 9). A similar case can be made for Muskoxen which continue to have a marginal existence in High-Arctic archipelagos but with marked fluctuations both in the population size and distribution.

During times of climatic adversity Muskoxen show a pronounced retreat to certain refugia which have better access for winter feeding. Refugia for Muskoxen in the High Arctic include lowlands on eastern Axel Heiberg Island in the Mokka Fiord region, the lowlands of north-eastern Devon Island, and the Bailey Point region of Melville Island (Figure 10.6). Historically, all these regions have supported high densities of Muskoxen from time to time, but the Bailey Point region on Melville Island has been considered as the best habitat for Muskoxen in the Canadian High Arctic. The Bailey Point refugium and its importance for Muskoxen was discovered only in 1961. It was not until aerial surveys were carried out between 1972 and 1974 that its importance for Muskoxen populations became apparent (Thomas *et al.*, 1981).

At present the total world population of Muskoxen is estimated to be about 100,000 and increasing. The greatest number for any one locality is on Banks Island, Canada (Figure 10.6) which in 1994 has a record number of 84,000 animals. In more recent surveys this has fallen to 58,000 which still accounts for about half the world total (Strutzik, 2000).

Although the distribution of Muskoxen is much reduced from what it was in the past, this species is not considered to be in any great danger of disappearing unless it succumbs to climatic change (Gunn and Forchhammer, 2008). The Muskox does have a low reproductive rate with single calves born annually, or else every 2–3 years, with cows normally bearing their first calf at an age of 3 years depending on environmental factors such as availability of food and weather severity.

Muskoxen in Alaska

Muskoxen were never a dominant component of the mega-mammal fauna in Alaska even in the Pleistocene period. After the end of the Pleistocene they became even less common. Inuit hunters apparently killed the last Muskoxen in north-western Alaska in the late 1850s (Lent, 1998). Despite the limited natural occurrence of Muskoxen in the past they have now been successfully reintroduced to their former ranges from the Seward Peninsula in the south and north across the Arctic Slope and east into the northern Yukon Territory.

From recent surveys it has been estimated that they now number over 3,000. The re-established herds either fluctuate or are increasing in size and range, and in some areas local people are concerned that they will compete for pasture with caribou (Gunn and Forchhammer, 2008).

Muskoxen in Canada

By the 1900s, muskox populations in Canada had dropped significantly due to an increased interest in the trade of hides, hunting with guns, and several icy and harsh winters. In 1917, the Canadian government put the Muskox under protection in the hope of saving nearly extinct herds. In 1927, the Thelon Game Sanctuary was established to protect Muskox on the Tundra mainland.

In the 1950s, muskox numbers dipped to a dangerously low level of approximately 1,000 animals. To give the herds a chance to repopulate, even local residents stopped harvesting them. By the early 1970s, the number of Muskox had increased significantly and the herds were once again viable.

As human activity increases in arctic Canada, monitoring the effects of disturbance on Muskoxen is clearly essential and is greatly facilitated through aerial surveys, tagging, and radio-tracking. There is therefore a degree of concern for the protection of Muskox populations and for this reason Polar Bear Pass on Bathurst Island has been declared a national wildlife area. muskox defence strategies have

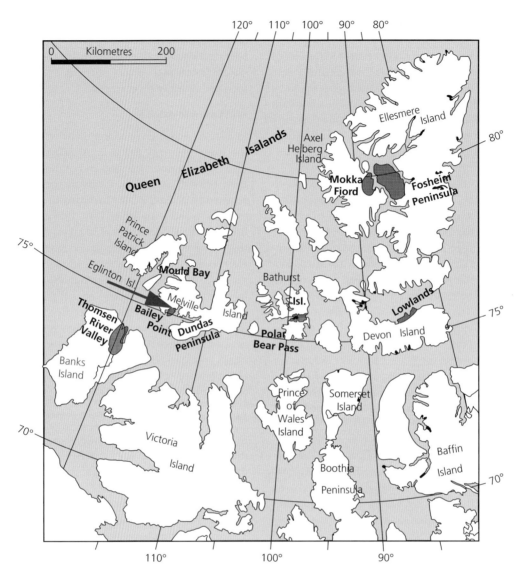

Figure 10.6 Locations on the Canadian Arctic Islands where the highest densities of Muskoxen have been recorded. The Bailey Point region on Melville Island is marked with a red arrow. The greatest number for any one locality is found on Banks Island. (Adapted from Thomas et al., 1981).

always been ineffective against people. When disturbed by low-flying aircraft or by human beings on foot, Muskoxen first stand their ground in a defensive ring and only when pressed closely do they stampede away from danger.

Muskoxen in Russia

Muskox populations survived in the Taymyr Peninsula until relatively recently. Remains of the species have been found which have been dated to between 2,000 and 4,000 BP. As a result of international co-operation there has been a reintroduction of Muskoxen to northern Russia ecosystems. For the pastures of the north, reintroduction is expected to improve nutrient cycling and thus increase the proportion of fodder vegetation and enhance the total productivity of the Tundra. It is hoped that the vast areas of the Russian Tundra might eventually be capable of supporting over 2 million Muskoxen.

The shoreline of the Polar Ocean was chosen as a site for reintroduction of the Muskoxen. According to the reintroduction plan, a chain of herds, located within 600–700 km from each other, could be established (Sipko and Gruzdev, 2006).

Initial reintroductions of Muskoxen took place in 1974 and 1975 when ten animals were delivered from Banks Island (Canada) and 20 animals from Nunivak Island (Alaska, USA) to the eastern part of the Taymyr Peninsula. These first introductions proved successful and have spread north, east, and south across the Putorana Plateau region of the Taymyr Peninsula. By 2002 number of Muskoxen increased to 2,500 animals and by 2010 reached 6,500. In 1975, 20 Muskoxen from Nunivak Island (Alaska, USA) were delivered to Wrangel Island. Here the population initially grew slowly as a number of the animals did not survive the period of introduction. Nevertheless, the population on the Island reached 750 by 2003 (Sipko and Gruzdev, 2006) and is now between 800 and 850 animals. The total population in northern Russia has now reached approximately 8,000 (Sipko, 2010).

Muskoxen in Norway

In Norway some success has been obtained with an introduction to the Dovrefjell (62°N) in southern Norway, which is considerably further south than their present natural distribution. Compared to the areas of origin for Muskoxen, the Dovrefjell is characterized by a relatively warmer and more humid climate. An initial introduction in 1932 appeared to be successful as the animals succeeded in breeding. However, by 1945 they had disappeared, not because of any intrinsic physiological mal-adaptation, but due to accidents on the railway line combined with poaching. Another attempt was therefore made between 1947 and 1953 with new introductions from east Greenland to Bardu in northern Norway (69°N) and again to the Dovrefjell. The Muskoxen in Bardu died out in the 1960s. The population in the Dovrefjell has persisted and has from time to time shown growth rates comparable to successful Muskox introductions elsewhere.

A population study (Asbjornsen *et al.*, 2005) has noted the existence of considerable annual population fluctuations. High growth rates have been found in years with mild weather during the period September–November and with minimal snow depths in May. It has been predicted that under these conditions the population is capable of increasing. Nevertheless there is an ever-present risk of infections and parasite infestations.

In recent years the Norwegian Muskoxen have been noted as suffering from a number of diseases together with parasite infestations. These conditions appear to be aggravated by warmer and more humid weather, which gives rise for concern for the long-term survival should there be significant climatic warming. In 2006 an outbreak of fatal *Pasteurellosis* occurred, causing the death of a large proportion of the animals. *Pasteurellosis* is the name given to a wide range of infections caused by bacteria in the genera *Pasteurella* and Mannheimia, which are opportunistic pathogens, prevalent in the pharynx of healthy mammals and birds, but which can lead to acute pneumonia and sepsis (Ytrehus *et al.*, 2008).

The Dovrefjell infection coincided with extraordinary warm and humid conditions that are often associated with outbreaks of *Pasteurellosis*. It has therefore been suggested that such instances may occur with increasing frequency among cold-adapted animals if global warming results in increased occurrence of heat waves, and could result in population declines and eventual extinctions. (Ytrehus *et al.*, 2008).

10.2.3 Muskox domestication

Muskox domestication projects were established at Unalakleet in Alaska and at Fort Chimo (Kuujjuaq) in northern Quebec. Now located at Palmer, the Alaskan domesticated herd has resulted in a native industry based on the manufacture of clothing and other woollen items. The Muskoxen from the Fort Chimo project have been released and now range freely in northern Quebec. In addition, Muskoxen from an introduced population in Alaska are regularly seen in the Yukon. Today, there are over 100,000 Muskoxen in the mainland Tundra and Arctic Islands of the North West Territories and Nunamove (the largest, northernmost territory of Canada). In the winter they roam in isolated herds of approximately 75 animals. During the mating season 15–20 cows will herd with one bull. These herds are so successful that the number of grizzly bears and wolves looking for Muskoxen as food are also increasing. For this reason, Muskox have altered their range, roaming nearer to the more northern human communities.

10.2.4 Wild Reindeer (Caribou)

Migratory wild Barren-Ground Caribou (*Rangifer tarandus groenlandicus*) is a sub-species that is found mainly in the Canadian Nunavut and North West Territories as well as western Greenland. The other sub-species found in Canada is the Woodland Caribou (*Rangifer tarandus caribou*) (Figure 9.22). Caribou numbers in general in Canada have dropped by about one-third since populations peaked in the 1990s and early 2000s (Festa-Bianchet *et al.*, 2011). Four of five major herds in Russia and one of two herds in Greenland are also declining. Several large Canadian herds have declined by 75–90%. Furthermore, it appears that the declines have recently accelerated, with the northerly Bathurst herd declining from over 450,000 caribou in 1986 to fewer than 50,000 in 2009 (Gunn and Russell, 2010).

Overall caribou ranges have shrunk substantially across North America due to the complex effects of human-habitat changes (Figure 10.7). Studies to define the reasons for this decline have been carried out on the *Little Rancheria Herd* of caribou, (Figure 10.8) which numbered about 1,000 in 1999 (Adamczewski *et al.*, 2003). The herd inhabits a lowland forested winter range with some merchantable pine and spruce stands just west of Watson Lake, Yukon. Up to 2003 the harvest of timber in this range had been been limited. Nevertheless, even at this time the potential for habitat fragmentation remained high. These examples show that there are both direct and indirect effects of development on woodland caribou. These will include loss of fragile, slow-growing lichens, an avoidance of disturbed areas, increased hunting, and collisions with vehicles.

In the remoter parts of the caribou range, improved access for predators, primarily wolves, is also a factor. It has been found that caribou are most likely to be killed by wolves in areas within 250 m of forestry cut-blocks and other human activities. This may account for the observation that caribou tend to use these areas less than the undisturbed forests.

The most northerly populations of North American Caribou are the Peary Caribou (*Rangifer tarandus pearyi*) which are found in the High Arctic islands of Canada's North West Territories. They are the smallest of the North American Caribou, and exist throughout this vast region in a number of distinct local herds. Recently their number has also diminished, from above 40,000 in 1961 to about 700 in 2009.

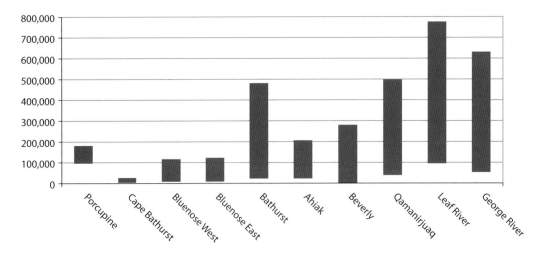

Figure 10.7 Range of maximum to minimum population estimates for Canadian *Rangifer* herds. (By courtesy of D. Russell and A. Gunn, *pers. comm*). 2012).

Figure 10.8 Caribou from the Little Rancheria Herd in the West Yukon crossing a frozen lake in search of grazing in an ice-encased landscape. (Photo courtesy of M. Pealow).

Climatic warming is creating an increased danger of winter starvation for caribou both in North America and Eurasia. Incursions of warm air with rain falling onto the frozen soils of the Arctic result in ice-encasement. Recently the number of days with above-freezing temperatures has increased significantly, resulting in partial melting of snow, only to be refrozen as ice layers in the snow pack. These ice layers hinder foraging and can cause dramatic drops in populations, as seen in the Spitsbergen reindeer population (Figures 4.21–4.23). This is a general phenomenon at northern latitudes as a result of climatic warming when intermittent periods of rain fall on frozen ground (see also section 4.6.3).

Further south in the Prince of Wales, Somerset, and Russell Islands in the south-central Canadian Arctic the number of caribou have been noted as having an equally catastrophic decline with their number declining by 98% in 15 years. In this case there was no evidence for large-scale winter mortality. It has been suggested that this last example was due to consequential reductions in survival rates, both of breeding females and of calves in their first year of life, associated with continued caribou harvesting and markedly increased wolf (*Canis lupus*) predation on the dwindling number of caribou through the 1980s and early 1990s (Gunn *et al.*, 2006).

Overall, for caribou in the Canadian Arctic, losses outweigh gains and not only have populations fallen from their historic highs, but also their ranges have shrunk. Canada's ten best-known herds of migratory Tundra caribou have all declined, in some cases to a shadow of their former selves. Elsewhere migratory wild reindeer numbers are also diminishing. Four of five major herds in Russia and one of two herds in Greenland are also declining.

There are, however, herds that are avoiding this general trend. East of the Hudson Bay, the George River herd is still close to 1 million animals. Further west in Alaska there are herds that are still growing, while in Norway and Finland, where herds are small, wild reindeer are either stable or increasing (Gunn et al., 2009).

There are several causes for this reduction in wild caribou and reindeer populations. Hunting, mining, and oil extraction and their pollutants are all negative factors in relation to the ability of reindeer to migrate, feed, and reproduce. In addition there are climatic factors. The *Arctic Oscillation* has been noted to have a prominent influence on caribou ecology and abundance (Zalatan et al., 2006). Climate sets the pattern at the annual and seasonal scale. Summer weather influences the timing and amount of plant growth as well as the levels of harassing insects and parasitic intestinal worms. Winter weather influences the availability of food through the amount of snow and the corresponding energy it takes the caribou to move and to dig for their forage. Interacting with the weather are the predators—wolves and grizzly bears, and less often wolverine, golden eagles, and lynx.

Predictions about the future of caribou populations need to take into account past population fluctuations. Such records are difficult to find and memories from native peoples are at the best only qualitative. Nevertheless in northern Quebec an ingenious proxy for measuring relative changes in the density has been made from recording frequency of tree-ring-dated trampling scars. These scars are produced by caribou hooves on tree roots of Black Spruce (*Picea mariana*) in the forest-tundra of North America in extensive trails distributed over the summer and winter ranges (Payette, 2004).

The caribou migrate annually (Payeette, 2004) for long distances from the forest to the open Tundra in late spring, and return to the forest in the autumn. The scar frequency distribution was determined with careful cross-dating and taking into account the influence of root age.

The technique was first demonstrated in a study to test the hypothesis that the long-term trends of caribou activity and population decline in caribou could be associated with native harvesting and fire in the Rivière-aux-Feuilles herd east of the Hudson Bay in northern Quebec (Payette et al., 2004). The age structure data of trampling scars from lichen woodlands distributed over the entire range of the herd confirmed the overall trends of caribou activity from the late 1700s to the present (Figures 10.9 and

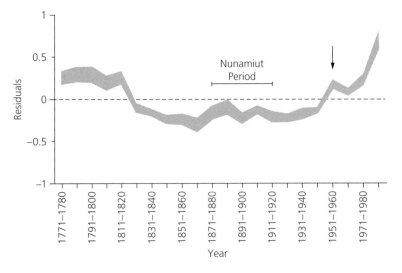

Figure 10.9 Standardized distribution of the number of trampling scars (residuals of regression curves ± SD) according to time (from the late 18th century to the late 20th century). The Nunamiut period is indicated. The arrow corresponds to caribou crowding during the 1950s likely associated with large wildfire conflagrations. (Reproduced with permission from Payette et al., 2004).

DISTURBANCE, POLLUTION, CONSERVATION, AND THE FUTURE 229

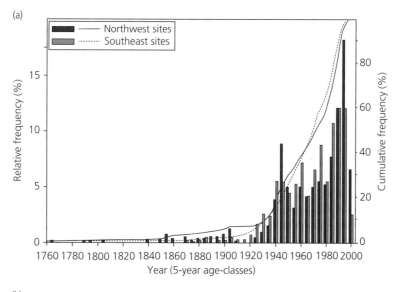

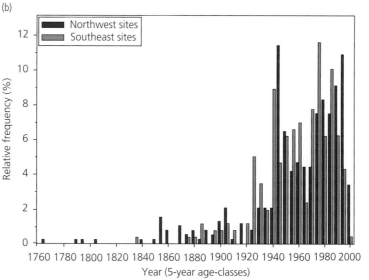

Figure 10.10 Scar frequency distribution (a) using all samples from sites and cumulative frequencies (plotted line) of the number of samples; (b) using only scars located on roots established before 1900. (Reproduced with permission from Zalatan et al., 2006).

10.10). Over the last 200 years, the herd has undergone two highs in the late 1700s and 1900s separated by a moderate activity pattern in the late 1800s. Native harvesting was possibly involved in the early 1900s decline, although at a moderate level (Figures 10.9 and 10.10).

The method was further applied to the Bathurst caribou herd in the Canadian North West Territories. The scar frequency distributions (dated from AD 1760–2000) showed similar abundance patterns through time. Caribou numbers were high during the mid-1940s and 1990s and were very low during the 1920s, 1950s–1970s, and at the turn of the 21st century.

These abundance patterns determined from scar frequencies correlate strongly with data obtained from traditional knowledge of Dogrib elders in the region and animal counts based on aerial photography.

The scar frequency distribution developed in this study is the longest proxy record of caribou abundance to date (Zalatan et al., 2006). It would appear that the populations levels of the caribou were very low during the 1920s and 1950s–1970s.

Taken as a whole, the data on arctic caribou and reindeer population sizes suggest that the 20th century witnessed a prolonged period of rising population levels but this has now given way to a general decline and in some cases a very severe reduction in herd sizes. Whether the present decline is merely a return to previous low population levels is not certain given the probability that the ancient root scars may not reflect population levels to the same extent as modern scars.

From the various studies, it would appear that both cold winters with heavy snowfall and warm winters with little sea ice can also be blamed for the recent decline in caribou populations. Winters with heavy snow cover can cause starvation, while warm winters causing thin sea ice can be disadvantageous for migration.

Reindeer are not now present on Franz Josef Land. However, on some of the constituent islands—Hooker, Scot, Keltie—as well as Alexandra Land, shed antlers can be found on unglaciated forelands. Radiocarbon dating of such antler remains have demonstrated the presence of reindeer between ca. > 6,400 and 1,300 BP. Reindeer therefore probably occupied Franz Josef Land for much of the Holocene, reflecting the existence of sufficient forage associated with summer air temperatures probably 1°C higher than today and accompanied by a significant retreat of glaciers and snowfields. A prominent neoglacial advance at ca. 1,000 BP was coincident with the inferred extinction of reindeer. It might be possible that late 20th-century warming and associated glacier retraction may provide environmental conditions for the return of reindeer to Franz Josef Land and other High Arctic areas, similar to earlier in the Holocene (Forman et al., 2000).

On Severnaya Zemlya (see Figure 3.6) the reindeer are seen only occasionally (Dekorte et al., 1995) but are found more frequently on the New Siberian Islands, having been recorded crossing the ice between the different islands.

Settlers introduced reindeer to Wrangel Island in 1950, although archaeological findings suggest that like Franz Josef Land they may have lived there in the past. Well adapted to the harsh conditions, the introduced reindeer quickly began to damage the delicate species-rich ecosystem, prompting the government to create the Wrangel Island Zapovednik (reserve), and to limit the reindeer numbers to a maximum of 1,500.

Although Spitsbergen has probably had a resident population of reindeer for 10,000 years or more, there are regions from which they have been absent in recent times. The isolated Brøgger Peninsula, Spitsbergen (78°55′N) is such a location where reindeer have been recently reintroduced. This region was chosen as due to its isolation the population remains relatively contained and therefore could be accurately monitored (see section 9.5.4). Fifteen animals were introduced to the Peninsula in 1978. The introduced population grew exponentially to 360 individuals by 1993. This number then declined to fewer than 80 individuals during the winter of 1993/94 when ice-encasement of the tundra caused mass starvation (Figure 4.22). Despite this 77% population reduction the reindeer have shown themselves able to recover gradually from this diminution in their numbers (see section 4.6.3) and by 1998 the Brøgger Peninsula population had regained approximately half its pre-1994 number (Aanes et al., 2000).

Reindeer were hunted very heavily throughout Svalbard from 1860–1925, and the population was markedly reduced. The hunt was stopped except for scientific sampling between 1925 and 1983. This period of protection resulted in recovery in numerical terms as well as the surviving reindeer spreading to occupy abandoned parts of their former range. Data gathered from numerous locations in the archipelago, and long-term monitoring from a few specific areas, suggest no indication of a decrease in the number of Svalbard reindeer during recent decades.

10.2.5 Rare plants

Every region of the world has its rare plant species and the Arctic is no exception. The inaccessibility and the vastness of the Arctic, combined with the belief that it is not a species-rich plant-hunter's ground, has protected its flora from excessive depletion by marauding botanists. In addition, arctic plants unlike those of alpine origins do not survive for long outside their native habitats which makes collection for horticulture usually unrewarding. Not only are the arctic species mal-adapted to the generally warmer conditions that they experience in cultivation but also their specialist habitat and reproductive requirements and mycorrhizal associations necessary for survival may not be present.

For many arctic species in cultivation, refrigerated glasshouses are a necessity. The Botanic Garden of the University of Copenhagen with its interest in the flora of Greenland was the first of very few gardens to create a refrigerated glasshouse where arctic plants could be kept alive by keeping them cool (Böcher, 1974).

Plant species differ fundamentally from animal species in their ability to hybridize and also in their capacity to persist without sexual reproduction. Consequently, the concept of rare species differs between plants and animals. For animals, rare species are usually taxonomically well-defined. Some may once have been relatively numerous and have now become rare. However, in plants many of the rare species that excite botanical interest at high latitudes are hybrids, apomicts, and polyploids, all of which are the product of the ever-continuing process of evolution. Typical examples of this ongoing process can be found in hybrid willows (*Salix* spp.) and the numerous species and sub-species of Whitlow Grasses (*Draba* spp.) (Figure 10.11), to list but two groups that have a facility to produce new and rare hybrids (Brochmann *et al.*, 1992).

The genus *Draba* has long been noted for its ability to evolve new species repeatedly from the same parent species, producing hybrid micro-species which are frequently considered sufficiently distinct to be given species status (Brochmann *et al.*, 1992). Rarity in these cases is therefore a natural phenomenon and not due to human disturbance, nor is it a sign of biological fragility. If some of the many rare hybrids disappeared it would not be a matter of any ecological consequence, as more would soon arise to fill their place.

Such phenomena do not mean that there are no rare sexually reproducing species in the Arctic. However, given the extent of the land areas in the Arctic the number of rare species is not particularly numerous.

By far the greatest number of these rare species is found in the Russian Arctic (Table 10.1). The next

Figure 10.11 *Draba oxycarpa* is an octoploid species, at present only known from the amphi-Atlantic regions (east Greenland, Spitsbergen, Scandinavia, north Russia, east to the Polar Urals, probably also Iceland, and Jan Mayen). A vicarious species, largely similar in morphology and also octoploid, is known from Wrangel Island, east Chukotka, and west Alaska, i.e. the central Beringian regions. It has provisionally been denoted as *Draba* 'pseudo-oxycarpa' (see Brochmann *et al.*, 1992).

Table 10.1 Number of rare, endemic angiosperm species as recorded for each country contributing to the Conservation of Arctic Flora and Fauna (CAFF) project (Talbot *et al.*, 1999).

Country	Number of taxa	Percentage
Russia	70	72.9
USA	11	11.4
Russia/USA	4	4.2
Greenland	4	4.2
Canada	3	3.1
Canada/USA	2	2.1
Norway	2	2.1
Total	**96**	**100.0**

highest national number is from the Beringian region of the USA (Figures 10.12 and 10.13). Ecologically, most of these species (excluding apomicts) are probably ancient species that have survived in refugial regions such as Wrangel Island (17%), and

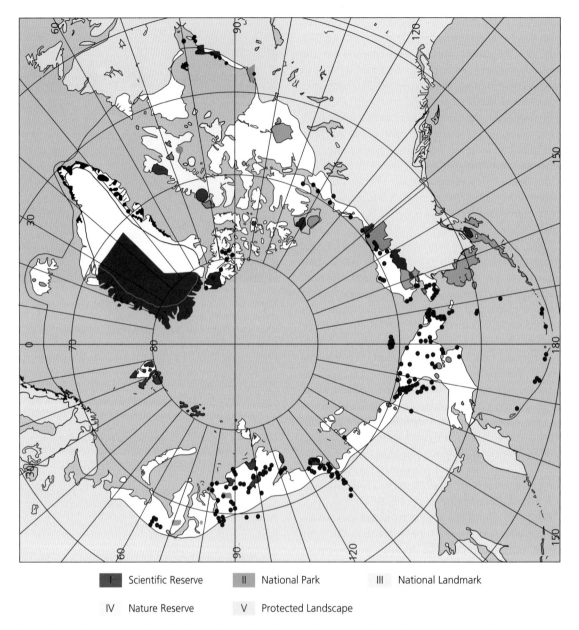

Figure 10.12 Circumpolar map of the distribution of rare endemic flowering plant species in the Arctic in relation to protected areas. (Reproduced with permission from Talbot *et al.*, 1999).

Beringia and Chukotka (14%). Overall only 1% of the 96 rare species within the protected regions of the Arctic are considered to be endangered. The rare species are found mostly in situations where competition is minimal and they are generally assumed to be relicts from the late-glacial steppes, hence their tendency to occur in xeric microhabitats in the modern tundra (Talbot *et al.*, 1999). Figure 10.13 illustrates the distribution of rare species of *Artemisia* which is an example of a genus with species that would have been widespread in the early Holocene Tundra-Steppe (see section 2.2).

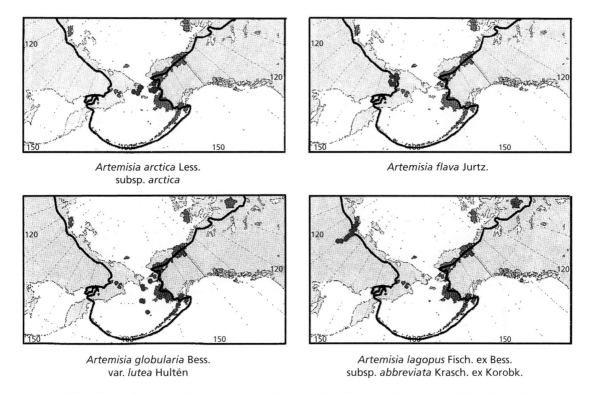

Figure 10.13 Distribution of four rare and isolated occurrences of ancient species of *Artemisia* with locations marked in red. (Reproduced with permission from Talbot *et al.*, 1999).

10.3 Pollution in the Arctic

Pollution can be defined as the contamination of air, water, or soil by substances that are harmful to living organisms. The remoteness of the Arctic does not render it immune to pollution irrespective of whether it occurs naturally, as arises with volcanic eruptions, or as a result of human activities such as oil or mining spillage. In the Arctic, pollution can have particularly severe and long-lasting effects due to the special nature of arctic habitats. This was made evident when the Russian nuclear reactor at the Ukrainian Chernobyl site exploded in 1986 with the resulting widespread pollution with radiocaesium (^{137}Cs).

Despite many well-documented examples of nutrient retention in unproductive habitats it was an astonishment to the scientific community that radioactive substances were retained for so long in arctic and sub-arctic soils. It had been popularly anticipated that rainwater, leaching, and snowmelt would remove the radioactive ions from shallow unproductive soils.

This erroneous belief was based in ignorance of the capacity for nutrient retention by oligotrophic vegetation in marginal areas. Where soils are shallow, plant roots extend through a greater portion of the soil than in areas where soils are deep. In the latter the radioactive substances are gradually removed to lower layers beneath the rooting zone. However, in shallow soils the nutrients are repeatedly recycled due to the mass of roots in a limited depth of soil. Thus, the radioactive materials persist longer in the vegetation and animals that graze on this vegetation still become contaminated, even many years after the deposition of the radioactive material.

The capacity for arctic soils to retain nutrients has long been known from 500-year-old abandoned Norse settlements in Greenland to hunters' camps in the High Arctic. These ancient sites usually possess a much enriched moss flora centuries after they

Figure 10.14 Site of long-abandoned High Arctic hunting station at Krossfjorden, Spitsbergen. This site was used by Russian trappers (Pomors) probably in the early 18th century. The encampment with its residual wooden posts is clearly distinguished by the lush bryophyte growth, demonstrating the retention of mineral nutrients by the flourishing moss community.

have been abandoned due to the capacity of the bryophytes to retain nutrients (Figure 10.14).

10.3.1 Heavy metal pollution in the Arctic

The consequences of early human activities in relation to the historical pollution by heavy metals can be traced in the ice cores of Greenland. The enthusiasm of the Roman Empire (27 BC to AD 476) for the luxuries of liberal supplies of water by means of lead pipes can be traced to noticeable increases in lead in arctic ice. Analysis of the Greenland ice core covering the period from 3,000–500 BP shows that lead was deposited in Greenland ice at concentrations four times greater than the natural values from about 500 BC to AD 300.

A continuous history of lead pollution has also been traced in a small lake in Finland (62°N) based on the accumulation rate of total lead and the ratio of the stable isotopes (Pb-^{206}/Pb-^{207}) in the annually accumulated laminated sediments (*varves*) beginning from the time of the Roman Empire to recent times (Figure 10.15). The rate of lead accumulation collapsed in the 5th century and remained at or close to the background level throughout the Dark Ages up to the 11th century. After this, the accumulation rate of lead pollution began to increase and reached 1.2 mg m^{-2} per annum in AD 1420–1439. During five centuries, from AD 1420–1895, the average accumulation of pollution Pb was 2.6 mg m^{-2} per annum. The lead accumulation rate as found in Greenland then started to increase exponentially in the early 20th century. These results show that both Greek and Roman lead and silver mining together with smelting activities polluted the middle troposphere on a hemispheric scale two millennia ago, long before the Industrial Revolution (Meriläinen *et al.*, 2010). Cumulative lead fallout onto the Greenland Ice Sheet during these past eight centuries was at times as high as 15% of that caused by the massive use of lead alkyl additives in gasoline from the 1930s onwards.

Although there is no evidence as to whether or not this pollution was toxic to the arctic biota it nevertheless puts into perspective the levels of heavy metal pollution that have been reached in modern times.

The continuing rise in the size of the world's human population makes it almost inevitable that

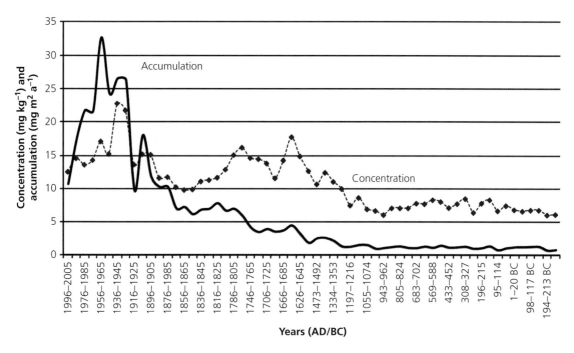

Figure 10.15 Concentration (mg kg^{-1}) and accumulation rate (mg m^{-2} a^{-1}) of total Pb (partial dissolution method) in the sediment of Lake Korttajärvi from 256 BC to AD 2005. Note the non-linear time scale. (Redrawn from Meriläinen et al., 2010).

there is hardly any area on Earth that is not disturbed in the search for resources. Even the remotest parts of the Arctic that were once considered uneconomic for exploitation are now being explored in the ever-growing demand for oil and minerals (Figure 10.16). The actual mining of the resources may not automatically lead to pollution, if properly carried out; nevertheless the transporting of oil and ore can interfere with migration routes and bring about considerable ecological disturbance.

When ore is being processed smelting can be highly damaging to the local vegetation as well as human health. In the Canadian and especially the Russian regions of the circumboreal zone, large-scale smelting of nickel and copper ores causes considerable damage to the vegetation by the emission of noxious gases, mainly SO_2 and nitrogen oxides, together with heavy metals. Adding to this, the dumping of mining waste and urbanization have inevitably caused a loss of landscape and habitat diversity (Toutoubalina and Rees, 1999).

A particular problem is the vastness of the Arctic and the enormous distances that are covered by oil pipes. Russia alone has 300,000 km of pipelines which makes checking for leaks a difficult process.

With the development of remote sensing techniques it is now possible to survey the extent of the damage that is being caused not only by toxic emission but also by the spillage of oil on frozen ground (Rees and Williams, 1997; Rees, 1999). The application of remote sensing to vegetation surveys uses multispectral visible and far-red radiation which has made it possible to develop an index that is suitable for sensing photosynthetic activity. Areas in which there is a noted absence of photosynthetic activity as measured by this index—the *Normalized Vegetation Index* (NDVI)—indicates vegetation damage.

From such surveys accompanied by subsequent ground studies it was found that in the vicinity of the mining and smelting complex at Noril'sk in north-central Siberia, the concentrations of heavy metals such as copper (Cu), nickel (Ni), and cobalt (Co) in the upper 10 cm of the soil layer exceeded the background level by a factor of 10–1,000 at distances of up to 40 km from the pollution source (Figure 10.17).

Figure 10.16 Site of a potential molybdenum mine near Mesters Vig, north-east Greenland 72 degrees N 20 degrees west. Greenland is rich in metal ores. This particular site due to its remoteness was not considered economic. Numerous other sites which would not have been economic in the past are now being actively mined for iron, lead, zinc, and other metal ores.

The sensitivity of tree species to pollutants was found to decrease from larch, to spruce, to birch, and then to willow. In the larch-dominant forests mortality was observed at 80–100 km distance from the source, with forest damage being observed at distances of up to 200 km from the smelters along the prevailing wind directions (Toutoubalina and Rees, 1999). Similar situations have been found for other heavy metals. Elements of highest concern include arsenic, cadmium, cobalt, chromium, copper, mercury, manganese, nickel, lead, tin, and thallium. Although some of these elements are metabolically necessary in minute amounts (cobalt, copper), others are carcinogenic or toxic.

The accumulation of heavy metals into arctic ecosystems is therefore a matter of concern for the well-being of local plant, animal, and human inhabitants and particularly those working and living in the mining areas. A circumpolar survey of heavy metals in the liver and kidneys of Willow Ptarmigan (*Lagopus lagopus*) revealed considerable variations in cadmium (Cd) content in Canada and Scandinavia. Some Canadian locations had exceptionally high levels, several birds having > 50 mg kg^{-1} in liver and > 400 mg kg^{-1} in kidneys (Pedersen *et al.*, 2006). Cadmium is also very injurious to human health when ingested, causing immediate poisoning and damage to liver and kidneys.

10.3.2 Oil and bitumen (tar sands) pollution

Fears have long been expressed in relation to the risks to arctic ecosystems from oil and tar sands extraction. Spillage is not infrequent as a result of shipping accidents with disastrous results for coastal regions. Accidents from oil on land, on the other hand, appear to be particularly common in the Russian Tundra. The greatest damage occurs from pipe leakage in the Russian North where permafrost is often present and pipelines are subject to corrosion and cryogenic processes with a consequent risk of rupture. A general lack of enforcement of environmental protection legislation is also a contributing factor. It has been estimated that every year, up to one-fifth of the total Russian oil production has been lost through leakage.

One particularly well-documented incident took place on the 26th September 1994 near the city of Usinsk, Komi Republic (66°30′N) in the Kolva River basin (Figure 10.18). Estimates of the amount of spillage vary, but a final assessment given by the Russian Roskomgidromet (Russian Mineralogical Society Centre) in December 1994 was 103,000–126,000 tonnes (Vilchek and Tishkov, 1997). Most of the leaked oil accumulated in bogs, but significant amounts also entered the Kolva and Usa Rivers.

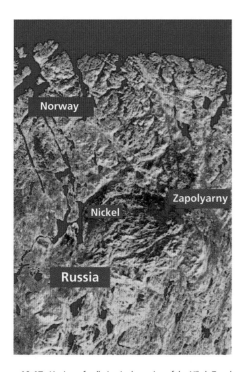

Figure 10.17 Air view of pollution in the region of the Nikel–Zapolyarny region as seen in 1991. Light yellow and brown indicates lichen-dominated heathland and woodland vegetation types; green, dwarf shrub dominated forests and heaths; violet, barren and damaged environments (air pollution) around the smelters in Nikel and Zapolyarny (Rees and Danks, 2007).

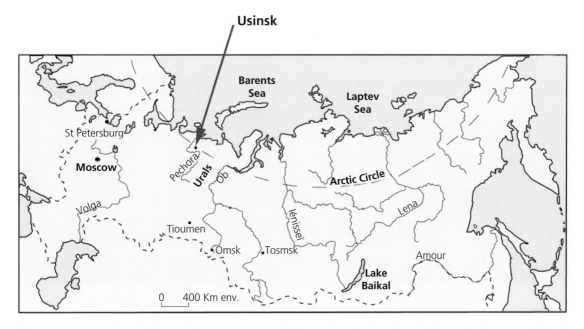

Figure 10.18 Location of Usinsk where possibly over 100,000 tonnes of oil were spilt in 1994. (Map adapted with permission from Gaye *et al.*, 2004).

This pipeline was no stranger to rupture. A list published in the Russian Newspaper *Izvestiya* on the 29th October 1994 reported a list of spillages and fires at pumping stations between 1986 and 1992. Arctic ecosystems have also been prone to pollution from other organic pollutants. Particularly toxic are the polycyclic aromatic hydrocarbons (PAHs) and polychlorinated biphenyls (PCBs) despite a ban on the production of the latter that was introduced in the USA in 1979. Electrical equipment is the most important source of PCBs due to leaks from operating installations.

An exacerbating cause of pollution comes from the incineration of these compounds. These and other persistent organic pollutants (POP) cause long-term adverse consequences for fish. The types of histopathological lesions observed in fish samples from the Pechora River indicate both a direct response to the toxic effects of contaminated water and sediments and secondary stress effects caused by factors such as parasitism (Hofgaard *et al.*, 2010; Lukin *et al.*, 2010).

The natural northern taiga landscapes of this region are normally rich in biological resources, including reindeer pasture and commercial fur animals such as sable, squirrel, muskrat, and elk. However, these have now been largely eliminated in the neighbourhood of the Vozei and Usinsk settlements. Nevertheless, rich hayfields and pastures are still to be found along the Kolva, Usa, and Pechora Rivers. The smaller tributary rivers have significant fish resources due to active management and preservation measures. However, the more valuable commercial fishing on the Pechora River itself was reduced to less than a third of its former levels (Vilchek and Tishkov, 1997).

Nearly two decades after the Usinsk oil spill, the Pechora River, one of the major rivers of the Arctic

Figure 10.19 The mouth of the Cree Creek Delta in the Peace-Athabasca Delta has become an area of active landscape evolution and vegetation change. Headwater capture by Cree Creek of a portion of the flow of the Embarras River has brought increased river discharge, flooding, and sediment deposition to the area since 1982. Date: 17th September 2005. (Photo courtesy of Dr K.P. Timoney).

Ocean drainage basin, had not yet fully recovered. The Pechora River originates from the western slope of the northern Urals. Its course extends over 1,800 km to its inland basin, covering two-thirds of the territory of the Komi Republic and a considerable part of the Nenets Autonomous District of the Archangelsk region. The Pechora River is very important both for terrestrial and aquatic biodiversity in the region, and for the reproduction of the most valuable fish species of the European North with 36 fish species from 13 families.

Tar sands pollution

Just as visible and damaging to the environment is the spread of pollutants from tar sands extraction. In Canada the development of a rapidly growing tar sands extraction industry in the northern boreal region is causing toxic pollution to rivers and lakes in regions rich in wildlife.

The Peace-Athabasca Delta (Figure 10.19) is a Ramsar wetland of international importance located at the west end of Lake Athabasca in north-eastern Alberta (Canada). It is the largest boreal delta in the world, and the varied habitats found in the delta include shallow lakes, mudflats, marshes, wet meadows, fens, and forest that support at least 219 species of birds and 364 species of flowering plants, and it is a staging migration site for large numbers of ducks, geese, swans, terns, gulls, and shorebirds (Figure 10.18). One-third of the existing population of the remaining free-roaming Wood Bison (*Bos bison athabaskae*) also lives in the Delta (Timoney, 2009).

The area is now threatened with pollution from the rapid development of open pit bitumen extraction

Figure 10.20 Extensive open pit bitumen mining is being conducted along the lower Athabasca River, upstream of the Peace-Athabasca Delta in northern Alberta, Canada. The environmental effects of bitumen development, mining, processing, transportation, and usage have become a focus of international concern. Date: 10th August 2006. (Photo courtesy of Dr K.P. Timoney).

(Figure 10.20). Mass mortality events take place among both resident and migratory birds when they land in the tailings ponds (Timoney and Ronconi, 2010). Athabasca River bitumen upgraders are also sources of dangerous airborne emissions, as found in the toxic concentrations of mercury and nickel which also pollute the Athabasca River (Kelly *et al.*, 2009).

10.3.3 Oil spill detection

Such is the frequency and often difficulty of immediate detection of oil spills that considerable attention has been given to the use of satellite remote sensing to monitor the state of the oil pipes on frozen ground.

In northern regions, where the days are short for a lengthy period of the year, normal optical surveillance by satellites is severely restricted. An alternative method of monitoring the state of the pipelines has, however, been tried using satellite radar in surveying the 1994 oil spill at Usinsk (Gaye *et al.*, 2004). Apart from being independent of the hours of daylight the methods appears to have the advantage of being able to detect the presence of oil both on the snow surface and also when it is covered by fresh dry snow (Figure 10.21). However, when covered by wet snow the oil spill is practically undetectable. Nevertheless, when conditions permit, radar imaging appears to be a method that is very well adapted for following the progressive spread of fresh oil spills.

10.3.4 Persistent organic pollutants

The groups of compounds described as persistent organic pollutants (POPs) are long-lasting, bioaccumulative, and toxic. In addition to persistent bioaccumulative and toxic pollutants (PBTs, *toxic organic micro pollutants* (TOMP) are long-lived organic compounds, usually lipids, that become concentrated as they pass upwards through the food chain. Field observations of population declines amongst the Tundra wildlife have long been suspected as being linked with POP exposure. Implicated compounds include aldrin, chlordane, DDT, dieldrin, endrin, heptachlor, polychlorinated biphenyls, and many more. In the human population they can cause a range of health disorders including injury to endocrine, reproductive, and immune systems. There are many well-known examples of bird populations that have been affected by POP exposure. Decreased or retarded egg production, increased embryo mortality, eggshell thinning, embryonic deformities, growth retardation, and reduced egg hatchability are among the effects reported. It is not just in Russia that the effects of pollution are found as a result of extraction of natural resources from the Arctic. An investigation carried out at Cape Vera, Devon Island (76°N) in the Canadian Arctic has traced the accumulation of pollutants from the sea and their transfer to the terrestrial environment (Choy *et al.*, 2010). It was shown that Northern Fulmars (*Fulmarus glacialis*) (Figure 8.6) concentrate POPs and transfer these contaminants from the marine to the terrestrial environment through their guano. The levels of the contaminants (Table 10.2) were measured in five species in this neighbouring land—Jewel Lichen (*Xanthoria elegans*), Worm Lichen (*Thamnolia vermicularis*), Northern Collared Lemmings (*Dicrostonyx groenlandicus*), Snow Buntings (*Plectrophenax nivalis*), and Ermine (*Mustela erminea arctica*). Snow Buntings (Figure 10.22) were the most contaminated species at Cape Vera by lipid and wet weight, followed by ermine, lichen, and lemmings.

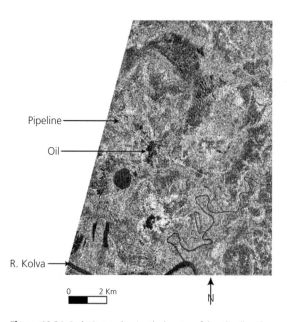

Figure 10.21 Radar image showing the location of the oil spill on the 29th September 1994 near Usinsk. (Redrawn from Gaye *et al.*, 2004).

Table 10.2 Summary of mean and standard errors of stable isotope and contaminant data for organisms collected from Cape Vera. (Reproduced with permission from Choy et al., 2010).

Species	Number	Percentage lipid	δ15N (‰)	ΣDDT (ng/g ww)	ΣPCB (ng/g ww)	PCB-153 (ng/g ww)
Worm Lichen	11	4.19	4.14 ± 0.88	0.31 ± 0.07	0.94 ± 0.14	0.12 ± 0.03
Jewel Lichen	14	1.69	10.37 ± 1.19	2.04 ± 0.60	3.28 ± 0.93	0.54 ± 0.21
Collared Lemming	5	2.36	13.7 ± 1.04	0.71 ± 0.32	0.76 ± 0.37	0.11 ± 0.07
Snow Bunting	18	2.91	18.1 ± 0.73	105.9 ± 210.2	168 ± 43.1	63.9 ± 16.8
Ermine	2	6.5	20.3 ± 0.18	37.3 ± 8.72	53.8 ± 12.9	18.4 ± 5.44

Figure 10.22 Adult Snow Bunting (*Plectrophenax nivalis*) in breeding plumage. This species is the most northerly breeding passerine. Even at northerly latitudes this terrestrial bird accumulates organic pollutants from the ocean through their transfer to land by fulmars. (Photo courtesy of Clare Kines).

Snow Buntings feed primarily on chironomids whose larval stages are aquatic. It was suggested therefore that it is the guano–sediment–chironomid pathway that is the main route for PCB contamination. This species is the most northerly breeding passerine. Even at such high latitudes this terrestrial bird accumulates organic pollutants from the ocean through their transfer to land by fulmars (Choy et al., 2010).

10.4 Radioactivity in the Arctic

Radioactive contamination of the Arctic has occurred at two different scales. The first is the widespread aerial contamination that resulted from global nuclear weapons testing, followed by the accident at the Russian nuclear reactor at Chernobyl. The second is oceanic contamination from the British Nuclear Energy site at Sellafield. For terrestrial ecosystems a major source of pollution is from aerial deposition. The rate of interception of aerially deposited radionuclides varies with surface characteristics and is particularly high for lichens and mosses. As discussed, in regions with deep soils precipitation can remove the radioactivity to below the main rooting zone of much of the vegetation. There has been detailed monitoring of ^{137}caesium and ^{90}strontium, as these radionuclides are important for determining the dose to human populations and have a half-life of approximately 30 years.

The two major sources of fallout in the Arctic region have been nuclear weapons testing and the Chernobyl accident. A total of 520 atmospheric nuclear weapons tests have been carried out, of which 88 took place in the Arctic on the Russian island of Novaya Zemlya. Radioactive contamination has also come from worn-out reactors leaking radioactive waste that has contaminated some northern Russian river basins, principally the Ob, Pechora, and northern Dvina. Following the cessation of widespread atmospheric weapons testing in the early 1960s, other sources, such as releases from European nuclear fuel reprocessing plants, increased in relative importance. However, in areas with shallow soils, because the entire soil profile is occupied by roots there is a continual recycling of the contaminants into the above-ground vegetation, which maintains the exposure of herbivores to the radioactivity. Thus in Arctic food chains the ability of lichen to intercept, absorb, and retain most of the deposited radiocaesium is particularly serious because of their utilization by reindeer (Figure 10.23).

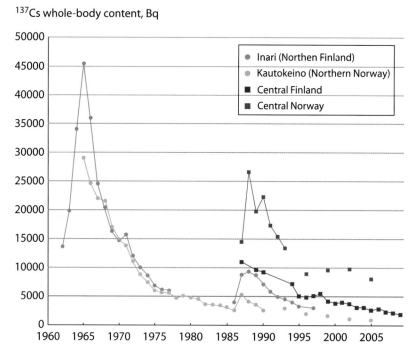

Figure 10.23 Radiocaesium (^{137}Cs) whole-body content averaged for male and female reindeer herders in northern Finland and northern and central Norway. (Reproduced with permission from Symon, 2010).

The levels of artificial radionuclides in the Arctic attained maximum values during the period 1950–1970, primarily as a consequence of atmospheric nuclear weapons testing. A second, but lower, peak in fission product radionuclides occurred in the Arctic marine environment in the early 1980s as a consequence of the peak in the rates of radionuclide discharge from Sellafield in the mid-1970s. Finally, fallout from the Chernobyl accident in 1986 made an additional contribution to radionuclide contamination of the Arctic. Since then, the levels of radionuclides have been in a general but not ubiquitous decline.

The major contribution to radiation doses for human Arctic residents comes from radionuclides originating from past nuclear weapons explosions in the atmosphere giving rise to global fallout. However, in some geographically limited but populated areas of the Arctic (Fennoscandia and western Russia), a substantial dose contribution has been made by additional fallout from the Chernobyl reactor accident (Figure 10.23). This contribution to the dose to Norwegian and Swedish Arctic residents was, and continues to be, reduced through the application of counter-measures based on diet regulation.

Arctic residents whose diets comprise a large proportion of traditional terrestrial and freshwater foodstuffs receive the highest radiation dosages, with the actual amounts depending on the rates of consumption of locally derived foodstuffs, including reindeer/caribou, freshwater fish, goat's cheese, berries, mushrooms, and lamb. In contrast, Arctic residents having diets largely comprising marine foodstuffs receive comparatively low radiation exposures because of the lower levels of contamination of marine organisms (Figure 10.23). For reindeer herders and others consuming comparatively large quantities of caribou/reindeer meat, the dominant pathway of natural radiation exposure is the intake of Polonium ($_{210}$Po, which has a half-life of 138 days) through caribou/reindeer meat consumption.

The highest time-integrated radiation exposures to the arctic human population from global fallout occurred in Canada, and the lowest in Greenland.

The variations in individual dose distributions are not primarily due to geographical heterogeneities in radionuclide fallout. Rather, they result from variations in diet among Arctic residents. Indigenous peoples comprise a relatively high proportion of the inhabitants of Arctic Canada, some of whom rely comparatively heavily on caribou as a source of food. In contrast, the population of Greenland is confined to coastal areas and has a diet containing a comparatively large proportion of marine foodstuffs having low radionuclide contamination (Strand, 1999).

Monitoring of radionuclides in the atmosphere in Finland and in seawater near Greenland and the Faroe Islands shows that traces of atmospheric weapons tests in the 1950s and 1960s are still detectable but have declined over time. Air monitoring data from Canada have highlighted how some of the fallout that has been incorporated into vegetation can be re-released into the environment through forest fires. Levels of ^{137}Cs exceeding detection limits have been shown to coincide with summer forest fires. Other data show that in spite of the peak of weapons testing having taken place over 50 years ago, the radiocaesium from the fallout is still retained in the top layer of soil in many areas, including notably Iceland. This is because processes that would normally favour mobility are slower in colder environments. Past fallout is thus likely to remain a source of radioactive contamination for grazing wildlife and for local human populations. The 1986 accident at the Chernobyl nuclear power plant added further radiocaesium to the environment, even though fallout in the Arctic was much less than further south in Fennoscandia and near Chernobyl. The additional contamination is seen as small peaks in the atmospheric record as well as in the monitoring data for deposition and levels in vegetation and food products such as milk and meat. Monitoring data have been used to estimate the effective ecological half-lives of radionuclides in different environments and food webs.

A food chain of major importance in the Arctic is from lichen to reindeer/caribou to people. Long-term studies in Scandinavia after the Chernobyl accident show that the effective ecological half-life for radioactive contamination in reindeer has increased from about 3 years shortly after the accident to 8–9 years. Internal human contamination shows the same trend. In general, the levels of anthropogenic radionuclides in the Arctic seawater are low, although they vary according to distance from sources and annual discharge rates. Some radionuclides may, however, concentrate in biota, as illustrated by the elevated levels of Technetium-99 (99Tc), an isotope of technetium which is common in seaweed along the Norwegian coast and decays with a half-life of 211,000 years. Concentrations of 99Tc in fish species are generally low and slowly decreasing with time. For seabirds, seals, and whales which prey mostly on fish, there is a higher concentration of radionuclides in kidney and liver compared to muscle for the natural radionuclides ^{210}Polonium and ^{210}lead (Symon, 2010).

10.5 Conclusions

The Arctic is sensitive to disturbance and the landscape easily scarred, as seen through human eyes. Nevertheless, this does not necessarily imply that it is automatically rendered inhospitable to the native flora and fauna. There is a case for considering the arctic biota as being well adapted to physical instability and climatic variation. Reproductive plasticity, gene exchange between adjacent populations, and certain morphological and physiological adaptations provide striking and visible evidence that many species in the Arctic are able to respond and even profit from the physical fragility and uncertainty of their environment. In plants clonal reproduction by ramet fractionation and movement is actually facilitated as a result of disturbance and cryoturbation. The high proportion of clonal species, combined with great longevity of individual plants, provides a biological answer to perturbation and produces a system that benefits from disturbance as it creates new opportunities for dispersal and colonization. As animal populations fluctuate, so are plant populations subjected to periods of intense grazing followed by periods of respite, when the animal populations eventually crash. The capacity of plant and animal populations to recover, often with great rapidity, is another testimony to the resilience of most arctic species to periods of adversity. Periods of over-grazing can be followed by renewed vigorous growth, as nutrients are recycled and the shade of old vegetation is removed and the summer sun is not impeded from melting the frozen soil. Being able to survive the uncertainties of life in the Arctic protects its adapted inhabitants from competition from more aggressive species that require a predictable environment and are unable to tolerate limited resources. The environment is also capable of eventually being cleansed from the worst effects of pollution. Whether this built-in resilience will be adequate to restore degraded ecosystems after persistent and severe human intervention is an unanswered problem that requires continued monitoring and investigation.

10.6 Coda—the future

Among the many doubts and uncertainties about the future of the Arctic, it is obvious that enormous differences exist between marine and terrestrial environments. Whatever fate may eventually overtake the Polar Bears, it is inevitable that the melting of the oceanic ice will make a fundamental difference to their way of life. The terrestrial environments of the tundra and the taiga are entirely different. Until very recently the sun was already declining in the sky before the snow and ice had retreated from its winter expanse. With climatic warming, spring is now coming earlier and the arctic lands are capturing significantly more energy (Figures 10.24–10.26). Plants begin to grow sooner, flowering takes place earlier, and the length of the growing season is extending (Goetz et al., 2005). Already results from a two-decade summer warming experiment in an Alaskan Tundra ecosystem have shown that warming has increased plant biomass and woody plant dominance (Sistla et al., 2013).

If these changes continue, these northern lands will provide more copious forage for lemmings, voles, and geese as well as improving grazing for

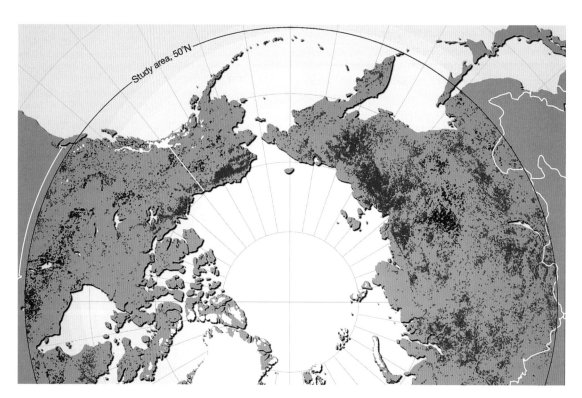

Figure 10.24 Trends in productivity derived from a 1982–2005 time series of vegetation indices (NDVI). Significant positive trends, shown as green, indicate an increase in both peak productivity and growing season length. Negative trends, showing as red, represent forested areas that declined in productivity and had not been recently disturbed by fire. (Reproduced with permission from CAFF, 2010).

Figure 10.25 Kolossen, a north-east Greenland nunatak (73°N)—a land awaiting new inhabitants.

Figure 10.26 Purple Saxifrage (*Saxifraga oppositifolia*), an ancient species with a bright future.

the larger herbivores such as muskoxen and reindeer. In the more southern parts of the Arctic, human disturbance and increasing parasite infections may cause problems. However, if suitably encouraged, reindeer might even recolonize Franz Josef Land and muskoxen should become more numerous. An improvement in conditions for herbivores could provide more prey for wolves, foxes, and raptors.

The brown bears of the taiga will have a greater supply of berries and other forage which may eventually enable them to move further north, where they may begin to hybridize and possibly have an effect in increasing bear diversity, thus aiding the Polar Bears to recover some of their ancestral characteristics and perhaps even the dentition needed to vary their present entirely carnivorous diet. In terms of past climatic history, an ice-free Arctic Ocean has been the norm. Such changes will not destroy the Arctic and, as long as the Earth rotates with approximately its present angle to the ecliptic, the polar night will persist and preserve the uniqueness of the polar scene.

If such views seem too optimistic and resemble *a first blast of the trumpet against the monstrous regiment of arctic pessimists* (cf John Knox, 1558) they should be read only as one possible outcome of climatic warming in relation to the terrestrial Arctic. In the more southerly regions of the planet our vested interests in the needs for human survival will meet political conflicts, which will create many other problems.

Glossary

Aapa mire (Finnish—open mire) A cold-climate mire with ridge and pool surfaces found in northern Scandinavia and boreal Canada. In North America more commonly called *string bogs*. They are soligenous mires with ridges arranged roughly normal to the slope, along the contours of the terrain. Water occupies the linear depressions between the ridges.

Acclimation A response by an animal or plant that enables it to tolerate a change in a single factor (e.g. temperature) in its environment. The term is applied most commonly to plants and animals used in laboratory experiments and normally implies a change in only one factor.

Acclimatization A reversible, adaptive response that enables animals or plants to tolerate environmental change (e.g. seasonal climatic change) involving several factors (e.g. temperature and availability of food). The response is physiological but may affect behaviour (e.g. when an animal responds physiologically to falling temperatures in ways that make hibernation possible, and behaviourally by seeking a nesting site, nesting materials, and food. With plants, acclimatization can involve gradual hardening to low temperatures by gradual exposure to falling temperatures.

Acrotelm/catotelm The upper layer of a peat bog, the *Acrotelm*, in which organic matter decomposes aerobically and much more rapidly than in the underlying, anaerobic *catotelm*. Derived from Greek *acro*—upper; *cato*—lower; *telm*—marsh.

Active layer In environments containing permafrost, the active layer is the top layer of soil that thaws during the summer and freezes again during the winter.

Albedo The proportion of incident light that an object reflects. From Latin *albedo*—whiteness. Commonly used in ecology to described the reflectivity of water, land, or vegetation.

Amplified fragment length polymorphism polymerase chain reaction (AFLP-PCR) Variation in DNA sequences detected by AFLP-PCR, involves digestion of DNA using restriction enzymes, followed by ligation of restriction fragments with specific adaptors to enable selective PCR amplification of certain fragments, and resolution of differences due to presence or absence of restriction site-specific sequences.

Autochthonous Describing an organism that is native to the place in which it is found.

Beringia A loosely defined region surrounding the Bering Strait, the Chukchi Sea, and the Bering Sea. It includes parts of Chukotka and Kamchatka as well as Alaska.

Biomolecular Pertaining to molecules that are produced by a living organism including large macromolecules such as proteins, polysaccharides, lipids, and nucleic acids.

Brown fat Brown adipose tissue (BAT) or *brown fat* is one of two types of fat or adipose tissue (the other being white adipose tissue) found in mammals. It is especially abundant in newborns and in hibernating mammals, where its primary function is to generate body heat without resorting to shivering. See also *thermogenesis*.

Caecotrophy The passing of food through the alimentary canal twice. Rabbits and some rodents take soft faecal pellets directly from the anus at night and store them in the stomach, to be mixed with food taken during the day, probably obtaining essential metabolites that are produced in the caecum. The animals will die if they are prevented from ingesting these pellets.

Calibrated carbon dates ^{14}C dates may be expressed as either uncalibrated or calibrated years. A raw ^{14}C BP date cannot be used directly as a calendar date, due to variations in cosmic radiation, the changing influences of substantial reservoirs of carbon in organic matter including the oceans, and sedimentary rocks. Changes in the Earth's climate can also affect the carbon flow between these reservoirs and the atmosphere, leading to changes in the atmosphere's ^{14}C fraction.

Cavitation This takes place in the xylem of vascular plants when the tension of water rises to an extent that liquid water (or sap) vapourizes locally from dissolved air within the water expanding to fill and create bubbles in the the vessel elements or tracheids. Plants can normally repair cavitated xylem provided that root pressure is sufficient to redissolve the air, or by the importation of solutes into the xylem from neighbouring tissues. In some plants the sound of the cavitation is clearly audible and can be heard as a series of clicks at times when the rate of evapotranspiration is high.

Cenozoic Era The third of the major eras of the Earth's history, beginning about 65.5 million years ago and extending to the present.

Clade A group consisting of an ancestor and all its descendants. The descendants can be shown as branches. From the Greek *Klados* meaning a twig or branch.

Coprophagy Eating of dung, or faeces, apparently instinctive among certain members of the order Lagomorpha (rabbits and hares) where it can be an essential part of their diet.

Corticular photosynthesis Photosynthetic activity in plant stems which contain chlorophyll. In illuminated stems and branches, CO_2 release is often reduced. Several light-triggered processes may contribute to this reduction, including photorespiration and photosynthesis.

Cosmogenic radionuclides Radionuclides are generated by the interaction of cosmic radiation with the atomic nuclei of the atmosphere or with extraterrestrial matter coming down to Earth as meteors or meteorites. Radionuclides are also generated by interaction of cosmic radiation with exposed rocks. In recent years this has provided a powerful tool for dating rock exposure and in determining when rock surfaces re-emerged from ice-cover with the passing of the Pleistocene era.

Cretaceous The third of the three periods that comprise the Mesozoic era. The Cretaceous lasted from 145.5–65.5 million years ago; its end is defined by the mass extinction of many invertebrate and vertebrate stocks associated with a bolide impact. The period is noted for the deposition of the chalk of the White Cliffs of Dover, England, much of the chalk being derived from the calcareous plates (coccoliths of marine algae). Angiosperms, which arose during the Jurassic period, came to dominance during the Cretaceous at the expense of such groups as cycads and pteridosperms. Woody species evolved during the Cretaceous period, and around 65 million years ago stratified forests appeared, with an increase in the variety of fruits.

Cryosols Soils that have a permanent ice-layer within 100 cm of the surface. Cryosols are a reference soil group in the FAO soil classification.

Cryoturbation Refers to the mixing of materials from various horizons of the soil down to the bedrock due to freezing and thawing. Also described as *frost churning*.

Eocene This epoch, lasting from about 56–34 million years ago, is a major division of the geological time and the second epoch of the Paleogene period in the Cenozoic era. The Eocene spans the time from the end of the Palaeocene to the beginning of the Oligocene epoch.

Fluviosols Young soils in an alluvial plain, of which there are many types. Frequently found in conjunction with Gley soils. Gley (Russian dialect *глей* for clay (standard Russian глина—*glina*—clay) is a wetland soil which, unless drained, is saturated with groundwater, developing a characteristic mottled-blue colour due to the presence of reduced ferrous iron.

Gelisoils An order defining cold climate soils with permafrost in the soil taxonomy system of the United States Department of Agriculture.

GISP2 ice core The Danish, Swiss, and American Greenland Ice Sheet Project (GISP) was a decade-long enterprise to drill ice cores in Greenland. On 1st July 1993, after 5 years of drilling, the Greenland Ice Sheet Project Two (GISP2) penetrated through the ice sheet and 1.55 m into bedrock, recovering an ice core 3,053.44 m in depth, the deepest ice core recovered in the world at the time. Together with the completion of this feat, a companion European ice coring venture (the Greenland Ice core Project (GRIP), took place 28 km to the east of GISP2. There are now two available records of environmental change covering *ca.* 110,000 years.

Hemicellulose A polysaccharide found in the cell walls of plants. The branched chains of this molecule bind to cellulose microfibrils, together with pectins, forming a network of cross-linked fibres.

Herbivore A general term for animals that consume living plant material. It can be more specifically subdivided. (1) *Browsing,* a form of herbivory in which a herbivore feeds on leaves, shoots, fruits, and twigs, as well as taller woody plants such as shrubs and trees. (2) *Grazing*, which is usually used to describe herbivores feeding on graminoids and other low-growing vegetation.

Holocene The geological epoch that began at the end of the Pleistocene around 12,000–11,600 or 11,200 BP and continues to the present.

Hypsithermal *Xerothermic* or *Climatic Optimum*. The warmest period so far of the Holocene. At high latitudes the summer temperature appears to have been 4°C above present temperatures, while in the tropics it was no more than 1°C. The timing was also variable. In south-western Saskatchewan the Hypsithermal period lay between 6,400 and 4,500 BP, while in north-west Montana the equivalent warm dry period lasted from 10,850–4,750 BP. In western Norway, the Hypsithermal is dated to between 8,000 and 6,000 BP when Hazel (*Corylus avellana*) reached its most northerly Scandinavian extension.

Iteroparity Condition of an organism that has more than one reproductive cycle in its lifetime.

Kilohertz Units for measuring sound frequency in terms of thousands of cycles per second. To the human ear this is sensed as pitch, which in adult human beings is normally audible between 20 and 16,000 Hertz.

King Alfred Alfred the Great (Old English Ælfrēd, Ælfrĕd, *'elf*—counsel' (AD 849–899), King of Wessex from AD 871–899.

Krummholz From German *krumm*—twisted, *holz*—wood, most commonly used to describe the misshapen trees at the timberline which result either from a phenological response to a harsh environment or from a genetic selection in favour of the crippled form.

Krüppelholz From the German *krüppel*—crippled. Hence *Krüppelholz* means crippled wood. Sometimes used to specifically denote stunted trees in which this form is genetically fixed.

Lesotundra—see Tundra-Taiga Interface.

Life strategies A general term used to describe ecological concepts originally developed in relation to reproduction and survival. An r-strategist (where r stands for maximum increase) responds swiftly to favourable conditions with most of their energies devoted to rapid maturity and reproduction. k-selected plants survive by putting their energies into persistence (Grime, 2001). Other strategies include pollen riskers and seed riskers (Molau, 1993) where plants are classified in relation to the production early in the growing season of large quantities of pollen (pollen riskers), while seed riskers maximize seed later in the growing season.

Neoendemic An endemic species having a limited geographical range due to its recent origin.

North Atlantic/Arctic Oscillation The North Atlantic Oscillation (NAO) is a climatic phenomenon in the North Atlantic Ocean caused by fluctuations in the difference of atmospheric pressure at sea level between the Icelandic Low and the Azores High Pressure Zones. The east–west oscillation motions of the Icelandic Low and the Azores High control the strength and direction of westerly winds and storm tracks across the North Atlantic.

Palaeo-Eskimo The Palaeo-Eskimos were the peoples who inhabited the Arctic region, from Chukotka in the east across North America to Greenland in the west, before the advent of the modern Inuit and/or Eskimo and related cultures. The first known Palaeo-Eskimo cultures developed by 2500 BC but were gradually displaced in most of the region, with the last one, the Dorset culture, disappearing around AD 1500.

Palaeolithic The Palaeolithic, or Old Stone Age, is the longest period of human history. Until recently, the earliest archaeological evidence for a Palaeolithic culture based on the use of stoned tools was placed at 2.5 million BP. New evidence now suggests that the use of stone tools began much earlier, possible 3.5 million BP, with the evolution of various hominine species. The end of the Palaeolithic period is traditionally dated to 11,000 BP, coinciding with the end of the Ice Age (the Pleistocene) and the onset of the Holocene.

Palsas Palsa mires are sub-arctic mire complexes with permanently frozen peat hummocks and are common in parts of Fennoscandia, Russia, Canada, and Alaska.

Paludification The growth of bogs, from Latin *palus*—marsh, swamp, bog, or fen.

Parapatric A term to describe populations that have become separated but remain contiguous with geographical ranges that do not overlap. Gene flow between the populations may still be possible.

Parthenogenesis A form of asexual reproduction in which growth and development of embryos occurs without fertilization. In plants, parthenogenesis takes place from the development of an embryo from an unfertilized egg cell, and is a component process of apomixis.

Phenology The study of climate, including photoperiodic effects of changes in day length on seasonal occurrences of flora and fauna.

Photoperiodism The response of organisms to changes in day length.

Pingo An ice-cored dome-shaped hill, oval in plan, and from 2–50 m high and 30–600 m in diameter.

Pleistocene The first of two periods of the Quaternary geological epoch which lasted from about 2,588,000–11,600 years ago, spanning the world's recent period of repeated glaciations.

Plio-Pleistocene Refers to the geological period more recent than *ca.* 5 million BP, incorporating both the formally defined epochs of the Pliocene and the Pleistocene. Opinion is divided as to whether it should include the Holocene.

Podzolization The process of mobilization and precipitation of dissolved organic matter and minerals within a soil profile. Acid leaching by organic acids originating from raw humus removes aluminium and iron from upper to lower soil horizons leaving an eluviated bleached zone. This bleached zone was believed by Russian peasants to be the ash remains from past forest fires (Russian *pod*—underneath, *zola*—ashes). Such soils are called Spodosols in China and the USA.

Polychlorinated biphenyls (PCBs) Halogenated aromatic compounds, typified by the polychlorinated dibenzo-p-dioxins (PCDDs), dibenzofurans (PCDFs), biphenyls (PCBs), and diphenylethers (PCDEs). All are potentially toxic industrial compounds or byproducts which have been widely identified in the environment and in chemical-waste dumpsites.

Polycyclic aromatic hydrocarbons (PAHs) Ubiquitous environmental pollutants formed from both natural and anthropogenic sources. The latter are by far the major contributors. Natural sources include forest fires, volcanic eruptions, and degradation of biological materials, which has led to the formation of these compounds in various sediments and fossil fuels. Major anthropogenic sources include the burning of coal refuse banks, coke production, automobiles, commercial incinerators, and wood gasifiers. PAHs are neutral, non-polar organic molecules that comprise two or more benzene rings arranged in various configurations. Members of this class of compounds have been identified as exhibiting toxic and hazardous properties.

Polyploidy Where an organism has three (triploid—3x), four (tetraploid—4x), or more complete sets of chromosomes, instead of two as in diploids (2x). Very common in plants, but rarer in animals, though examples are known in insects, crustaceans, molluscs, fish, amphibians, reptiles, and mammals. *Allopolyploidy* is where the sets of chromosomes are derived from different parent species due to duplication of a hybrid's chromosome number. *Autopolyploidy* is where the organism has more than two sets of chromosomes, all of which were derived from the same species. *Endopolyploidy* is a process by which chromosomes replicate without the division of the cell nucleus, resulting in a polyploid nucleus. Also called *endomitosis*. *Neopolyploidy* Occurrence of a recently evolved polyploid species. *Palaeopolyploidy* is occurrence of an ancient polyploid species.

Pomors The Pomors (Russian *po more*—on the sea) were people who lived on the coast in the north of Russia and had a long tradition of hunting in the Arctic, including Spitsbergen. Whether or not their arrival in in Spitsbergen preceded that of the Dutchman Willem Barentsz in 1596 is still a matter of historical and archaeological argument.

Pre-adaptation This term is used to describe a situation where a species evolves to use a pre-existing structure or trait inherited from an ancestor for a potentially unrelated function.

Reactive oxygen species (ROS) Chemically reactive molecules containing oxygen. Examples include oxygen ions and peroxides. ROS are produced as a byproduct of normal aerobic metabolism. Under periods of physiological stress ROS levels can increase sufficiently to cause significant damage to cell structures. Cumulatively, this is known as oxidative stress. ROS are also generated by exogenous sources such as ionizing radiation.

Refugia In biology a *refugium* (plural *refugia*) is a location of an isolated or relict population of a once more widespread species. In anthropology, refugia often refer specifically to the Last Glacial Maximum, when some ancestral human populations may have been forced back to glacial refugia, forming small isolated survival pockets in the face of the continental ice sheets during the last ice age. Isolation by refugia and the resulting ice evolution can be due to climatic changes, geography, or human activities such as deforestation and over-hunting. Recent genetic studies have demonstrated that refugial areas have helped to preserve biodiversity during previous periods of major climate change, and therefore could do so again in the future.

Taiga The northern coniferous forest and the coldest terrestrial biome after the Tundra and the ice caps. The word has been taken from the Russian тайга and is of Turkik or Mongolian origin (cf Turkish *dağ*—mountain, Altai *tayya*—a forest-covered mountain).

Thermogenesis The process of heat production in organisms. It occurs mostly in warm-blooded animals and a few species of flowering plants particularly in the Araceae. In warm-blooded mammals futile respiration cycles generate heat by non-shivering thermogenesis from *brown fat*.

Thermokarst A feature of periglacial lands with irregular surfaces of marshy hollows and small hummocks formed as permafrost ice-rich soils thaw. Slumping terrain can also give rise to thermokarst lakes sometimes described as *thaw-lakes*.

Thúfur Cryogenic earth hummocks in periglacial environments. The name is derived from Icelandic *þufur* (cognate with English *tuft*).

Trypsin A serine protease produced by the pancreas and found in the digestive system of many vertebrates, where it hydrolyses proteins.

Tundra Arctic lands with frozen soils and devoid of trees. The word is most probably derived from an ancient Turkish word for frozen ground. The root exists in modern Turkish as *dondurma*—frozen.

Tundra-Steppe An arctic plant community sharing a codominance of both steppe and tundra species including prostrate shrubs, which was once widespread in the late Pleistocene and is now found only as an occasional relict in eastern Siberia and most notably on Wrangel Island.

Tundra-Taiga Interface The transition zone between Tundra and Taiga. In North America the *Tundra-Taiga Interface* is commonly referred to as the *Forest-Tundra Ecotone*, while in Russia, where this biome is particularly extensive latitudinally, it is called the *lesotundra* (Russian *les*—forest).

Vitrification In the biological sense this refers to the ultrarapid freezing of water in plant and animal tissues, giving the ice a glass-like structure with minimal growth of crystals.

(Sources—various, including Allaby, M. (2004) *The Oxford Dictionaries of Plant Sciences and Ecology*. Oxford University Press, Oxford).

References

Aanes, R., Saether, B.-E., & Øritsland, N.A. (2000) Fluctuations of an introduced population of Svalbard reindeer: the effects of density dependence and climatic variation. *Ecography*, **23**, 437–43.

Aanes, R., Saether, B.E., Smith, F.M., Cooper, E.J., Wookey, P.A., & Oritsland, N.A. (2002) The Arctic Oscillation predicts effects of climate change in two trophic levels in a high-arctic ecosystem. *Ecology Letters*, **5**, 445–53.

Abbott, R.J. & Comes, H.P. (2004) Evolution in the Arctic: a phylogeographic analysis of the circumarctic plant *Saxifraga oppositifolia* (Purple saxifrage). *New Phytologist*, **161**, 211–24.

Abbott, R.J., Chapman, H.M., Crawford, R.M.M., & Forbes, D.G. (1995) Molecular diversity and derivations of populations of *Silene acaulis* and *Saxifraga oppositifolia* from the high Arctic and more southerly latitudes. *Molecular Ecology*, **4**, 199–207.

Abbott, R.J., Smith, L.C., Milne, R.I., Crawford, R.M.M., Wolff, K., & Balfour, J. (2000) Molecular analysis of plant migration and refugia in the Arctic. *Science*, **289**, 1343–6.

Adamczewski, J.Z., Florkiewicz, R.F., & Loewen, V. (2003) *Habitat Management in the Yukon Winter Range of the Little Rancheria Caribou Herd*. Department of Environment, Fish and Wildlife Branch, Yukon Territory.

Aiken, S.G., Dallwitz, M.J., L.L. Consaul, L.L., et al. (1999, onwards) *Flora of the Canadian Arctic Archipelago: Descriptions, Illustrations, Identification, and Information Retrieval*. Version 29 April 2003. http://www.mun.ca/biology/delta/arcticf/.

Alexandrovna, V.D. (1980) *The Arctic and Antarctic: Their Division Into Geobotanical Areas*. Cambridge University Press, Cambridge.

Alm, T. (1993) Ovre Aeråsvatn—palynostratigraphy of a 22,000 to 10,000 BP lacustrine record on Andøya, northern Norway. *Boreas*, **22**, 171–88.

Alm, T. & Birks, H.H. (1991) Late Weichselian flora and vegetation of Andøya, northern Norway—macrofossil (seed and fruit) evidence from Nedre Aeråsvatn. *Nordic Journal of Botany*, **11**, 465–76.

Alsos, I.G., Eidesen, P.B., Ehrich, D., *et al.* (2007) Frequent long-distance plant colonization in the changing Arctic. *Science*, **316**, 1606–9.

Alsos, I.G., Ehrich, D., Thuiller, W., *et al.* (2012) Genetic consequences of climate change for northern plants. *Proceedings of the Royal Society B-Biological Sciences*, **279**, 2042–51.

Andersson, T., Forman, S.L., Ingolfsson, O., & Manley, W.F. (2000) Stratigraphic and morphologic constraints on the Weichselian glacial history of northern Prins Karls Forland, western Svalbard. *Geografiska Annaler Series A—Physical Geography*, **82A**, 455–70.

Andreev, A.V. (1991) Winter adaptations in the willow ptarmigan. *Arctic*, **44**, 106–14.

Angerbjorn, A., Tannerfeldt, M., & Erlinge, S. (1999) Predator–prey relationships: Arctic foxes and lemmings. *Journal of Animal Ecology*, **68**, 34–49.

Armstrong, W. & Armstrong, J. (2005) Stem photosynthesis not pressurized ventilation is responsible for light-enhanced oxygen supply to submerged roots of alder (*Alnus glutinosa*). *Annals of Botany*, **96**, 591–612.

Arneborg, J., Heinemeier, J., Lynnerup, N., Nielsen, H.L., Rud, N., & Sveinbjornsdottir, A.E. (1999) Change of diet of the Greenland Vikings determined from stable carbon isotope analysis and C-14 dating of their bones. *Radiocarbon*, **41**, 157–68.

Asbjornsen, E.J., Saether, B.E., Linnell, J.D.C., Engen, S., Andersen, R., & Bretten, T. (2005) Predicting the growth of a small introduced muskox population using population prediction intervals. *Journal of Animal Ecology*, **74**, 612–18.

Bairlein, F., Norris, D.R., Nagel, R., *et al.* (2012) Cross-hemisphere migration of a 25 g songbird. *Biology Letters*, **8**, 505–7.

Balloux, F., Handley, L.J.L., Jombart, T., Liu, H., & Manica, A. (2009) Climate shaped the worldwide distribution of human mitochondrial DNA sequence variation. *Proceedings of the Royal Society B-Biological Sciences*, **276**, 3447–55.

Bannikova, A.A., Dokuchaev, N.E., Yudina, E.V., Bobretzov, A.V., Sheftel, B.I. & Lebedev, V.S. (2010) Holarctic phylogeography of the tundra shrew (*Sorex tundrensis*) based on mitochondrial genes. *Biological Journal of the Linnean Society*, **101**, 721–46.

Barber, D.C., Dyke, A., Hillaire-Marcel, C., et al. (1999) Forcing of the cold event of 8,200 years ago by catastrophic drainage of Laurentide lakes. *Nature*, **400**, 344–8.

Barboza, P.S., Peltier, T.C., & Forster, R.J. (2006) Ruminal fermentation and fill change with season in an arctic grazer: responses to hyperphagia and hypophagia in muskoxen (*Ovibos moschatus*). *Physiological and Biochemical Zoology*, **79**, 497–513.

Barlow, L.K., Sadler, J.P., Ogilvie, A.E.J., et al. (1997) Interdisciplinary investigations of the end of the Norse western settlement in Greenland. *Holocene*, **7**, 489–99.

Barnes, B.V. (1966) The clonal growth of the American aspen. *Ecology*, **47**, 439–47.

Barnosky, A.D., Koch, P.L., Feranec, R.S., Wing, S.L., & Shabel, A.B. (2004) Assessing the causes of Late Pleistocene extinctions on the continents. *Science*, **306**, 70–5.

Battley, P.F., Warnock, N., Tibbitts, T.L., et al. (2012) Contrasting extreme long-distance migration patterns in bar-tailed godwits *Limosa lapponica*. *Journal of Avian Biology*, **43**, 21–32.

Batzli, G.O., White, R.G., Maclean Jr, S.F., Pitelka, F.A., & Collier, B.D. (1980) The herbivore-based tropic system. In: *An Arctic Ecosystem* (eds J. Brown, P.C. Miller, L.L. Tieszen, & F.L. Bunnell), pp. 335–410. Dowden, Hutchinson & Ross, Stroudsburg, Pennsylvania.

Bauert, M.R. (1996) Genetic diversity and ecotypic differentiation in arctic and alpine populations of *Polygonum viviparum*. *Arctic and Alpine Research*, **28**, 190–5.

Beerling, D.J. (2007) *The Emerald Planet*. Oxford University Press, Oxford.

Bennike, O. (1998) Late Cenozoic wood from Washington Land, North Greenland. *Geology of Greenland Survey Bulletin*, **189**, 155–8.

Bennike, O. & Bocher, J. (1990) Forest-tundra neighboring the North-Pole—plant and insect remains from the Pliopleistocene Kap Kobenhavn Formation, North Greenland. *Arctic*, **43**, 331–8.

Bennington, C.C., McGraw, J.B., & Vavrek, M.C. (1991) Ecological genetic variation in seed banks. II. Phenotypic and genetic differences between young and old populations of *Luzula parviflora*. *Journal of Ecology*, **79**, 627–43.

Beringia Conservation Program (1999) *Wrangel Island: Russian Stronghold for Beringian Biodiversity*. Beringia Conservation Program, Anchorage, Alaska.

Bianchi, G.G. & McCave, I.N. (1999) Holocene periodicity in North Atlantic climate and deep-ocean flow south of Iceland. *Nature*, **397**, 515–17.

Bierzychudek, P. (1985) Patterns in plant parthogenesis. *Experimentia*, **41**, 1255–63.

Binford, L.R. (1978) *Nunamiut: Ethnoarchaeology*. Academic Press, New York.

Birch, S. & Young, I. (2009) Uamh an Claoaite. *Discovery and Excavation in Scotland*, **10**, 89–90.

Birkhead, T. (2012) *Bird Sense: What it's Like to Be a Bird*. Bloomsbury, London.

Birks, H.H., Giesecke, T., Hewitt, G.M., Tzedakis, P., Bakke, J., & Birks, J.B. (2012) Comment on 'Glacial Survival of Boreal Trees in Northern Scandinavia'. *Science*, **338**, 742.

Bliss, L.C. (1988) Arctic tundra and polar desert biome. In: *North American Terrestrial Vegeation* (eds M.G. Barbour & W.D. Billings), pp. 1–32. Cambridge University Press, New York.

Bliss, L.C. (1997) Arctic ecosystems of North America. In: *Polar and Alpine Tundra* (ed. F.E. Wielgolaski), pp. 551–683. Elsevier, Amsterdam.

Bliss, L.C. & Matveyeva, N.V. (1992) Circumpolar arctic vegetation. In: *Arctic Ecosystems in a Changing Climate* (eds F.S. Chapin III, R.L. Jefferies, J.F. Reynolds, G.R. Shaver, J. Svoboda, & E.W. Chu), pp. 59–89. Academic Press, San Diego.

Blix, A.S. (2005) *Arctic Animals: And Their Adaptations to Life on the Edge*. Tapir Academic Press, Trondheim.

Blix, A.S., Grav, H.J., Markussen, K.A., & White, R.G. (1984) Modes of thermal protection in newborn muskoxen (*Ovibos-moschatus*). *Acta Physiologica Scandinavica*, **122**, 443–53.

Böcher, T.W. (1974) The arctic greenhouse. *Botanisk Tidsskrift*, **69**, 121–9.

Bockstoce, J.R. (2009) *Furs and Frontiers in the Far North*. Yale University Press, New Haven.

Boertje, R.D., Keech, M.A., Young, D.D., Kellie, K.A., & Seaton, C.T. (2009) Managing for elevated yield of moose in interior Alaska. *Journal of Wildlife Management*, **73**, 314–27.

Bond, G., Kromer, B., Beer, J., et al. (2001) Persistent solar influence on north Atlantic climate during the Holocene. *Science*, **294**, 2130–6.

Bowsher, W., Steer, M., & Tobin, A. (2008) *Plant Biochemistry*. Garland Science, New York.

Briffa, K.R., Shishov, V.V., Melvin, T.M., et al. (2008) Trends in recent temperature and radial tree growth spanning 2000 years across northwest Eurasia. *Philosophical Transactions of the Royal Society B-Biological Sciences*, **363**, 2271–84.

Brinkhuis, H., Schouten, S., Collinson, M.E., et al. (2006) Episodic fresh surface waters in the Eocene Arctic Ocean. *Nature*, **441**, 606–9.

Brochmann, C. & Brysting, A.K. (2008) The Arctic—an evolutionary freezer? *Plant Ecology & Diversity*, **1**, 181–95.

Brochmann, C., Soltis, P.S., & Soltis, D.E. (1992) Recurrent formation and polyphyly of nordic polyploids in *Draba* (Brassicaceae). *American Journal of Botany*, **79**, 673–88.

Brochmann, C., Gabrielsen, T.M., Nordal, I., Landvik, J.Y., & Elven, R. (2003) Glacial survival or tabula rasa? The history of North Atlantic biota revisited. *Taxon,* **52**, 417–50.

Brochmann, C., Brysting, A.K., Alsos, I.G., et al. (2004) Polyploidy in arctic plants. *Biological Journal of the Linnean Society,* **82**, 521–36.

Brockmann-Jerosch, M. (1908) Die Geschichte der Schweizerischen Alpen Flora. In: *Das Pflanzenleben der Alpen* (ed. C. Schröter), pp. 743–77. Raustein, Zurich.

Broecker, W.S. (2003) Does the trigger for abrupt climate change reside in the ocean or in the atmosphere? *Science,* **300**, 1519–22.

Brook, E. (2012) The ice age carbon puzzle. *Science,* **336**, 682–3.

Brouwers, E.M., Clemens, W.A., Spicer, R.A., Ager, T.A., Carter, L.D., & Sliter, W.V. (1987) Dinosaurs on the North Slope, Alaska: high latitude, latest Cretaceous environments. *Science,* **237**, 1608–10.

Brysting, A.K., Mathiesen, C., & Marcussen, T. (2011) Challenges in polyploid phylogenetic reconstruction: a case story from the arctic-alpine *Cerastium alpinum* complex. *Taxon,* **60**, 333–47.

Buckland, P. & Dugmore, A.J. (1991) 'If this is a refugium, why are my feet so bloody cold?' The origins of the Icelandic biota in the light of recent research. In: *Environmental Change in Iceland: Past and Present* (eds J.K. Maizels & C. Caseldine), pp. 107–25. Kluwer, Dordrecht.

Burnham, K.K. & Newton, I. (2011) Seasonal movements of Gyrfalcons Falco rusticolus include extensive periods at sea. *Ibis,* **153**, 468–84.

Burnham, K.K., Burnham, W.A., & Newton, I. (2009) Gyrfalcon Falco rusticolus post-glacial colonization and extreme long-term use of nest-sites in Greenland. *Ibis,* **151**, 514–22.

CAFF (Conservation of Arctic Fauna and Flora) (2010) *Arctic Biodiversity: Selected Indicators of Change* (ed. C.I. Secretariat). CAFF, Akureyri, Iceland.

Cajander, A.K. (1913) Studien über die Moore Finnlands. *Acta Forestalia Fennica,* **2**, 1–208.

Callaghan, T.V., Crawford, R.M.M., Eronen, M., et al. (2002) The dynamics of the tundra-taiga boundary: an overview and suggested coordinated and integrated approach to research. *Ambio,* **Special Issue 12**, 3–5.70.

Campos, P.F., Willerslev, E., Sher, A., et al. (2010) Ancient DNA analyses exclude humans as the driving force behind late Pleistocene musk ox (*Ovibos moschatus*) population dynamics. *Proceedings of the National Academy of Sciences of the United States of America,* **107**, 5675–80.

Chernov, Y.I. (1988) *The Living Tundra.* Cambridge University Press, Cambridge.

Choat, B., Cobb, A.R., & Jansen, S. (2008) Structure and function of bordered pits: new discoveries and impacts on whole-plant hydraulic function. *New Phytologist,* **177**, 608–25.

Choy, E.S., Kimpe, L.E., Mallory, M.L., Smol, J.P., & Blais, J.M. (2010) Contamination of an arctic terrestrial food web with marine-derived persistent organic pollutants transported by breeding seabirds. *Environmental Pollution,* **158**, 3431–8.

Christian, M. & Parker, M.G. (2010) The engineering of brown fat. *Journal of Molecular Cell Biology,* **2**, 23–5.

Church, J.M., Arge, S.M., Edwards, K.J., Ascough, P.L., Bond, J.M. & et al. (2013) The Vikings were not the first colonizers of the Faroe Islands. *Quaternary Science Reviews,* **80 (In Press)**.

Constantine, A., Chinsamy, A., Vickers-Rich, P., & Rich, T.H. (1998) Periglacial environments and polar dinosaurs. *South African Journal of Science,* **94**, 137–41.

Conway, S.M. (1906) *No Man' Lands.* (Reprinted 1995, Damms Antikvariat 1995.) Cambridge University Press, Cambridge/Oslo.

Cooper, E.J. (2011) Polar desert vegetation and plant recruitment in Murchisonfjord, Nordaustlandet, Svalbard. *Geografiska Annaler Series A-Physical Geography,* **93A**, 243–52.

Cooper, E.J. & Wookey, P.A. (2003) Floral herbivory of *Dryas octopetala* by Svalbard reindeer. *Arctic Antarctic and Alpine Research,* **35**, 369–76.

Cooper, E.J., Alsos, I.G., Hagen, D., Smith, F.M., Coulson, S.J., & Hodkinson, I.D. (2004) Plant recruitment in the High Arctic: seed bank and seedling emergence on Svalbard. *Journal of Vegetation Science,* **15**, 115–24.

Cote, S.D., Dallas, J.F., Marshall, F., Irvine, R.J., Langvatn, R., & Albon, S.D. (2002) Microsatellite DNA evidence for genetic drift and philopatry in Svalbard reindeer. *Molecular Ecology,* **11**, 1923–30.

Crater, A.R. & Barboza, P.S. (2007) The rumen in winter: cold shocks in naturally feeding muskoxen (*Ovibos moschatus*). *Journal of Mammalogy,* **88**, 625–31.

Crawford, R.M.M. (1993) Plant survival without oxygen. *Biologist,* **40**, 110–14.

Crawford, R.M.M. (1995) Plant survival in the High Arctic. *Biologist,* **42**, 101–5.

Crawford, R.M.M. (1997) *Disturbance and Recovery in Arctic Lands: An Ecological Perspective.* Kluwer, Dordrecht.

Crawford, R.M.M. (2000) Ecological hazards of oceanic environments. *New Phytologist,* **147**, 257–81.

Crawford, R.M.M. (2004) Long-term plant survival at high latitudes. *Botanical Journal of Scotland,* **56**, 1–23.

Crawford, R.M.M. (2008) *Plants at the Margin—Ecological Limits and Climate Change.* Cambridge University Press, Cambridge.

Crawford, R.M.M. (2012) Long-term anoxia—tolerance in flowering plants. In: *Anoxia: Paleontological Strategies and Evidence for Eukaryote Survival* (eds A.V. Alltenbach, J.M. Bernard, & J. Seckbach), pp. 219–46. Springer, Heidelberg.

Crawford, R.M.M., Chapman, H.M., & Hodge, H. (1994) Anoxia tolerance in high Arctic vegetation. *Arctic and Alpine Research,* **26**, 308–12.

Crawford, R.M.M., Chapman, H.M., & Smith, L.C. (1995) Adaptation to variation in growing season length in

Arctic populations of *Saxifraga oppositifolia* L. *Botanical Journal of Scotland*, **41**, 177–92.

Danilov, P.I. (2005) *Game Animals of Karelia: Ecology, Resources, Management and Conservation*. Nauka, Moscow.

Danilov, P., Kan'shiev, V.Y., & Fedorov, F.V. (2008) European (*Castor fiber*) and Canadian (*Castor canadensis*) beavers from the Russian north-west. *Zoologichesky Zhurnal*, **87**, 348–60.

Dawson, A.G., Hickey, K., Mayewski, P.A., & Nesje, A. (2007) Greenland (GISP2) ice core and historical indicators of complex North Atlantic climate changes during the fourteenth century. *Holocene*, **17**, 427–34.

de Witte, L.C., Armbruster, G.F., Gielly, L., Taberlet, P., & Stöcklin, J. (2012) AFLP markers reveal high clonal diversity and extreme longevity in four key arctic-alpine species. *Molecular Ecology* **21**(5), 1081–97.

DeConto, R.M. & Pollard, D. (2003) Rapid Cenozoic glaciation of Antarctica induced by declining atmospheric CO_2. *Nature*, **421**, 245–9.

Dekorte, J., Volkov, A.E., & Gavrilo, M.V. (1995) Bird observations in Severnaya-Zemlya, Siberia. *Arctic*, **48**, 222–34.

Denton, G.H., Anderson, R.F., Toggweiler, J.R., Edwards, R.L., Schaefer, J.M., & Putnam, A.E. (2010) The Last Glacial Termination. *Science*, **328**, 1652–6.

Despres, L., Loriot, S., & Gaudeul, M. (2002) Geographic pattern of genetic variation in the European globeflower *Trollius europaeus* L. (Ranunculaceae) inferred from amplified fragment length polymorphism markers. *Molecular Ecology*, **11**, 2337–47.

DeVries, A.L. & Wohlschlag, D.E. (1969) Freezing resistance in some Antarctic fishes. *Science*, **163**, 1073–5.

Diggle, P.K. (1997) Extreme preformation in alpine *Polygonum viviparum*: an architectural and developmental analysis. *American Journal of Botany*, **84**, 154–69.

Diggle, P.K., Lower, S., & Ranker, T.A. (1998) Clonal diversity in alpine populations of *Polygonum viviparum* (Polygonaceae). *International Journal Of Plant Sciences*, **159**, 606–15.

Dillehay, T.D., Ramirez, C., Pino, M., Collins, M.B., Rossen, J., & Pino-Navarro, J.D. (2008) Monte Verde: seaweed, food, medicine, and the peopling of South America. *Science*, **320**, 784–6.

Dugmore, A.J., Church, M.J., Buckland, P.C., et al. (2005) The Norse landnám on the North Atlantic islands: an environmental impact assessment. *Polar Record*, **41**, 21–37.

Dugmore, A.J., Keller, C., & McGovern, T.H. (2007) Norse Greenland settlement: reflections on climate change, trade, and the contrasting fates of human settlements in the North Atlantic islands. *Arctic Anthropology*, **44**, 12–36.

Dunnett, G.M. (1992) A forty-three year study of the fulmars on Eynhallow, Orkney. *Scottish Birds*, **16**, 155–9.

Dyck, A.P. & Macarthur, R.A. (1993) Seasonal variation in the microclimate and gas composition of beaver lodges in a boreal environment. *Journal of Mammalogy*, **74**, 180–8.

Edwards, C.J., Suchard, M.A., Lemey, P., et al. (2011) Ancient hybridization and an Irish origin for the modern polar bear matriline. *Current Biology*, **21**, 1251–8.

Edwards, M.E., Anderson, P.M., Brubaker, L.B., et al. (2000) Pollen-based biomes for Beringia 18,000, 6000 and 0 ^{14}C yr BP. *Journal of Biogeography*, **27**, 521–54.

Egevang, C., Stenhouse, I.J., Phillips, R.A., Petersen, A., Fox, J.W., & Silk, J.R.D. (2010) Tracking of Arctic terns *Sterna paradisaea* reveals longest animal migration. *Proceedings of the National Academy of Sciences of the United States of America*, **107**, 2078–81.

Eide, N.E., Stien, A., Prestrud, P., Yoccoz, N.G., & Fuglei, E. (2012) Reproductive responses to spatial and temporal prey availability in a coastal Arctic fox population. *Journal of Animal Ecology*, **81**, 640–8.

Eidesen, P.B., Carlsen, T., Molau, U., & Brochmann, C. (2007) Repeatedly out of Beringia: *Cassiope tetragona* embraces the arctic. *Journal of Biogeography*, **34**, 1559–74.

Elias, P.M., Menon, G., Wetzel, B.K., & Williams, J.W. (2010) Barrier requirements as the evolutionary 'driver' of epidermal pigmentation in humans. *American Journal of Human Biology*, **22**, 526–37.

Elton, C. (1927) *Animal Ecology*. Sidgwick & Jackson, London.

Elvebakk, A. (2005) 'Arctic hotspot complexes'—proposed priority sites for studying and monitoring effects of climatic change on arctic biodiversity. *Phytocoenologia*, **35**, 1067–79.

Enk, J.M., Yesner, D.R., Crossen, K.J., Veltre, D.W., & O'Rourke, D.H. (2009) Phylogeographic analysis of the mid-Holocene mammoth from Qagnax Cave, St Paul Island, Alaska. *Palaeogeography Palaeoclimatology Palaeoecology*, **273**, 184–90.

Erlendsson, E. & Edwards, K.J. (2009) The timing and causes of the final pre-settlement expansion of *Betula pubescens* in Iceland. *Holocene*, **19**, 1083–91.

Eskelinen, O. (2002) Diet of the wood lemming *Myopus schisticolor*. *Annales Zoologici Fennici*, **39**, 49–57.

Festa-Bianchet, M., Ray, J.C., Boutin, S., Cote, S., & Gunn, A. (2011) Conservation of caribou (*Rangifer tarandus*) in Canada: an uncertain future. *Canadian Journal of Zoology-Revue Canadienne De Zoologie*, **89**, 419–34.

Fickert, T., Friend, D., Gruninger, F., Molnia, B., & Richter, M. (2007) Did debris-covered glaciers serve as Pleistocene refugia for plants? A new hypothesis derived from observations of recent plant growth on glacier surfaces. *Arctic Antarctic and Alpine Research*, **39**, 245–57.

Flagstad, O. & Røed, K.H. (2003) Refugial origins of reindeer (*Rangifer tarandus* L.) inferred from mitochondrial DNA sequences. *Evolution*, **57**, 658–70.

Forbes, B.C. & Kumpula, T. (2009) The ecological role and geography of reindeer (*Rangifer tarandus*) in northern Eurasia. *Geography Compass*, **3/4**, 1356–80.

Forbes, B.C. & Stammler, F. (2009) Arctic climate change discourse: the contrasting politic of research agendas in the West and Russia. *Polar Research,* 28(1), 28–42.

Forman, S.L., Lubinski, D., & Weihe, R.R. (2000) The Holocene occurrence of reindeer on Franz Josef Land, Russia. *Holocene,* 10, 763–8.

Forsyth, J. (1992) *A History of the Peoples of Siberia.* Cambridge University Press, Cambridge.

Fox, A.D., Bergersen, E., Tombre, I.M., & Madsen, J. (2007) Minimal intra-seasonal dietary overlap of barnacle and pink-footed geese on their breeding grounds in Svalbard. *Polar Biology,* 30, 759–68.

Friend, A.D. & Woodward, F.I. (1990) Evolutionary and ecophysiological responses of mountain plants to the growing season environment. *Advances in Ecological Research,* 20, 59–124.

Fuglei, E. & Oritsland, N.A. (1999) Seasonal trends in body mass, food intake and resting metabolic rate, and induction of metabolic depression in arctic foxes (*Alopex lagopus*) at Svalbard. *Journal of Comparative Physiology B–Biochemical Systemic and Environmental Physiology,* 169, 361–9.

Fuglei, E., Oritsland, N.A., & Prestrud, P. (2003) Local variation in arctic fox abundance on Svalbard, Norway. *Polar Biology,* 26, 93–8.

Gabrielsen, T.M. & Brochmann, C. (1998) Sex after all: high levels of diversity detected in the arctic clonal plant *Saxifraga cernua* using RAPD markers. *Molecular Ecology,* 7, 1701–8.

Galbreath, K.E., Cook, J.A., & Hoberg, E.P. (2012) Climate's role in polar bear past. *Science,* 336, 1230.

Gamache, I. & Payette, S. (2005) Latitudinal response of subarctic tree lines to recent climate change in eastern Canada. *Journal of Biogeography,* 32, 849–62.

Gaye, E., Marchand, Y., & Rees, G. (2004) Etude par imagerie radar des pollutions pétrolières. *Cybergeo: Revue Européenne de Géographie,* 273, 1–14.

Gerloff, L.M., Hills, L.V., & Osborn, G.D. (1995) Postglacial vegetation history of the Mission Mountains, Montana. *Journal of Paleolimnology,* 14, 269–79.

Gilbert, M.T.P., Jenkins, D.L., Gotherstrom, A., *et al.* (2008) DNA from pre-Clovis human coprolites in Oregon, North America. *Science,* 320, 786–9.

Gilg, O., Sittler, B., & Hanski, I. (2009) Climate change and cyclic predator–prey population dynamics in the high Arctic. *Global Change Biology,* 15, 2634–52.

Gill, R.E., Piersma, T., Hufford, G., Servranckx, R., & Riegen, A. (2005) Crossing the ultimate ecological barrier: evidence for an 11000-km-long nonstop flight from Alaska to New Zealand and eastern Australia by Bar-tailed Godwits. *Condor,* 107, 1–20.

Girardin, M.P., Tardif, J., & Bergeron, Y. (2001) Radial growth analysis of *Larix laricina* from the Lake Duparquet area, Quebec, in relation to climate and larch sawfly outbreaks. *Ecoscience,* 8, 127–38.

Glahder, C.M., Fox, T.A., Hubner, C.E., Madsen, J., & Tombre, I.M. (2006) Pre-nesting site use of satellite transmitter tagged Svalbard Pink-footed Geese *Anser brachyrhynchus. Ardea,* 94, 679–90.

Goebel, T., Waters, M.R., & O'Rourke, D.H. (2008) The Late Pleistocene dispersal of modern humans in the Americas. *Science,* 319, 1497–502.

Goetcheus, V.G. & Birks, H.H. (2001) Full-glacial upland tundra vegetation preserved under tephra in the Beringia National Park, Seward Peninsula, Alaska. *Quaternary Science Reviews,* 20, 135–47.

Goetz, S.J., Bunn, A.G., Fiske, G.J., & Houghton, R.A. (2005) Satellite-observed photosynthetic trends across boreal North America associated with climate and fire disturbance. *Proceedings of the National Academy of Sciences of the United States of America,* 102, 13521–5.

Gore, A.J.P. (1983) Mires: swamp, bog, fen and moor. In: *Ecosystems of the World* (ed. D.W. Goodall). Elsevier, Amsterdam.

Gornall, J.L., Woodin, S.J., Jonsdottir, I.S., & van der Wal, R. (2011) Balancing positive and negative plant interactions: how mosses structure vascular plant communities. *Oecologia,* 166, 769–82.

Grab, S. (2005) Aspects of the geomorphology, genesis and environmental significance of earth hummocks (thufur, pounus): miniature cryogenic mounds. *Progress in Physical Geography,* 29, 139–55.

Grace, J., Berninger, F., & Nagy, L. (2002) Impacts of climate change on the tree line. *Annals of Botany,* 90, 537–44.

Gratton, C., Donaldson, J., & van der Zanden, M.J. (2008) Ecosystem linkages between lakes and the surrounding terrestrial landscape in northeast Iceland. *Ecosystems,* 11, 764–74.

Gravlund, P., Meldgaard, M., Paabo, S., & Arctander, P. (1998) Polyphyletic origin of the small-bodied, high-arctic subspecies of tundra reindeer (*Rangifer tarandus*). *Molecular Phylogenetics and Evolution,* 10, 151–9.

Grime, J.P. (2001) *Plant Strategies, Vegetation Processes and Ecosystem Properties.* John Wiley, Chichester.

Grove, J.M. (1988) *The Little Ice-Age.* Methuen, London.

Groves, P. (1997) Intraspecific variation in mitochondrial DNA of muskoxen, based on control-region sequences. *Canadian Journal of Zoology–Revue Canadienne De Zoologie,* 75, 568–75.

Gudmundson, A.T. (2007) *Living Earth: Outline of the Geology of Iceland.* Mál og Menning, Reykjavik.

Guglielmo, C.G. (2010) Move that fatty acid: fuel selection and transport in migratory birds and bats. *Integrative and Comparative Biology,* 50, 336–45.

Guiliazov, A.S. (1998) Causes of reindeer (*Rangifer tarandus*) and moose (*Alces alces*) mortality in the Lapland Reserve and its surroundings. *Alces,* 34, 319–27.

Gunn, A. & Forchhammer, M. (2008) *Ovibos moschatus*. IUCN Red List of Threatened Species, Version 2010.4. http://www.iucnredlist.org.

Gunn, A. & Russell, D. (2010) The dwindling numbers of caribou in Canada's Arctic. *Meridian*, **Fall/Winter**, 1–6.

Gunn, A., Miller, F.L., Barry, S.J., & Buchan, A. (2006) A near-total decline in caribou on Prince of Wales, Somerset, and Russell islands, Canadian Arctic. *Arctic*, **59**, 1–13.

Gunn, A., Russell, D., White, R.G., & Kofinas, G. (2009) Facing a future of change: wild migratory caribou and reindeer. *Arctic*, **62**, III–VI.

Gusta, L.V., Trischuk, R., & Weiser, C.J. (2005) Plant cold acclimation: the role of abscisic acid. *Journal of Plant Growth Regulation*, **24**, 308–18.

Gustine, D.D., Barboza, P.S., Adams, L.G., Farnell, R.G., & Parker, K.L. (2011) An isotopic approach to measuring nitrogen balance in caribou. *Journal of Wildlife Management*, **75**, 178–88.

Gwinner, E. & Brandstätter, R. (2001) Complex bird clocks. *Philosophical Transactions of the Royal Society of London Series B–Biological Sciences*, **356**, 1801–10.

Hailer, F., Kutschera, V.E., Hallstroem, B.M., *et al*. (2012) Nuclear genomic sequences reveal that polar bears are an old and distinct bear lineage. *Science*, **336**, 344–7.

Hakala, A.V.K., Staaland, H., Pulliainen, E., & Roed, K.H. (1985) Taxonomy and history of arctic island reindeer with special reference to Svalbard (Norway) reindeer. *Aquilo Ser Zoologica*, **23**, 1–12.

Halliday, G. (2013) The flora of the inland mountains of south-east Greenland between 66° and 69°N with particular reference to altitude limits. *Meddelelser om Grønland*, in press.

Hancock, A.M., Witonsky, D.B., Ehler, E., *et al*. (2010) Human adaptations to diet, subsistence, and ecoregion are due to subtle shifts in allele frequency. *Proceedings of the National Academy of Sciences of the United States of America*, **107**, 8924–30.

Hannon, G.E. & Bradshaw, R.H.W. (2000) Impacts and timing of the first human settlement on vegetation of the Faroe Islands. *Quaternary Research*, **54**, 404–13.

Hansen, B.B., Aanes, R., Herfindal, I., Kohler, J., & Saether, B.-E. (2011) Climate, icing, and wild arctic reindeer: past relationships and future prospects. *Ecology*, **92**, 1917–23.

Hay, W.W., Soeding, E., DeConto, R.M., & Wold, C.N. (2002) The Late Cenozoic uplift—climate change paradox. *International Journal of Earth Sciences*, **91**, 746–74.

Hearne, S. (1795) *A Journey to the Northern Ocean*. Republished 2007 with Foreword by K. McGoogan. TouchWood Editions, Victoria, British Columbia.

Hegarty, M.J. & Hiscock, S.J. (2008) Genomic clues to the evolutionary success of review polyploid plants. *Current Biology*, **18**, R435–44.

Heide, O.M. (1992) Flowering strategies of the high-arctic and high-alpine snow bed grass species *Phippsia algida*. *Physiologia Plantarum*, **85**, 606–10.

Heide, O.M. (1997) Environmental control of flowering in some northern *Carex* species. *Annals of Botany*, **79**, 319–27.

Henry, G.H.R. (1998) Environmental influences on the structure of sedge meadows in the Canadian High Arctic. *Plant Ecology*, **134**, 119–29.

Hijma, M.P. & Cohen, K.M. (2010) Timing and magnitude of the sea-level jump preluding the 8200 yr event. *Geology*, **38**, 275–8.

Hobson, K.A., Stirling, I., & Andriashek, D.S. (2009) Isotopic homogeneity of breath CO_2 from fasting and berry-eating polar bears: implications for tracing reliance on terrestrial foods in a changing Arctic. *Canadian Journal of Zoology–Revue Canadienne De Zoologie*, **87**, 50–5.

Hoch, G. & Körner, C. (2005) Growth, demography and carbon relations of *Polylepis* trees at the world's highest treeline. *Functional Ecology*, **19**, 941–51.

Hofgaard, A., Rees, G., Tommervik, H., *et al*. (2010) Role of disturbed vegetation in mapping the boreal zone in northern Eurasia. *Applied Vegetation Science*, **13**, 460–72.

Holm, L.E., Forchhammer, M.C., & Boomsma, J.J. (1999) Low genetic variation in muskoxen (*Ovibos moschatus*) from western Greenland using microsatellites. *Molecular Ecology*, **8**, 675–9.

Holsten, E.H., Werner, R.A., & Develice, R.L. (1995) Effects of a spruce beetle (Coleoptera: Scolytidae) outbreak and fire on Lutz spruce in Alaska. *Environmental Entomology*, **24**, 1539–47.

Holtmeier, F.-K. (2003) *Mountain Timberlines*. Kluwer, Dordrecht.

Holtmeier, F.K. & Broll, G. (2010) Wind as an ecological agent at treelines in North America, the Alps, and the European Subarctic. *Physical Geography*, **31**, 203–33.

Hong, S.M., Candelone, J.P., Patterson, C.C., & Boutron, C.F. (1994) Greenland ice evidence of hemispheric lead pollution 2-millennia ago by Greek and Roman civilizations. *Science*, **265**, 1841–3.

Høye, T.T. & Forchhammer, M.C. (2008) The influence of weather conditions on the activity of high-arctic arthropods inferred from long-term observations. *BMC Ecology*, **8:8**, Epub.

Hübner, C.E. (2006) The importance of pre-breeding areas for the arctic barnacle goose *Branta leucopsis*. *Ardea*, **94**, 701–13.

Hulbert, R.C. & Harington, C.R. (1999) An early Pliocene hipparionine horse from the Canadian Arctic. *Palaeontology*, **42**, 1017–25.

Hultén, E. & Fries, M. (1986) *Atlas of North European Vascular Plants North of the Tropic of Cancer*. Koeltz, Königstein.

Huntley, B. & Birks, H.J.B. (1983) *An Atlas of Past and Present Pollen Maps for Europe: 0–13000 Years Ago*. Cambridge University Press, Cambridge.

Ims, R.A., Henden, J.A., & Killengreen, S.T. (2008) Collapsing population cycles. *Trends in Ecology & Evolution*, **23**, 79–86.

Ineson, S., Scaife, A.A., Knight, J.R., *et al.* (2011) Solar forcing of winter climate variability in the northern hemisphere. *Nature Geoscience*, **4**, 753–7.

Ingold, T. (1982) The significance of storage in hunting societies. *Man*, **18**, 553–71.

Ingram, H.A.P. (1978) Soil layers in mires—function and terminology. *Journal of Soil Science*, **29**, 224–7.

Ingstad, H. (1951) *Nunamiut: Among Alaska's Inland Eskimos*. Allen and Unwin, London.

Ingstad, H. (1992) *The Land of Feast and Famine*. McGill-Queens University Press, Montreal.

IUCN (International Union for Conservation of Nature) (2010) *Polar Bear Specialists Group Report*. IUCN, Gland, Switzerland.

Jablonski, N.G. (2004) The evolution of human skin and skin color. *Annual Review of Anthropology*, **33**, 585–623.

Jacoby, G.C., Workman, K.W., & D'Arrigo, R.D. (1999) Laki eruption of 1783, tree rings, and disaster for northwest Alaska Inuit. *Quaternary Science Reviews*, **18**, 1365–71.

Jahren, A.H. (2007) The Arctic Forest of the Middle Eocene. *Annual Review of Earth and Planetary Sciences*, **35**, 509–40.

Jasinski, M.E. & Søreide, F. (2003) The Norse settlements in Greenland from a maritime perspective. In: *Vinland Revisited: The Norse World at the Turn of the First Millennium* (ed. L. Lewis-Simpson), pp. 123–32. Historic Sites Association of Newfoundland and Labrador, St John's, Newfoundland.

Jayen, K., Leduc, A., & Bergeron, Y. (2006) Effect of fire severity on regeneration success in the boreal forest of northwest Quebec, Canada. *Ecoscience*, **13**, 143–51.

Jefferies, R.L. (1997) Long-term damage to sub-arctic coastal ecosystems by geese: ecological indicators and easures of ecosystem dysfunction. In: *Disturbance and Recovery in Arctic Lands* (ed. R.M.M. Crawford), pp. 151–65. Kluwer, Dordrecht.

Jefferies, R.L. & Gottlieb, L.D. (1983) Genetic variation within and between populations of the asexual plant *Puccinellia x phryganodes*. *Canadian Journal of Botany*, **61**, 774–9.

Jenkins, D.L., Davis, L.G., Stafford, T.W., Jr, *et al.* (2012) Clovis age western stemmed projectile points and human coprolites at the Paisley Caves. *Science*, **337**, 223–8.

Jóhansen, J. (1971) A palaeobotanical study indicating a previking settlement in Tjørnuvik, Faroe Islands. *Fróðskaparrit*, **19**, 147–57.

Jóhansen, J. (1978) Cereal cultivation in Mykines, Faroe Islands AD 600. *Geological Survey of Denmark, Yearbook*, 93–103.

Jóhansen, J. (1985) Studies in the vegetational history of the Faroe and Shetland Isles. *Anales Societatis Scientiarium Faeroensis*, Supplement XI.

Johansson, A., Ingman, M., Mack, S.J., Erlich, H., & Gyllensten, U. (2008) Genetic origin of the Swedish Sami inferred from HLA class I and class II allele frequencies. *European Journal of Human Genetics*, **16**, 1341–9.

Jones, P.D. & Mann, M.E. (2004) Climate over past millennia. *Reviews of Geophysics*, **42**(2), RG2002.

Jonsdottir, I.S., Virtanen, R., & Karnefelt, I. (1999) Large-scale differentiation and dynamics in tundra plant populations and vegetation. *Ambio*, **28**, 230–8.

Jonsdottir, I.S., Augner, M., Fagerstrom, T., Persson, H., & Stenström, A. (2000) Genet age in marginal populations of two clonal *Carex* species in the Siberian Arctic. *Ecography*, **23**, 402–12.

Jonsson, B.O., Jónsdóttir, I.S., & Cronberg, N. (1996) Clonal diversity and allozyme variation in populations of the arctic sedge *Carex bigelowii* (Cyperaceae). *Journal of Ecology*, **84**, 449–59.

Jorgenson, M.T., Racine, C.H., Walters, J.C., & Osterkamp, T.E. (2001) Permafrost degradation and ecological changes associated with a warming climate in central Alaska. *Climatic Change*, **48**, 551–79.

Kadereit, J.W. & Comes, H.P. (2005) The temporal course of alpine plant diversification in the Quaternary. *Plant Species—Level Systematics: New Perspectives on Pattern & Process*, **143**, 117–30.

Kadereit, J.W., Licht, W., & Uhink, C.H. (2008) Asian relationships of the flora of the European Alps. *Plant Ecology & Diversity*, **1**, 171–9.

Kahlke, R.-D. (2013) The origin of the Eurasian mammoth faunas (*Mammuthus–Coelodonta* faunal complex). *Quaternary Science Reviews* (in press).

Kasuga, J., Arakawa, K., & Fujikawa, S. (2007) High accumulation of soluble sugars in deep supercooling Japanese white birch xylem parenchyma cells. *New Phytologist*, **174**, 569–79.

Kelly, E.N., Short, J.W., Schindler, D.W., *et al.* (2009) Oil sands development contributes polycyclic aromatic compounds to the Athabasca River and its tributaries. *Proceedings of the National Academy of Sciences of the United States of America*, **106**, 22346–51.

Kelly, E.N., Schindler, D.W., Hodson, P.V., Short, J.W., Radmanovich, R., & Nielsen, C.C. (2010) Oil sands development contributes elements toxic at low concentrations to the Athabasca River and its tributaries. *Proceedings of the National Academy of Sciences of the United States of America*, **107**, 16178–83.

Khar'kov, V.N., Stepanov, V.A., Medvedev, O.F., *et al.* (2008) The origin of Yakuts: analysis of Y-chromosome haplotypes. *Molekuliarnaia Biologiia (Mosk)*, **42**, 226–37.

Kielland, K., J. Bryant, et al. (2006). Mammalian herbivory, ecosystem engineering, and ecological cascades in Alaskan boreal forests. *Alaska's changing Boreal forest* Eds F. S. Chapin III, M. W. Oswood, K. Van Kleve, L. A. Viereck and D. L. Verbyla. Oxford, Oxford University Press: 211–226.

Kiklman, A.O. (1892) Pflanzenbiologische Studien aus Russsisch Lappland. Ein Beitrag zur Kenntnis der regionalen Gliederung an der polaren Waldgrenze. *Acta Societatis pro Fauna et Flora Fennica*, **16**(3).

Kirpotin, S.N., Berezin, A., Bazanov, V., *et al.* (2009) Western Siberian wetlands as indicator and regulator of climate change on a global scale. *International Journal of Environmental Studies*, **66**(4), 409–21.

Klinger, L.F. (1996) Coupling of soils and vegetation in peatland succession. *Arctic and Alpine Research*, **28**, 380–7.

Knox, J. (1558) *The Monstrous Regiment of Women*. J. Crespin, Geneva.

Kohler, J. & Aanes, R. (2004) Effect of winter snow and ground-icing on a Svalbard reindeer population: results of a simple snowpack model. *Arctic Antarctic and Alpine Research*, **36**, 333–41.

Körner, C. (2003) *Alpine Plant Life*. Springer, Berlin.

Körner, C., Neumayer, M., Pelaez Menendez-Riedl, S., & Smeets-Scheel, A. (1989) Functional morphology of mountain plants. *Flora*, **182**, 353–83.

Korsten, M., Ho, S.Y.W., Davison, J., *et al.* (2009) Sudden expansion of a single brown bear maternal lineage across northern continental Eurasia after the last ice age: a general demographic model for mammals? *Molecular Ecology*, **18**, 1963–79.

Koutaniemi, L. (1999) Twenty-one years of string movements on the Liippasuo aapa mire, Finland. *Boreas*, **28**, 521–30.

Kremenetski, C.V., Chichagova, O.A., & Shishlina, N.I. (1999) Palaeoecological evidence for Holocene vegetation, climate and landuse change in the low Don basin and Kalmuk area, southern Russia. *Vegetation History and Archaeobotany*, **8**, 233–46.

Krupnik, I. (1993) *Arctic Adaptations*. University Press of New England, Hanover.

Kuijper, D.P.J., Bakker, J.P., Cooper, E.J., Ubels, R., Jonsdottir, I.S., & Loonen, M.J.J.E. (2006) Intensive grazing by Barnacle geese depletes High Arctic seed bank. *Canadian Journal of Botany–Revue Canadienne de Botanique*, **84**, 995–1004.

Kukal, O. & Kevan, P.G. (1987) The influence of parasitism on the life-history of a high arctic insect, *Gynaephora groenlandica* (Wocke) (Lepidoptera, Lymantriidae). *Canadian Journal of Zoology–Revue Canadienne de Zoologie*, **65**, 156–63.

Kurten, B. (1995) *The Cave Bear Story: Life and Death of a Vanished Animal*. University of Columbia Press, New York.

Kuzmin, Y.V. (2010) Extinction of the woolly mammoth (*Mammuthus primigenius*) and woolly rhinoceros (*Coelodonta antiquitatis*) in Eurasia: review of chronological and environmental issues. *Boreas*, **39**, 247–61.

Kuzmin, Y.V., Orlova, L.A., Zol'nikov, I.D., & Igol'nikov, A.E. (2000) The history of mammoth (*Mammuthus primigenius* Blum.) population in Siberia and adjacent areas (based on radiocarbon data). *Geologiya I Geofizika*, **41**, 746–54.

Laaksonen, S., Solismaa, M., Kortet, R., Kuusela, J., & Oksanen, A. (2009) Vectors and transmission dynamics for *Setaria tundra* (Filarioidea; Onchocercidae), a parasite of reindeer in Finland. *Parasites & Vectors*, **2**(3), Epub.

Labutin, Y.V. & Ellis, D.H. (2006) Gyrfalcon (*Falco rusticolus*) in Yakutia: distribution, nesting areas, and features of nutrition. *Zoologichesky Zhurnal*, **85**, 1354–61.

Laitinen, J., Rehell, S., Huttunen, A., Tahvanainen, T., Heikkilä, R., & Lindholm, T. (2007) Mire systems in Finland—special view to aapa mires and their water-flow pattern. *Suo*, **58**, 1–26.

Landvik, J.Y., Brook, E.J., Gualtieri, L., Raisbeck, G., Salvigsen, O., & Yiou, F. (2003) Northwest Svalbard during the last glaciation: ice-free areas existed. *Geology*, **31**, 905–8.

Lefebvre, C. & Vekemans, X. (1995) A numerical taxonomic study of *Armeria maritima* (Plumbaginaceae) in North-America and Greenland. *Canadian Journal of Botany*, **73**, 1583–95.

Lent, P.C. (1998) Alaska's indigenous muskoxen: a history. *Rangifer*, **18**, 133–44.

Lescop-Sinclair, K. & Payette, S. (1995) Recent advance of the arctic treeline along the eastern coast of Hudson Bay. *Journal of Ecology*, **83**, 929–36.

Lewis, P.O. & Crawford, D.J. (1995) Pleistocene refugium endemics exhibit greater allozymic diversity than widespread congeners in the genus *Polygonella* (Polygonaceae). *American Journal of Botany*, **82**, 141–9.

Lid, J. (1964) The flora of Jan Mayen. *Norsk Polar Institutt Skr*, **130**, 1–107.

Lid, J. & Lid, D.T. (1994) *Norsk Flora*. Det Norske Samlaget, Oslo.

Lindgard, K., Stokkan, K.A., Lemaho, Y., & Groscolas, R. (1992) Protein-utilization during starvation in fat and lean Svalbard ptarmigan (*Lagopus-mutus-hyperboreus*). *Journal of Comparative Physiology B-Biochemical Systemic and Environmental Physiology*, **162**, 607–13.

Lindqvist, C., Schuster, S.C., Sun, Y.Z., *et al.* (2010) Complete mitochondrial genome of a Pleistocene jawbone unveils the origin of polar bear. *Proceedings of the National Academy of Sciences of the United States of America*, **107**, 5053–7.

Lloyd, C.R. (2001) On the physical controls of the carbon dioxide balance at a high Arctic site in Svalbard. *Theoretical and Applied Climatology*, **70**, 167–82.

Lomonosov, M.V. (1763) *Layers of the Earth and Other Works of Geology. The Complete Published Works*. Akademiya Nauk, Moscow.

Longton, R.E. (1988a) *Biology of Polar Bryophytes and Lichens*. Cambridge University Press, Cambridge.

Longton, R.E. (1988b) Life-history strategies among bryophytes of arid regions. *Journal of the Hattori Botanical Laboratory*, **64**, 15–28.

Lukin, A., Sharova, J., & Belicheva, L. (2010) Assessment of fish health status in the Pechora River: effects of contamination. *Ecotoxicology and Environmental Safety*, **74**, 355–65.

Lynch, J.A., Clark, J.S., Bigelow, N.H., Edwards, M.E., & Finney, B.P. (2002) Geographic and temporal variations in fire history in boreal ecosystems of Alaska. *Journal of Geophysical Research-Atmospheres*, **108**(D1), 8152–69.

MacArthur, R.H. & Wilson, E.O. (1967) *The Theory of Island Biogeography*. Princeton University Press, Princeton, New Jersey.

Mackenzie, G.S. (1811) *Travels in the Islands of Iceland During the Summer of the Year MDCCCX*. Republished 2011 by Cambridge University Press, Cambridge.

MacPhee, R.D.E., Tikhonov, A.N., Mol, D., et al. (2002) Radiocarbon chronologies and extinction dynamics of the Late Quaternary mammalian megafauna of the Taimyr Peninsula, Russian Federation. *Journal of Archaeological Science*, **29**, 1017–42.

MacPhee, R.D.E., Tikhonov, A.N., Mol, D., & Greenwood, A.D. (2005) Late Quaternary loss of genetic diversity in muskox (*Ovibos*). *BMC Evolutionary Biology*, **5**, 49, Epub.

Maessen, O., Freedman, B., Nams, M.L.N., & Svoboda, J. (1983) Resource allocation in high-arctic vascular plants of differing growth form. *Canadian Journal of Botany*, **61**, 1680–91.

Mallik, A.U., Gimingham, C.H., & Rahman, A.A. (1984) Ecological effects of heather burning. 1. Water infiltration, moisture retention and porosity of surface soil. *Journal of Ecology*, **72**, 767–76.

Mann, M.E. (2007) Climate over the past two millennia. *Annual Review of Earth and Planetary Sciences*, **35**, 111–36.

Mark, A.F., Fetcher, N., Shaver, G.R., & Chapin III, F.S. (1985) Estimated ages of mature tussocks of *Eriophorum vaginatum* along a latitudinal gradient in central Alaska, U.S.A. *Arctic and Alpine Research*, **17**, 1–5.

Marthinsen, G., Wennerberg, L., Solheim, R., & Lifjeld, J.T. (2009) No phylogeographic structure in the circumpolar snowy owl (*Bubo scandiacus*). *Conservation Genetics*, **10**, 923–33.

Mathiesen, S.D., Haga, O.E., Kaino, T., & Tyler, N.J.C. (2000) Diet composition, rumen papillation and maintenance of carcass mass in female Norwegian reindeer (*Rangifer tarandus tarandus*) in winter. *Journal of Zoology*, **251**, 129–38.

Mayrose, I., Zhan, S.H., Rothfels, C.J., et al. (2011) Recently formed polyploid plants diversify at lower rates. *Science*, **333**, 1257.

McGraw, J.B. & Antonovics, J. (1983) Experimental ecology of *Dryas octopetala* ecotypes. I. Ecotypic differentiation and life cycle stages of selection. *Journal of Ecology*, **71**, 879–97.

Mechnikova, S.A., Romanov, M.S., Kalyakin, V.N., & Kudryavtsev, N.V. (2010) The Gyrfalcon, *Falco rusticolus*, in the Yamal Peninsula: dynamics of brood size and nest size over the years 1973–2008. *Russian Journal of Ecology*, **41**, 256–63.

Meeker, L.D. & Mayewski, P.A. (2002) A 1400-year high-resolution record of atmospheric circulation over the North Atlantic and Asia. *Holocene*, **12**, 257–66.

Merilainen, J.J., Kustula, V., & Witick, A. (2011) Lead pollution history from 256 BC to AD 2005 inferred from the Pb isotope ratio (Pb-206/Pb-207) in a varve record of Lake Korttajarvi in Finland. *Journal of Paleolimnology*, **45**, 1–8.

Meriot, C. (1984) The Saami peoples from the time of the voyage of Ottar to Vonwesten, Thomas. *Arctic*, **37**, 373–84.

Meusel, H. & Jäger, E.J. (1992) *Vergleichende Chorologie der Zentral Europäischen Fora*. Gustav Fischer, Jena.

Mitchell, W.W. (1992) Cytogeographic races of *Arctagrostis latifolia* (Poaceae) in Alaska. *Canadian Journal of Botany*, **70**, 80–3.

Molau, U. (1992) On the occurrence of sexual reproduction in *Saxifraga-cernua* and *S-foliolosa* (Saxifragaceae). *Nordic Journal of Botany*, **12**, 197–203.

Molau, U. (1993) Reproductive ecology of the three Nordic *Pinguicula* species (Lentibularieaceae). *Nordic Journal of Botany*, **13**, 149–57.

Möller, P., Bolshiyanov, D., & Bersten, H. (1999) Weichselian geology and palaeoenvironmental history of the central Taymyr Peninsula, Siberia, indicating no glaciation during the last global glacial maximum. *Boreas*, **28**, 92–114.

Monnett, C. & Gleason, J.S. (2006) Observations of mortality associated with extended open-water swimming by polar bears in the Alaskan Beaufort Sea. *Polar Biology*, **29**, 681–7.

Morewood, W.D. & Ring, R.A. (1998) Revision of the life history of the High Arctic moth *Gynaephora groenlandica* (Wocke) (Lepidoptera: Lymantriidae). *Canadian Journal of Zoology–Revue Canadienne de Zoologie*, **76**, 1371–81.

Morrison, I. (1990) Climatic changes and human geography: Scotland in a North Atlantic context. *Northern Studies*, **27**, 1–11.

Mudar, K. & Speaker, S. (2003) Natural catastrophes in Arctic populations: the 1878–1880 famine on Saint Lawrence Island, Alaska. *Journal of Anthropological Archaeology*, **22**, 75–104.

Munn, A.J., Barboza, P.S., & Dehn, J. (2009) Sensible heat loss from muskoxen (*Ovibos moschatus*) feeding in winter: small calves are not at a thermal disadvantage compared with adult cows. *Physiological and Biochemical Zoology*, **82**, 455–67.

Murray, D.G. & Lipkin, R. (1987) *Candidate Threatened and Endangered Plants of Alaska*. University of Alaska, Fairbanks.

Murthy, U.M.N., Kumar, P.P., & Sun, W.Q. (2003) Mechanisms of seed ageing under different storage conditions for *Vigna radiata* (L.) Wilczek: lipid peroxidation, sugar hydrolysis, Maillard reactions and their relationship to glass state transition. *Journal of Experimental Botany*, **54**, 1057–67.

Myrberget, S. (1987) Introduction of mammals and birds in Norway including Svalbard. *Meddelelser fra Norsk Viltforskning*, **3**, 1–26.

Natcheva, R. & Cronberg, N. (2007) Recombination and introgression of nuclear and chloroplast genomes between the peat mosses, *Sphagnum capillifolium* and *Sphagnum quinquefarium*. *Molecular Ecology*, **16**, 811–18.

Nei, N. (1978) Estimation of average heterozygosity and genetic distance from a small number of individuals. *Genetics*, **89**, 583 90.

Neilson, R.E., Ludlow, M.M., & Jarvis, P.G. (1972) Photosynthesis in Sitka spruce (*Picea sitchensis* (Bong.) Carr.). *Journal of Applied Ecology*, **9**, 721–45.

Nesje, A. & Kvamme, M. (1991) Holocene glacier and climate variations in western Norway—evidence for early Holocene glacier demise and multiple neoglacial events. *Geology*, **19**, 610–12.

Nesje, A., Dahl, S.O., Linge, H., *et al.* (2007) The surface geometry of the Last Glacial Maximum ice sheet in the Andoya-Skanland region, northern Norway, constrained by surface exposure dating and clay mineralogy. *Boreas*, **36**, 227–39.

Neuvonen, S., Ruohomäki, K., Bylund, H., & Kaitaniemi, P. (2001) Insect herbivores and herbivory effects on mountain birch dynamics. In: *Nordic Mountain Birch Ecosystems* (ed. F.E. Wiellgolaski), pp. 207–22. Parthenon, New York.

Nielsen, O.K. (2010) Rock Ptarmigan and Gyrfalcon. *Natturufraedingurinn*, **79**, 8–18.

Nummi, P. & Hahtola, A. (2008) The beaver as an ecosystem engineer facilitates teal breeding. *Ecography*, **31**, 519–24.

Nydal, R. (1989) A critical review of radiocarbon dating of a Norse settlement at L'Anse aux Meadows, Newfoundland, Canada. *Radiocarbon*, **31**, 976–85.

Nystrom, J., Ekenstedt, J., Engstrom, J., & Angerbjorn, A. (2005) Gyr falcons, ptarmigan and microtine rodents in northern Sweden. *Ibis*, **147**, 587–97.

Nystrom, V., Dalen, L., Vartanyan, S., Liden, K., Ryman, N., & Angerbjorn, A. (2010) Temporal genetic change in the last remaining population of woolly mammoth. *Proceedings of the Royal Society B-Biological Sciences*, **277**, 2331–7.

Oksanen, L. & Virtanen, R. (1997) Adaptation to disturbance as a part of the strategy of Arctic and Alpine plants. In: *Disturbance and Recovery in Arctic Lands* (ed. R.M.M. Crawford), pp. 91–113. Kluwer, Dordrecht.

Oksanen, T., Oksanen, L., Dahlgren, J., & Olofsson, J. (2008) Arctic lemmings, *Lemmus* spp. and *Dicrostonyx* spp.: integrating ecological and evolutionary perspectives. *Evolutionary Ecology Research*, **10**, 415–34.

Oli, M.K. (1999) The Chitty effect: a consequence of dynamic energy allocation in a fluctuating environment. *Theoretical Population Biology*, **56**, 293–300.

Olson, V.A., Davies, R.G., Orme, C.D.L., *et al.* (2009) Global biogeography and ecology of body size in birds. *Ecology Letters*, **12**, 249–59.

Paetkau, D., Amstrup, S.C., Born, E.W., *et al.* (1999) Genetic structure of the world's polar bear populations. *Molecular Ecology*, **8**, 1571–84.

Parducci, L., Jorgensen, T., Tollefsrud, M.M., *et al.* (2012) Glacial survival of boreal trees in northern Scandinavia. *Science*, **335**, 1083–6.

Patterson, W.P., Dietrich, K.A., Holmden, C., & Andrews, J.T. (2010) Two millennia of North Atlantic seasonality and implications for Norse colonies. *Proceedings of the National Academy of Sciences of the United States of America*, **107**, 5306–10.

Payette, S. & Delwaide, A. (2004) Dynamics of subarctic wetland forests over the past 1500 years. *Ecological Monographs*, **74**, 373–91.

Payette, S. & Rochefort, L. (2001) *Ecologies des Tourbièes du Québec-Labrador*. Les Presses de l'Université Laval, Saint-Nicolaus.

Payette, S., Boudreau, S., Morneau, C., & Pitre, N. (2004) Long-term interactions between migratory caribou, wildfires and nunavik hunters inferred from tree rings. *Ambio*, **33**, 482–6.

Payette, S., Filion, L., & Delwaide, A. (2008) Spatially explicit fire-climate history of the boreal forest-tundra (eastern Canada) over the last 2000 years. *Philosophical Transactions of the Royal Society B-Biological Sciences*, **363**, 2301–16.

Pearson, P.N. & Palmer, M.R. (2000) Atmospheric carbon dioxide concentrations over the past 60 million years. *Nature*, **406**, 695–9.

Pedersen, A.O., Lier, M., Routti, H., Christiansen, H.H., & Fuglei, E. (2006) Co-feeding between Svalbard rock ptarmigan (*Lagopus muta hyperborea*) and Svalbard reindeer (*Rangifer tarandus platyrhynchus*). *Arctic*, **59**, 61–4.

Pedersen, H.C., Fossoy, F., Kalas, J.A., & Lierhagen, S. (2006) Accumulation of heavy metals in circumpolar willow ptarmigan (*Lagopus l. lagopus*) populations. *Science of the Total Environment*, **371**, 176–89.

Pfanz, H. (1999) Photosynthetic performance of twigs and stems of trees with and without stress. *Phyton-Annales Rei Botanicae*, **39**, 29–33.

Philipp, M. (1998) Genetic variation in four species of *Pedicularis* (Scrophulariaceae) within a limited area in West Greenland. *Arctic and Alpine Research*, **30**, 396–9.

Philipp, M., Böcher, J., Mattsson, O., & Woodell, S.R.J. (1990) A quantitative approach to the sexual reproductive biology and population structure in some arctic flowering plants: *Dryas integrifolia, Silene acaulis* and *Ranunculus nivalis. Meddelelser om Grønland: Bioscience*, **34**, 3–60.

Pielou, E.C. (1994) *A Naturalist's Guide to the Arctic.* University of Chicago Press, Chicago.

Pitulko, V.V. (2011) The Berelekh Quest: a review of forty years of research in the mammoth graveyard in northeast Siberia. *Geoarchaeology—an International Journal*, **26**, 5–32.

Pitulko, V.V., Nikolsky, P.A., Girya, E.Y., *et al.* (2004) The Yana RHS site: humans in the Arctic before the Last Glacial Maximum. *Science*, **303**, 52–6.

Polunin, N. (1959) *Circumpolar Arctic Flora.* Clarendon Press, Oxford.

Porter, S.C., Sauchyn, D.J., & Delorme, L.D. (1999) The ostracode record from Harris Lake, southwestern Saskatchewan: 9200 years of local environmental change. *Journal of Paleolimnology*, **21**, 35–44.

Powell, J.A. & Engelhardt, K.A.M. (2000) Optimal trajectories for the short-distance foraging flights of swans. *Journal of Theoretical Biology*, **204**, 415–30.

Prentice, I.C. & Webb, T. (1998) BIOME 6000: reconstructing global mid-Holocene vegetation patterns from palaeoecological records. *Journal of Biogeography*, **25**, 997–1005.

Prins, H.H.T. (1982) Why are mosses eaten in cold environments only? *Oikos*, **38**, 374–80.

Raunkiaer, C. (1904) Biological types with reference to the adaptation of plants to survive the unfavourable season. In: *History of Ecology: Life Form of Plants and Statistical Plant Geography* (ed. F.N. Egerton). Arno Press, New York.

Rees, W.G. (1999) Remote sensing of oil spills on frozen ground. *Polar Record*, **35**, 19–24.

Rees, W.G. & Danks, F.S. (2007) Derivation and assessment of vegetation maps for reindeer pasture analysis in Arctic European Russia. *Polar Record*, **43**, 290–304.

Rees, W.G. & Williams, M. (1997) Satellite remote sensing of the impact of industrial pollution on tundra biodiversity. In: *Disturbance and Recovery in Arctic Lands: An Ecological Perspective* (ed. R.M.M. Crawford), pp. 253–82. Kluwer, Dordrecht.

Reierth, E., Van't Hof, T.J., & Stokkan, K.A. (1999) Seasonal and daily variations in plasma melatonin in the high-arctic Svalbard ptarmigan (*Lagopus mutus hyperboreus*). *Journal of Biological Rhythms*, **14**, 314–19.

Rexstad, E. & Kielland, K. (2006) Mammalian herbivore population dynamics in the Alaskan boreal forest. In: *Alaska's Changing Boreal Forest* (eds F.S. Chapin III, M.W. Oswood, K. Van Kleve, L.A. Viereck, & D.L. Verbyla), pp. 121–32. Oxford Univiersity Press, Oxford.

Rich, T.H., Vickers-Rich, P., & Gangloff, R.A. (2002) Polar dinosaurs. *Science*, **295**, 979–80.

Rickaby, R.E.M. & Elderfield, H. (2005) Evidence from the high-latitude North Atlantic for variations in Antarctic Intermediate water flow during the last deglaciation. *Geochemistry Geophysics Geosystems*, **6**, Q05001, Epub.

Robinson, M. & Becker, C.D. (1986) Owls on Fetlar. *British Birds*, **79**, 228–42.

Roed, K.H. (2007) Taxonomy and origin of reindeer. *Rangifer Report*, **27**(3), 17–20.

Roed, K.H., Flagstad, O., Bjornstad, G., & Hufthammer, A.K. (2011) Elucidating the ancestry of domestic reindeer from ancient DNA approaches. *Quaternary International*, **238**, 83–8.

Roesdal, E. (2005) Walrus ivory—demand, supply, workshops, and Greenland. In: *Viking and Norse in the North Atlantic* (eds A. Mortensen & S.V. Arge), pp. 182–91. Annales Societatis Scientiarum Faeroensis, Torshavn.

Rosell, F., Bozser, O., Collen, P., & Parker, H. (2005) Ecological impact of beavers *Castor fiber* and *Castor canadensis* and their ability to modify ecosystems. *Mammal Review*, **35**, 248–76.

Roy, V., Bernier, P.Y., Plamondon, A.P., & Ruel, J.C. (1999) Effect of drainage and microtopography in forested wetlands on the microenvironment and growth of planted black spruce seedlings. *Canadian Journal of Forest Research–Revue Canadienne de Recherche Forestiere*, **29**, 563–74.

Rozema, J., Boelen, P., Doorenbosch, M., *et al.* (2006) A vegetation, climate and environment reconstruction based on palynological analyses of high arctic tundra peat cores (5000–6000 years BP) from Svalbard. *Plant Ecology*, **182**, 155–73.

Rundgren, M. & Ingolfsson, O. (1999) Plant survival in Iceland during periods of glaciation? *Journal of Biogeography*, **26**, 387–96.

Safriel, U.N., Volis, S., & Kark, S. (1994) Core and peripheral populations and global climate change. *Israel Journal of Plant Sciences*, **42**, 331–45.

Sala, A., Piper, F., & Hoch, G. (2010) Physiological mechanisms of drought-induced tree mortality are far from being resolved. *New Phytologist*, **186**, 274–81.

Sale, R. (1988) *A Complete Guide to Arctic Wild Life.* Christoher Helm, London.

Salewski, V. & Bruderer, B. (2007) The evolution of bird migration—a synthesis. *Naturwissenschaften*, **94**, 268–79.

Sand-Jensen, K., Riis, T., Markager, S., & Vincent, W.F. (1999) Slow growth and decomposition of mosses in Arctic lakes. *Canadian Journal of Fisheries and Aquatic Sciences*, **56**, 388–93.

Sarmaja-Korjonen, K. & Seppa, H. (2007) Abrupt and consistent responses of aquatic and terrestrial ecosystems to the 8200 cal. yr cold event: a lacustrine record from Lake Arapisto, Finland. *Holocene,* **17**, 457–67.

Savolainen, O. (1996) Pines beyond the polar circle: adaptation to stress conditions. *Euphytica,* **92**, 139–45.

Schmitt, J., Schneider, R., Elsig, J., *et al.* (2012) Carbon isotope constraints on the deglacial CO_2 rise from ice cores. *Science,* **336**, 711–14.

Schofield, J.E. & Edwards, K.J. (2011) Grazing impacts and woodland management in Eriksfjord: Betula, coprophilous fungi and the Norse settlement of Greenland. *Vegetation History and Archaeobotany,* **20**, 181–97.

Schönswetter, P., Paun, O., Tribsch, A., & Niklfeld, H. (2003) Out of the Alps: colonization of northern Europe by East Alpine populations of the Glacier Buttercup *Ranunculus glacialis* L. (Ranunculaceae). *Molecular Ecology,* **12**, 3373–81.

Schönswetter, P., Popp, M., & Brochmann, C. (2006) Rare arctic-alpine plants of the European Alps have different immigration histories: the snow bed species *Minuartia biflora* and *Ranunculus pygmaeus*. *Molecular Ecology,* **15**, 709–20.

Schöter, C. (1898) Über die Vielgestaltigkeit der Fichte (Pice excelsa Link). *Vierteljahresschrift der Naturforschenden Gesellschaft in Zürich,* **43**, 125–252.

Schurr, T.G. (2004) The peopling of the New World: perspectives from molecular anthropology. *Annual Review of Anthropology,* **33**, 551–83.

Schwarzenbach, F.H. (2000) Altitude distribution of vascular plants in mountains of East and Northeast Greenland *Meddelelser øm Grønland, Bioscience,* **50**, 1–192.

Seel, W.E. & Press, M.C. (1993) Influence of the host on 3 sub-arctic annual facultative root hemiparasites. 1. Growth, mineral accumulation and aboveground drymatter partitioning. *New Phytologist,* **125**, 131–8.

Seldal, T., Andersen, K.-J., & Högstedt, G. (1994) Grazing-induced proteinase inhibitors: a possible cause for lemming population cycles. *Oikos,* **70**, 3–11.

Sformo, T., Walters, K., Jeannet, K., *et al.* (2010) Deep supercooling, vitrification and limited survival to −100 degrees C in the Alaskan beetle *Cucujus clavipes puniceus* (Coleoptera: Cucujidae) larvae. *Journal of Experimental Biology,* **213**, 502–9.

Sher, A.V. (1996) Late-Quaternary extinction of large mammals in northern Eurasia: a new look at the Siberian contribution. In: *Past and Future Rapid Environmental Changes* (eds B. Huntley, W. Cramer, A.V. Morgan, H.C. Prentice, & J.R.M. Allen), pp. 319–39. Springer, Berlin.

Shi, T., Reeves, R.H., Gilichinsky, D.A., & Friedmann, E.I. (1997) Characterization of viable bacteria from Siberian permafrost by 16S rDNA sequencing. *Microbial Ecology,* **33**, 169–79.

Sigman, D.M. & Boyle, E.A. (2000) Glacial/interglacial variations in atmospheric carbon dioxide. *Nature,* **407**, 859–69.

Sigourney, D.B., Letcher, B.H., & Cunjak, R.A. (2006) Influence of beaver activity on summer growth and condition of age-2 Atlantic salmon parr. *Transactions of the American Fisheries Society,* **135**, 1068–75.

Simard, M., Lecomte, N., Bergeron, Y., Bernier, P.Y., & Pare, D. (2007) Forest productivity decline caused by successional paludification of boreal soils. *Ecological Applications,* **17**, 1619–37.

Sipko, T.P. (2010) Reintroduction of musk ox in the northern Russia. *Large Herbivore Network,* http://www.large-herbivore.org/interesting-articles.

Sipko, T.P. & Gruzdev, A.R. (2006) Re-introduction of muskoxen in northern Russia. *Re-introduction News,* **25**, 25–6.

Sistla, S.A., Moore, J.C., Simpson, R.T., Gough, L., Shaver, G., & Schimel, J.P. (2013) Long-term warming restructures Arctic tundra without changing net soil carbon storage. *Nature,* Epub, 15 May 2013.

Smith, L.C., Sheng, Y., MacDonald, G.M., & Hinzman, L.D. (2005) Disappearing Arctic lakes. *Science,* **308**, 1429–9.

Smith, T.M., Tafforeauc, P., Reid, D.J., *et al.* (2010) Dental evidence for ontogenetic differences between modern humans and Neanderthals. *Proceedings of the National Academy of Sciences of the United States of America,* **107**, 20923–8.

Soltis, D.E., Buggs, R.J.A., Doyle, J.J., & Soltis, P.S. (2010) What we still don't know about polyploidy. *Taxon,* **59**, 1387–403.

Sonstebo, J.H., Gielly, L., Brysting, A.K., *et al.* (2010) Using next-generation sequencing for molecular reconstruction of past Arctic vegetation and climate. *Molecular Ecology Resources,* **10**, 1009–18.

Southwood, T.R.E. (1988) Tactics, strategies and templates. *Oikos,* **52**, 3–18.

Sperber, I., Bjornhag, G., & Ridderstrale, Y. (1983) Function of proximal colon in lemming and rat. *Swedish Journal of Agricultural Research,* **13**, 243–56.

Staaland, H. & Punsvik, T. (1979) Reindeer grazing on Nordaustlandet, Svalbard. *Proceedings of the Second International Reindeer/Caribou Symposium* (eds E. Reimers, E. Gaare, & S. Skjenneberg), 17–21 September, Roros, Norway.

Stebbins, G.L. (1971) *Chromosomal Evolution in Higher Plants.* Edward Arnold, London.

Stokkan, K.A. (1992) Energetics and adaptations to cold in ptarmigan in winter. *Ornis Scandinavica,* **23**, 366–70.

Stokkan, K.A., Mortensen, A., & Blix, A.S. (1986) Food-intake, feeding rhythm, and body-mass regulation in Svalbard rock ptarmigan. *American Journal of Physiology,* **251**, R264–7.

Stone, N. (2011) *Turkey: A Short History*. Thames and Hudson, London.

Strakhovenko, V.D., Shcherbov, B.L., & Khozhina, E.I. (2005) Distribution of radionuclides and trace elements in the lichen cover of West Siberian regions. *Geologiya I Geofizika*, **46**, 206–16.

Strand, P. (1999) Radioactivity. In: *AMAP Assessment Report: Arctic Pollution Issues. Arctic Monitoring and Assessment Programme* (eds S.J. Wilson, J.L. Murray, & H.P. Huntington), pp. 525–619. Arctic Monitoring and Assessment Programme (AMAP), Oslo.

Strutzik, E. (2000) And then there were 84,000: the return of musk-oxen to Canada's Banks Island in recent decades is just one chapter of a beguiling Arctic mystery. *International Wildlife*, **30**(1), 28.

Stuart, A.J. (2005) The extinction of woolly mammoth (*Mammuthus primigenius*) and straight-tusked elephant (*Palaeoloxodon antiquus*) in Europe. *Quaternary International*, **126**, 171–7.

Sundell, J. & Norrdahl, K. (2002) Body size-dependent refuges in voles: an alternative explanation of the Chitty effect. *Annales Zoologici Fennici*, **39**, 325–33.

Sundset, M.A., Barboza, P.S., Green, T.K., Folkow, L.P., Blix, A.S., & Mathiesen, S.D. (2010) Microbial degradation of usnic acid in the reindeer rumen. *Naturwissenschaften*, **97**, 273–8.

Symon, C. (2010) *Radioactivity in the Arctic*. Arctic Monitoring and Assessment Programme (AMAP), Oslo.

Taberlet, P., Dawson, A.G., Hickey, K., Mayewsk, P.A., & Nesje, A. (2007) Greenland (GISP2) ice core and historical indicators of complex North Atlantic climate changes during the fourteenth century. *The Holocene*, **17**, 425–32.

Talbot, S.L. & Shields, G.F. (1996) A phylogeny of the bears (Ursidae) inferred from complete sequences of three mitochondrial genes. *Molecular Phylogenetics and Evolution*, **5**, 567–75.

Talbot, S.S., Yurtsev, B.A., Murray, D.F., Argus, G.W., Bay, C., & Elvebakk, A. (1999) Atlas of rare endemic vascular plants of the Arctic. In: *Conservation of Arctic Flora and Fauna (CAFF) Technical Report No. 3*, US Fish and Wildlife Service, pp. iv + 73.

Tambets, K., Rootsi, S., Kivisild, T., *et al.* (2004) The western and eastern roots of the Saami—the story of genetic 'outliers' told by mitochondrial DNA and Y chromosomes. *American Journal of Human Genetics*, **74**, 661–82.

Tenow, O., Bylund, H., & Holmgren, B. (2001) Impact on mountain birch forests in the past and future of outbreaks of two geometrid insects. In: *Nordic Mountain Birch Ecosystems* (ed. F.E. Wielgolaski), pp. 223–39. Parthenon, New York.

Teskey, R.O., Saveyn, A., Steppe, K., & McGuire, M.A. (2008) Origin, fate and significance of CO_2 in tree stems. *New Phytologist*, **177**, 17–32.

Therrien, J.-F., Gauthier, G., & Bety, J. (2011) An avian terrestrial predator of the Arctic relies on the marine ecosystem during winter. *Journal of Avian Biology*, **42**, 363–9.

Thomas, D.C., Miller, F.L., Russell, R.H., & Parker, G.R. (1981) The Bailey Point region and other muskox refugia in the Canadian Arctic: a short review. *Arctic*, **34**, 34–6.

Timoney, K.P. (2009) Three centuries of change in the Peace-Athabasca Delta, Canada. *Climatic Change*, **93**, 485–515.

Timoney, K.P. & Ronconi, R.A. (2010) Annual bird mortality in the bitumen tailings ponds in northeastern Alberta, Canada. *Wilson Journal of Ornithology*, **122**, 569–76.

Tkach, N.V., Hoffmann, M.H., Roeser, M., Korobkov, A.A., & von Hagen, K.B. (2008) Parallel evolutionary patterns in multiple lineages of arctic *Artemisia* L. (Asteraceae). *Evolution*, **62**, 184–98.

Tollefsrud, M.M., Bachmann, K., Jakobsen, K.S., & Brochmann, C. (1998) Glacial survival does not matter—II: RAPD phylogeography of Nordic *Saxifraga cespitosa*. *Molecular Ecology*, **7**, 1217–32.

Tolmatchev, A.I. (1960) Der autochthone Grundstock der arktischen Flora und ihre Beziehungen zu den Hochgebirgsfloren Nord- und Zentralasiens. *Botanisk Tidsskrift*, **55**, 269–76.

Toutoubalina, O.V. & Rees, W.G. (1999) Remote sensing of industrial impact on Arctic vegetation around Noril'sk, northern Siberia: preliminary results. *International Journal of Remote Sensing*, **20**, 2979–90.

Valentini, R., Dore, S., Marchi, G., *et al.* (2000) Carbon and water exchanges of two contrasting central Siberia landscape types: regenerating forest and bog. *Functional Ecology*, **14**, 87–96.

van Altena, C., van Logtestijn, R.S.P., Cornwell, W.K., & Cornelissen, J.H.C. (2012) Species composition and fire: non-additive mixture effects on ground fuel flammability. *Frontiers in Plant Science*, **3**, 63, Epub.

Van Bogaert, R., Jonasson, C., De Dapper, M., & Callaghan, T.V. (2009) Competitive interaction between aspen and birch moderated by invertebrate and vertebrate herbivores and climate warming. *Plant Ecology & Diversity*, **2**, 221–32.

van der Wal, R. (2006) Do herbivores cause habitat degradation or vegetation state transition? Evidence from the tundra. *Oikos*, **114**, 177–86.

Väre, H. (2001) Mountain birch taxonomy and floristics of mountain birch woodlands. In: *Nordic Mountain Birch Ecosystems* (ed. F.E. Wielgolaski). Parthenon, New York.

Varricchio, D.J., Martin, A.J., & Katsura, Y. (2007) First trace and body fossil evidence of a burrowing, denning dinosaur. *Proceedings of the Royal Society B-Biological Sciences*, **274**, 1361–8.

Vartanyan, S.L., Arslanov, K.A., Karhu, J.A., Possnert, G., & Sulerzhitsky, L.D. (2008) Collection of radiocarbon dates on the mammoths (*Mammuthus primigenius*) and other genera of Wrangel Island, northeast Siberia, Russia. *Quaternary Research*, **70**, 51–9.

Vavrek, M.C., McGraw, J.B., & Bennington, C.C. (1991) Ecological genetic-variation in seed banks. 3. Phenotypic and genetic differences between young and old seed populations of *Carex bigelowii*. *Journal of Ecology*, **79**, 645–62.

Vera, F.W.M. (2000) *Grazing Ecology and Forest History*. CABI, Wallingford.

Vilchek, G.E. & Tishkov, A.A. (1997) Usinsk oil spill. In: *Disturbance and Recovery in Arctic Lands: An Ecological Perspective* (ed. R.M.M. Crawford), pp. 411–20. Kluwer, Dordrecht.

Vorren, T.O., Vorren, K.D., Alm, T., Gulliksen, S., & Lovlie, R. (1988) The last deglaciation (20,000 to 11,000 BP) on Andøya, northern Norway. *Boreas*, **17**, 41–77.

Vors, L.S. & Boyce, M.S. (2009) Global declines of caribou and reindeer. *Global Change Biology*, **15**, 2626–33.

Wagner, D. & Liebnerz, S. (2010) Methanogenesis in Arctic permafrost habitats. In: *Handbook of Hydrocarbonvand Lipid Microbiology* (ed. K.N. Timmis), pp. 655–66. Springer, Berlin.

Walker, D.A., Raynolds, M.K., Daniels, F.J.A., et al. (2005) The Circumpolar Arctic vegetation map. *Journal of Vegetation Science*, **16**, 267–82.

Wallace, B. (2011) L'Anse aux Meadows: different disciplines, divergent views. In: *Viking Settlements and Viking Society* (eds A. Holt, G. Sigurðsson, G. Olafsson, & O. Vesteinsson), pp. 448–68. University of Iceland Press, Rekkjavik.

Wallace, B.L. (2003) Vinland and the death of Thorvalldr. In: *Vinland Revisited: The Norse World at the Turn of the First Millennium* (ed. S. Lewis-Simpson), pp. 377–90. Historic Sites Association of Newfoundland and Labrador, St John's, Newfoundland.

Waters, M.R., Stafford, T.W. Jr, McDonald, H.G., et al. (2011) Pre-Clovis mastodon hunting 13,800 years ago at the Manis Site, Washington. *Science*, **334**, 351–3.

Weber, J.-M. (2009) The physiology of long-distance migration: extending the limits of endurance metabolism. *Journal of Experimental Biology*, **212**, 593–7.

Weladji, R.B., Gaillard, J.M., Yoccoz, N.G., et al. (2006) Good reindeer mothers live longer and become better in raising offspring. *Proceedings of the Royal Society B-Biological Sciences*, **273**, 1239–44.

Westergaard, K.B., Alsos, I.G., Popp, M., Engelskjon, T., Flatberg, K.I., & Brochmann, C. (2011) Glacial survival may matter after all: nunatak signatures in the rare European populations of two west-arctic species. *Molecular Ecology*, **20**, 376–93.

Wielgolaski, F.E. & Sonesson, M. (2001) Nordic mountain birch ecosystems—a conceptual overview. In: *Nordic Mountain Birch Ecosystems* (ed. F.E. Wielgolaski), pp. 377–84. Parthenon, New York.

Wieser, G., Gigele, T., & Pausch, H. (2005) The carbon budget of an adult *Pinus cembra* tree at the alpine timberline in the Central Austrian Alps. *European Journal of Forest Research*, **124**, 1–8.

Wilkins, K.A., Malecki, R.A., Sullivan, P.J., et al. (2010) Migration routes and bird conservation regions used by eastern population tundra swans *Cygnus columbianus columbianus* in North America. *Wildfowl*, **60**, 20–37.

Willerslev, E., Cappellini, E., Boomsma, W., et al. (2007) Ancient biomolecules from deep ice cores reveal a forested southern Greenland. *Science*, **317**, 111–14.

Williams, C.T., Barnes, B.M., & Buck, C.L. (2012) Daily body temperature rhythms persist under the midnight sun but are absent during hibernation in free-living arctic ground squirrels. *Biology Letters*, **8**, 31–4.

Winkler, M., Tribsch, A., Schneeweiss, G.M., et al. (2012) Tales of the unexpected: phylogeography of the arctic-alpine model plant *Saxifraga oppositifolia* (Saxifragaceae) revisited. *Molecular Ecology*, **21**, 4618–30.

Wolff, J.O. (1980) Social organization of the taiga vole *Microtus xanthognathus*. *Biologist (Charleston)*, **62**, 34–45.

Yan, J., Burman, A., Nichols, C., et al. (2006) Detection of differential gene expression in brown adipose tissue of hibernating arctic ground squirrels with mouse microarrays. *Physiological Genomics*, **25**, 346–53.

Young, S.B. (1989) *To the Arctic*. John Wiley, New York.

Ytrehus, B., Bretten, T., Bergsjo, B., & Isaksen, K. (2008) Fatal pneumonia epizootic in musk ox (*Ovibos moschatus*) in a period of extraordinary weather conditions. *Ecohealth*, **5**, 213–23.

Zalatan, R., Gunn, A., & Henry, G.H.R. (2006) Long-term abundance patterns of barren-ground caribou using trampling scars on roots of *Picea mariana* in the Northwest Territories, Canada. *Arctic Antarctic and Alpine Research*, **38**, 624–30.

Zhang, T., Barry, R.G., Knowles, K., Heginbottomb, J.A., & Brown, J. (1999) Statistics and characteristics of permafrost and ground-ice distribution in the northern hemisphere. *Polar Geography*, **31**, 47–68.

Index

8,200 BP event, 22

A
aapa mire, 101, 102, 246
Abies balsamea, 118
abscisic acid, 152, 255
acclimation, 151, 246
acclimatization, 129, 151, 246
  acclimatization and metabolic rates, 127
acrotelm, 101, 102, 105, *see also* catotelm
active oxygen species, 155
adaptation
  antifreeze proteins, 153
  capacity adaptation, 149
  functional adaptation, 151
  cold, 127–129
  diet, 214
  fat accumulation, 145
  fatty acids, 198
  freezing tolerance, 130
  low growth rates, 191
  metabolic depression, 138
  pigmentation, 126
  pre-adaptation, 190
adenosine triphosphate (ATP), 127
Aedes spp. *see* mosquitoes
agamospermy, 189 *see* apomixis
agriculture, 1, 214
Alaska
  Agrostis latifolia – chromosome races, 198
  bar-tailed godwits, 197
  beetle injury, 118
  climate and volcanoes, 167
  climate reconstruction, 166
  Eskimo encounters, 68
  forest fires, 97
  ground squirrels, 127
  moose, 110
  moss flora, 79
  muskoxen, 91
  North Slope 9, 84
  Northern Alaska, 61–63
  palsas, 103
  paludification, 104
  Pliocene refugia, 204
  refugia, 26, 30
  seed survival, 149
  semipalmated sandpipers, 195
  swans, 194
  taiga, 93
  tall shrubs, 76
  Tanana Flats, 105
  thermal microsites, 25
  treeline, 2, 32, 33
  tundra shrew clades, 206
  tussock ages, 168
  wheatears, 193
  wolf predation, 119
albedo, 5, 8, 15, 17, 246
Alces alces, 37, 76, 108, 110, 119, 254
  see Elk (Europe) or Moose (North America)
alcohol dehydrogenase, 215
alcohol, 215
Alder, *see Alnus*
Aleuts, 53, 59
Allen's rule, 124, 137
Alnus, 76, 84, 93, 96, 97, 250, *see* Alder
Alopex lagopus, see Arctic Fox
Alpine Butterwort (*Pinguicula alpina*), 176
Alpine Meadow Grass (*Poa alpina* var. *vivipara*), 188
Altai Mountains, 202
American beaver, 110 *see Castor canadensis*
amphipods, 196
amplified fragment length polymorphism, 28
Anabaena azollae, 15
anaerobic environments, 15, 43, 44, 101, 102, 149, 155, 156, 189, 246
anaerobiosis, 100, 154, 162
ancient phenomena
  DNA, 33, 39, 54, 56, 221, 222, 260
  forests, 10
  megafauna, 218
  Polar Bears, 208
  root scars, 230
  species, 231, 233, 245
  trees, 98
  Turkish, 70
Andøya, 26, 28, 29, 250
annual plants, 184
anoxia, 155, 156, 187, 189, 252
Anthoxanthum odoratum, 76
  sweet vernal grass
anti-oxidants, 155
antifreeze proteins, 153
apical buds, 152
Apodeumus sylvaticus, 133 *see* wood mouse
Apomixis, 180, *see* agamospermy
apoplast, 153
Archangel, 111, *see* White Sea
arctic avifauna, 192
Arctic Barrens, 71–2 *see* Polar desert
Arctic Bearberry, *see Arctostaphylos alpinus*
Arctic beetle, *see Pterostichus brevicornis*
arctic bryophytes, 79
Arctic Cotton Grass, 204 *see Eriophorum scheuchzeri*
arctic forests, *see* taiga
Arctic Fox, 87, 125, 136, 137, 138, 168, 172, 173, *see Alopex lagopus*
Arctic Hare (*Lepus arcticus*), 102
Arctic Heather, *see Cassiope tetragona*
Arctic herbivores, 130
Arctic insects, 129
Arctic islands
  Axel Heiberg Island, 10, 11, 12, 13, 223
  Baffin Island, 66, 139, 159, 202
  Franz Josef Land, 142

New Siberian Islands, 137, 230
Novaya Zemlya, 26, 51, 89, 137, 210, 241
Wrangel Island, 36, 37, 38, 39, 40, 56, 57, 58, 72, 225, 230, 231
Arctic Norway, 5
Arctic Ocean
circulation changes, 23
climatic influences, 4–8
coastal migration routes, 203–204
effect on paludification, 100–104
geographical history, 15–16, 42
ice status, 245
Nansen's voyage, 137
Palaeolithic shore settlements, 190
Arctic Poppy, 146 see *Papaver radicatum*
arctic ruminants, 131
Arctic Tern, 138 see *Sterna paradiseae*
arctic terrestrial birds, 138
arctic treeline, 1, 2, 257
Arctic Willow see *Salix arctica*,
Arctic Wintergreen see *Pyrola grandiflora*, 191
arctic-alpine species, 150
arctic-altai element, 205
Arctostaphylos alpinus, 75 see Arctic bearberry
Arenaria humifusa, 27, 28 see Low Sandwort
asexual reproduction, 180, 181, 182, 198, 248, see clones
Aspen see *Populus tremula* Europe *P. tremuloides* N. America
Atlantic meridional overturning circulation (AMOC), 20
Autumn moth see *Epirrita autumnata*
Avena sativa, 65 see oats
avian circadian pacemaker, 175
Azolla event, 16
Azolla spp, 15, 16

B

baby mammoth, 128
Barents Sea, 5, 42, 137, 219, 237
Barnacle Geese, 172, 173, 174, 192 see *Branta leucopsis*
Basal metabolic rates, 127
bears, 51, 110, 116, 118, 127, 168
see also
Brown Bear
Grizzly Bear
Polar Bear
adipose metabolism, 219
behaviour, 123

Beavers, 110–113, 124 see European and American beavers
Bell Sound, 218
Beluga Whale (*Delphinapterus leucas*), 213
Bergmann's rule, 124, 125
Beringia, 26, 29, 32, 33, 40, 41, 51, 53, 119, 184, 186, 204, 209, 210, 232, 246, 251, 253
Betula nana, 79, 158, 159, 160 see Dwarf Birch
Betula pendula, 159 see Silver Birch
Betula pubescens var. *czerpanovii*, 159 see Mountain Birch
Betula pubescens see common birch
Betula, 12, 13, 33, 42, 47, 76, 79, 93, 94, 96, 97, 109, 117, 158, 159, 160, 161, 162, 253, 261
Big Freeze see Younger Dryas period
biodiversity, 100, 169, 217, 253
birch see *Betula*
bird migration, 191–8, 260
birth rate, 164
Bison
Bison bison, 34, 108
Black Spruce see *Picea mariana*,
Blomsterstrand Peninsula, Kongsfjorden, Spitsbergen, 13, 19, 261
Blue Heath see *Phyllodoce caerulea*
body core temperature, 137
body mass, 108, 123, 127, 130, 131, 138, 143, 194, 196, 254
Bog bilberry, 199, see *Vaccinium uliginosum*
bog growth, see paludification,
bogs, 54, 93, 100, 101, 102, 103, 104, 105, 106, 107, 113, 119, 120, 168, 237, 246, 248
Bølling-Allerød interstadial, 21
boreal forests, 93, 97, 101, 104, 106, 110, 113, 215 see Taiga
Bos bison atha-baskae, 239 see Woodland Caribou
Branta leucopsis, 173, see Barnacle geese
Brattahlið, Eric the Red's Greenland base, 65
breeding
breeding life/success, 167–8, 172–3, 174, 193, 212, 225, 241, 252, 254–5, 259
breeding seasons 135, 136
breeding territoiries 57, 104, 111, 122, 140–2, 167, 173, 192, 194–7, 202, 241
Brøgger Peninsula, 81, 89, 90, 108, 172, 230

brown adaptive tissue, 127
Brown Bear, 30, 118–9, 165, 125, 165, 206–9, 216, 228, 242, see *Ursus arctos*
brown fat, 127–8
Bubo scandiacus see Snowy Owl
bulbils, 181, 183
Bupalus piniarius, 118 see Pine-Looper
Butternuts (*Juglans cinerea*), 66
C. *buccinator*, 194
C. *magellanica*, 77
C. *marina*, 77
C. *membranacea*, 77
C. *rostrata*, 77
C. *sylvatica*, 131

C

Calidris pusilla, 195, 196, 197, see Semipalmated Sandpiper
Calliergon giganteum, 80 Giant water moss
Canadian Arctic Islands with the highest densities of Muskoxen, 224
Canadian Barren Ground Caribou (*R. t. groen- landicus*), 209
Canis lupus arctos, 119 see White Wolf
Canis lupus, 119, 227 see Grey Wolf
Carbon dioxide
atmospheric carbon dioxide, 13, 19, 20, 44, 82, 105, 124, 178
fixation, 105, 150, 151, 162
from soil respiration, 82, 105
in beaver lodges, 124
Carex spp.
C. *buccinator*, 194
C. *magellanica*, 77
C. *marina*, 77
C. *membranacea*, 77
C. *rostrata*, 77
C. *sylvatica*, 131
Caribou 41, 59, 102, 131, 209, 210, 211, 218, 223, 226, 227, 229, 250, 261, see *Rangifer tarandus*
Cassiope tetragona, 75, 102, 200, 201 see Arctic heather
Castor canadensis, 253 see American Beaver
Castor fiber, 110, 253, 260 see European Beaver
catotelm, 101, 102, 246 see also acrotelm
Cenozoic climatic decline/cooling, 204
Cephenemyia trompe, 108
Cerastium alpinum, 201, 252

Cerastium arcticum, 146, 147, 198, 201 see Arctic Mouse-ear Chickweed
Cerastium regelii, 133
Cervus canadensis, 41
Cervus elaphus, 108, see Red Deer
Cetraria islandica, 131 see Iceland Moss (a lichen)
Changes in obliquity, 17
Chen caeruluscens, 102, see Snow Geese
Chernobyl reactor accident, 242–3
Chitty effect, 126, 259, 262
chloroplast DNA haplotypes, 27, 29, 223
chloroplast DNA, 29
Choristoneura fumiferana, 118
Chukotka, 33, 61, 231
climatic oscillations, 14, 21, 199
climatic warming, 8, 14, 23, 43, 69, 77, 87, 88, 104, 105, 106, 115, 116, 154, 182, 183, 193, 227, 244, 245
Clonal genetic diversity, 181
clones, 82, 149, 169
Cochlearia officinalis, 133 see Scurvygrass
Coelodonta antiquitatis, 34 37, 38, 51 see Woolly Rhinoceros
cold climates, 213
cold phase of the Last Glacial, 206
cold stages of the Pleistocene, 204
cold winters, 34, 48, 129, 230
cold-adapted animals, 225
Colesdalen, 77, 78, 211
Collapsing vole population cycles, 86
Collared Lemming (*Dicrstonyx groenlandicus*), 87
Colville River Delta, 84
Common Butterwort (*Pinguicula vulgaris*, 176
Common Cotton Grass see Eriophorum vaginatum
Copenhagen University Botanic Garden, 231
Corophium volutator, 196, see mud shrimp
Corticular photosynthesis, 77, 161–2
Corvus corvex, 125, see Raven
cosmic-ray-produced isotopes, 26
cosmogenic nuclides, 48 counts 220
Cree Creek Delta in the Peace-Athabasca Delta, 238
 fixation, 150–1
 from soil respiration, 82
 inbreeding, 299
 isotope composition in Polar Bear breath, 219
 molecular ratio fixed: respired, 178
 outbreeding, 176, 200
 refixation by corticular photosynthesis, 162
 wetlands sinks, 105
Creeping Glow Wort see *Sibbaldia procumbens*
Creeping saltmarsh grass see *Puccinellia phryganodes*
Croll-Milankovitch theory, 16
Crowberry see *Empetrum nigrum*
Cucujus clavipes puniceus Flat Bark Beetle, 129
Croll, James, 16, 17
cryoprotection in ectothermic species, 129
cryoprotection, 129, 147
cryosols, 215, 247
cryoturbation and patterned ground, 82
cryoturbation, 43, 72, 80, 82, 94, 102, 217, 243
cyanobacteria, 82
cyclic vole populations, 126
Cygnus spp.
 C. columbianis bewickii, 194, Bewick's Swan
 C. columbianis columbianis, 194 Whistling (or Tundra) Swan
 C. cygnus cygnus, 194, Whooper Swan
 C. buccinator, 194 Trumpeter Swan

D
dating the ice retreat, 29
deep body temperature, 130
 defence strategies, 223
defining the Arctic, 1
defining the Tundra, 70
demography, 85, 86, 164, 255
Dendroctonus rufipennis, 118
Deschampsia flexuosa, 76
diapause, 129
Dicranum, 136
Dicrostonyx groenlandicus, 240
Disjunct plant distributions, 204–5
disruption by ice crystals, 129
distribution of rare endemic flowering plant species, 232
Ditrichum, 74
DNA, 12, 27, 28, 29, 35, 52, 54, 56, 119, 175, 201, 202, 204, 206, 208, 210, 217, 219, 221, 223, 246, 250, 252, 253, 254, 262
 human dated DNA, 53
 dating of mammoth survival, 39
 methylation, 201
 sequences in skeletons of Holocene muskoxen, 221
 vegetation history, 33
domestication, 225
Dovrefjell, 221, 225
Draba spp, 155, 198, 231, 251
 D. oxycarpa, 231
 D. pseudo-oxycarpa, 251
dropstones, 15
Dryas integrifolia, 102, 169, 182, 260
Dryas octopetala, 16, 21, 22, 155, 169, 170, 198, 200, 252, 258
Dwarf Birch see *Betula nana*
Dwarf Siberian Pine see *Pinus pumila*, 44

E
early-flowering outbreeder, 176
ectothermic species, 129
effects of permafrost, 43
Empetrum nigrum, 98 see Crowberry
endemism, 25
endotherm, 123
Eocene wood, 10
Eocene, 10, 11, 13, 14, 15, 16, 247, 251, 256
Epirrita autumnata, 117 see Autumn Moth
Equus caballus, 34, 37, 41 see Horses
Erect/tall shrub Tundra, 75
Erik the Red, 65
Eriophorum angustifolium. see common cotton grass
Eriophorum scheuchzeri, 77, 78, 102
Eriophorum vaginatum, 78 see Hare's-tail cotton grass
Erling Porsild, 84
Ermine, 87, 88, 126, 240, 241 see *Mustella erminea*
Eskimo, 54, 56, 57, 59, 63, 84, 248
ethanol, 215
Eurasian Reindeer herding, 64
Eurasian Tundra Reindeer (*R. t. tarandus*), 209
European beaver, 110–1 see *Castor fiber*
European Red Deer. see *Cervus elaphus*
Euthermic, 123
evidence for glacial refugia, 27
evolution of migration, 193
Evolution, 190, 191, 250, 253, 254, 256, 261, 262
 evolutionary lineages of northern mammals, 206
 evolutionary novelty, 201
 evolutionary success of polyploids, 200
extinctions, 34, 68, 225, 251

F

Falco rusticolus, 138, 139, Gyr falcon
Faroe Islands, 64, 243, 255, 256
Fatty acid composition of diet, 196
Festuca baffinensis, 147
Festuca hyperborea, 147
Festuca vivipara, 156
fixed-heterozygosity, 200
Flat Bark Beetle, 129 see *Cucujus clavipes puniceus*
floral induction, 181
Flowering in the Arctic, 182
Forest fire, 77, 94, 97–98
  bottom burning fires, 98
  flame speed, 98
Forest-Tundra, 1, 93, 95, 249, see Tundra-Taiga Interface
Forests, 209
frost damage, 94, 129, 153
frost-heave, 217
fur insulation, 130, 137
fur traders, 68
futile cycle, 128, 130

G

gene silencing, 201
genetic diversity, 27, 41, 56, 119, 168, 181, 198, 199, 204, 208, 211, 221, 258
genetic history and plant migration, 202
genetic isolation in arctic island reindeer, 211
glacial refugia, 23–33, 202
  bryophyte refufia, 80
  genetic evidence for glacial refugia, 27
  muskoxen refugia, 40, 223
Gleys and Fluviosols, 215
Gloger's rule, 126
Glycerol, 128–9
glycerol concentrations, 129
Godwits, 195, 197, 254
Graminoid Tundras, 71, 77
Great Bear, 1 see Ursa major
Greenland
  abandonment, 67
  ice core, 46, 234
  lemming cycles, 88
  polar oasis, 23
  settlement, 65
  treeline, 76
Grey Wolf, see *Canis lupus*
Grizzly Bear, 119, 125, 206–7, 228, 245, see *Ursus arctos* ssp. *horribilis*

H

haplotypes, 202, 203, 204, 209, 223, 256
Hare's Tail Grass (*Eriophorum vaginatum*), 78
Heavy metal pollution, 109, 234–6, 259
Hekla, 49
herbivores, 34, 41, 80, 82, 85, 89, 102, 108, 121, 130, 131, 132, 133, 135, 170, 190, 217, 241, 245, 247, 262, see herbivory
herbivory, 170, 247, 252, 259
hibernation, 126, 127
high-altitude record, 147, 150, 201
Holocene presence of plants at high latitudes, 30
homoiotherm, 123
horses in the Arctic, 41
  Equus caballus—sensu latu, 37
  North Asian Horse, 37 *Equus caballus—sensu latu—ssp. Leniencies*
  Przewalski's horse *E. caballus przewalskii*, 41
  Tarpan (*Equus caballus ferus*)
huddling and burrowing, 123
Hudson Bay, 99, 101, 171, 257
human contamination, 243
human immigrations, 53
Hylocomium splendens, 98
Hyperboreans, 1
hypothalamic oscillator, 175
hypothermia, 123

I

ice cores, 12, 18, 31, 46, 234, 247, 261, 1
ice-encased landscape, 227
Ice Grass See *Phippsia algida*
Iceland, 31, 44–7, 49, 64, 65, 67, 84, 108, 113, 114, 115, 116, 140, 156, 158, 159, 160, 166, 167, 172, 184, 186, 192, 194, 204, 231
Iceland Moss, see *Cetrarioa icelandica* 131
Incidence of fire, 98, 99
indigenous peoples of Arctic America, 61
Indigenous peoples of Arctic Europe, 58
Ingstad, Anne-Stine, 62, 66, 67
Ingstad, Helge, 62, 63, 66, 67
Insects, see arctic insects, 129
inter-specific hybridization, 201
Inupiaq Nunamiut, 63
Irish Brown Bears, 208
Islands, 5, 39, 40, 64, 65, 79, 89, 117, 137, 151, 167, 184, 210, 224, 225, 227, 230, 244, 255, 256, 258

K

Kamchatka, 208, 246
King Alfred, 64, 248
Krüppelholz, 95, 257, 160–1, 248
Krummholz, 95–6, 98, 99, 106, 157, 159–161, 248

L

L'Anse aux Meadows, 62, 66, 67, 259, 1
Labrador, 98, 103, 259
landnam, 64
language, 58
Larix gmelenii, 2
Larix laricina, 105, 254
Larix. sibirica, 2
Last Gla- cial Maximum, 51, 206, 260
Last Glacial Maximum, 23, 25, 26, 27, 28, 30, 31, 32, 40, 53, 206, 209, 210, 249, 259
late-flowering inbreeder, 176
Lauge Koch, 10
lead deposited in Greenland ice, 234
Least Weasel (*Mustela nivalis nivalis*), 126
Ledum groenlandicum, 79
Lemming population cycles, 85
lemmings, 82, 85, 86, 92, 123, 133, 135, 136, 138, 140, 142, 145, 164, 171, 172, 173, 188, 240, 244, 250, 259
Lesotundra, 72, 248
lichenin, 131
Little Ice Age, 44, 47, 48, 108
Little Rancheria Herd, 226, 227
Lodgepole Pine (*Pinus contorta*), 99
long migration journeys, 193
long-distance flying, 196
Long-tailed Skua, 87–8, 172 see *Stercorarius longicaudus*,
longevity, 80, 82, 149, 164, 167, 168, 169, 244, 253
Low Sandwort, see *Arenaria humifusa*

M

Mackenzie Delta, 32, 77, 111
Mackenzie River 56, 101, 171
macrofossil assemblages, 29
magmatism, 14
Mammoth, 34, 51, 129 see also Woolly Mammoth
mammoth calf, known as *Lyuba*, 128
Mammoth Graveyard, 51, 53
Martes zibeline, 68, see Sable
Mastodon, 52, 55
Mastodon Plant *Tephroseris palustris* var. *congesta*, 185
Mat-grass, see *Nardus stricta*
Maunder Minimum, 47, 48

Medieval Warm Period, 44, 45, 46, 47, 64, 65
megafauna decline, 34
Mesters Vig, 41, 75, 158, 236
metabolic adaptations to diet, 214
Metasequoia sp. 12
methanogenic activity, 43
migration phenomena, 14, 56
migration theory, 9
migrations of arctic species, 27
Mikhail Vasilyevich Lomonosov, 70
Milutin Milankovitch, 17
mining and smelting complex at Noril'sk, 235
Minuartia rubella, 147 see Mountain Sandwort
Minuartia biflora, 204 see Tufted Sandwort
mitochondria, 56, 127, 128
mitochondrial DNA, 56
modern Muskox mtDNA, 222
mollusks, 46
Mongol Hordes, 214
Mongolia, 38, 205
mosquitoes, 62, 115, see *Aedes spp.*
mosses as facilitators, 80
Mountain Avens, 21, 22, 170
Mountain Birch, 94, 109, 117, 158, 159, 160, 162, 259, 262
Mountain Sandwort, *see Minuartia rubella*
  Tufted Sandwort see *Minuartia biflora*, 204
muktu, 213
Mountain Sandwort see *Minuartia biflora*
Mud shrimp, see *Corophium volutator*
Muskoxen, 34, 37, 41, 51, 59, 89, 91, 130, 168, 218–221, 220, 223–226, 245, see *Ovibos moschatus*
  conservation, 220
  genetic diversity, 221
  in Alaska, 223
  in Canada, 223
  in Russia, 224
  mtDNA, 222
  Norway, 225
  numbers, 223
Mustella erminea Ermine, 87, 88, 126, 240, 241 see Ermine
Mustela nivalis, 126, see Least Weasel
Myodes glareolus, 133 see voles
Myopus schisticolor, 136, 253, see voles

N
Nardus stricta, 76 see Mat-grass
Nelumbo nucifera, 149 see Water Lily

Nenets, 58, 64, 76, 107, 239
neo-endemic species, 25
non- shivering thermogenesis, 127
non-equilibrium, 92
Nordaustlandet, 132, 147, 149, 252, 261
norepinephrine, 127
Norse arrival in America, 65
Norse settlements across the North Atlantic, 64
North American Woodland Caribou (*R. t. caribou*), 210
Northern European Shield, 58
Northern Indians, 68
northern limits for tree survival, 42
Northern mires, 100
Northern Wheatear, 193
Northern Wheatears, 192, 193, 194
'northern-home-theory', 191
nuclear weapons testing, 241, 242
Numbers of rare, endemic angiosperm species, 231
Nunamiut, 61, 62, 63, 213, 228, 251, 256
Nutritional deprivation in winter, 131

O
Oats, 65 see *Avena sativa*
obliquity of the ecliptic, 5
oceanic deep-water flow, 44
oil and bitumen (tar sands) pollution, 237
Oil spill detection, 240
Oldest Dryas, 21, 117
Ottar from Halogoland, 64
over-grazing, 170, 171, 244
Ovibos moschatus, 37, 41, 89, 91, 130, 168, 220, 251, 252, 255, 258
oxygen, 10, 14, 43, 78, 100, 105, 147, 150, 153, 154, 155, 156, 162, 187, 189, 249, 250, 252

P
Pale Whitlow-grass, see *Draba oxycarpa*
Palaeo-lithic cave artists, 128
Palaeo-Siberians, 58
Palaeocene–Eocene thermal maximum, 13
paludification, 5, 43, 93, 100, 104, 106, 120, 214, 261
Panolis flammea, 118 see Pine Beauty Moth
Papaver dahlianum, 133, see Svalbard Poppy
Papaver radicatum, 146, 147, see Arctic Poppy
pasteurellosis, 225
persistent organic pollutants, 240

Petuniabukta, 22, 79, 80, 84
phenology and reproduction, 173
Phippsia algida, 182, 184, 255 see Ice Grass
photoperiodic response, 173
Phyllodoce caerulea, 18 see Blue Heath
phylogeny, 207
phylogenetic relationships, 207, 210
physiological evolution, 196
physical and biological fragility, 217
Picea, 12, 29, 33, 42, 93, 96, 97, 100, 104, 105, 118, 146, 157, 228
  P. abies, 42, 100, 118, 157
  P. glauca, 97, 118, 157 see White Spruce
  P. glauca x lutzii, 118
  P. mariana, 96, 97, 100, 104, 105, 157, 228 see Black spruce
Pigmentation in the Arctic, 126
Pine Beauty Moth see *Panolis flammea*
Pine looper see *Bupalus piniarius*
Pingos, 79, 83
Pinguicula alpina, 176–7 see Alpine Butterwort
Pinus spp., 29, 44, 93, 96, 99
  P. banksiana, 100
  Pinus cembra, 13
  Picea sitchensis, 11
  Pinguicula villosa, 176–7
  Pinus nigra, 118
  Pinus pumila, 2, 4, 44, 96
  Pinus sibirica, 2
  Pinus sylvestris, 109, 118, 157, 198, 214
plant life-forms, 156
plant phenology, 176
Plectrophenax nivalis, see Snow Bunting
Pleistocene glaciations, 22, 29
Pleistocene climate changes, 20
Pleistocene overkill, 54
Pleurozium, 136
Pliocene, 11, 12, 16, 41, 203, 204, 248, 255
Ploidy
  diploids, 179–80, 200, 202, 249
  autotetraploids, 201
  polyploids, 180, 199, 200, 201, 202, 231, 251199
  polyploidy, 180, 199, 200, 201–2, 202, 231, 249, 251–2, 261
Poa abbreviata, 147
Poa arctica, 147
Poa nemoralis, 76
podzol, 214, 215
podzolization, 214
poikilotherm, 123

Polar Bear, 39, 57, 61, 124–8, 136, 140, 168, 206, 207, 208, 216, 218, 219, 223, 244–5
  counts, 220
  fasting, 219 see *Ursus maritimus*
polar desert, 29, 71, 72, 75, 80, 132
polar night, 5, 6, 8, 9, 12, 13, 62, 122, 131, 144, 190, 245
pollen analysis, 32, 33
pollen riskers, 176, 248
pollution in the Arctic, 233
polonium ($_{210}$Po), 242–3
polygons, 8304
Polygonum vivipara, 155, 181, 185, 198
polymorphism, 28, 204, 217, 246, 253
polyploidization, 201
Polytrichum sp., 136
Poolepynten, 207, 208
population estimates for Canadian *Rangifer spp* herds, 226
Populus tremula, 161 see Aspen
Populus tremuloides, 169 see Aspen
Post-hypsithermal climatic cooling, 42
Potentilla hyparctica, 180
Potentilla spp, 35
pre-Clovis hunter-gatherers, 55
Prince Charles Foreland, Spitsbergen, 25, 122
prostrate shrub Tundra, 74
Pterostichus brevicornis, 129 see Arctic beetle
Puccinellia phryganodes, 170, 186, 187, 188, 198 see Creeping Saltmarsh Grass
Puccinellia vahliana, 155
Pyrola grandiflora, 191 see Arctic Wintergreen

Q
Qiviut, 89, 130
Quebec, 91, 97, 98, 101, 104, 106, 160, 221, 225, 228, 254, 256

R
Rangifer tarandus ssp
  Rangifer tarandus caribou, 210
  R. t. fennicus, 210
  R. t. tarandus, 209
  R.t. granti, 210
  R.t. groenlandicus, 210
  R.t. platyrhunchus, 210
Racomitrium, 74
radiation exposure, 242
radioactivity in the Arctic, 241, 262
Rangifer tarandus, 34, 37, 41, 88, 102, 119, 132, 172, 226, 253, 254, 258, 259

Ranunculus nivalis, 169, 260 snow buttercup
Ranunculus pygmaeus, 155, 180, 200, 204, 205, 261 see Pygmy Buttercup
rare arctic plants, 230
Raven see *Corvus corvus*
reconstructing past plant distributions, 28
Red Deer, 108, see *Cervus elaphus*
regional heterothermy, 123
Reindeer, 56, 60–64, 69, 77, 80, 88–9, 91–4, 107–9, 113, 115, 117, 119, 122, 124, 125, 128, 131–3, 143, 168, 170, 171, 173, 209–12, 227–30, 238, 241–3, 250–4, 254–5, 257–62, see *Rangifer tarandus*
Reindeer brigades, 69
Reintroductions
  Muskoxen, 224
resources, 59, 61, 63, 64, 69, 77, 82, 85, 86, 92, 112, 134, 142, 147, 148, 149, 157, 165, 173, 191, 217, 235, 238, 240, 244
resting metabolic rate, 127, 138, 254
Rhododendron palustre formerly *Ledum groenlandicum*, 81
Role of staging areas, 173
Roman Warm Period, 44, 45, 46
Russian trappers (Pomors), 234

S
Salix, see willows
Saami, 58, 64, 70, 258, 262
Sable, 68 see *Martes zibeline*
Sagina caespitosa, 27, 28
Saiga tatarica, 34, 108
Salix, 12, 31, 32, 35, 48, 49, 76, 77, 82, 87, 91, 131, 146, 157, 169, 180, 231
  S. arctica, 49, 75, 148 see Arctic Willow
  S. polaris see Polar Willow
  *S. reticula*ta, 75 see Reticulate Willow
salt marshes, 77, 92, 121, 170, 186, 191
Samoyeds, 58, 60
Samuel Hearne, 56, 68, 171
Sandworts, see *Minuartia spp.*
Saxifrages
  Saxifraga caespitosa, 155, tufted saxifrage
  Saxifraga cernua, 171, 181, 198 Drooping saxifrage
  Saxifraga foliosa, 155, see Foliose Saxifrage
  Saxifraga cespitosa, 262
  Tufted Saxifrage

Saxifraga oppositifolia, 48, 74, 75, 82, 133, 146, 147, 156, 176, 177, 178, 179, 180, 183, 198, 200, 201, 202, 203, 223, 245, 250, 253 see Purple Saxifrage
Scurvygrass, see *Cochlearia officinalis*, 133
sea ice, 8, 59, 61, 68, 140, 184, 206, 208, 217, 219, 220, 230
seasonal bird migrations, 191
sedge tundra, 77, 78
sedges see *Carex spp.*
seed bank, 98, 99, 100, 149, 168, 252, 257
seed survival, 149
Sellafield, British Nuclear Energy site, 241
Semipalmated Sandpiper, 195, 196, 197
sequestration of carbon, 15
serotinus cones, 99
Sibbaldia procumbens, 180 see Creeping Glow Wort
Siberian Tundra, 104, 128
Silene acaulis, 169, 182, 198, 250, 260
Skin colour, 212
sleep hormone, 174, 175
smelting of nickel and copper ores, 235
Snow Bunting (*Plectrophenax nivalis*), 85, 240, 241
Snow geese, 102
Snowy Owl see *Bubo scandiacus*
Solar activity and climate, 48
solifluction, 149, 217
southern/northern-home-theory, 191
Species under threat, 218
Spermophilus citellus, 127 European Ground Squirrel
Spermophilus parryii, 127 see Arctic Ground Squirrel
Spitsbergen, see Svalbard
  anoxia tolerance in plants, 155–6
  arctic fox, 138, 159, 172–3
  bear fossils, 207
  bird cliffs, 81–82, 192
  bird migration, 173–4, 192
  climate, 3, 5, 6, 48–9, 122, 147, 157, 165
  earth mounds, 22, 84, see also thúfur
  fulmar longevity, 167
  hunting stations, 234
  ice-free refugia, 23–8, 30
  plant colonization, 217
  plant growth, 157, 159, 177–8, 183, 180, 186–8, 201, 204
  polar deserts, 74–75

Spitsbergen, *see Svalbard (cont.)*
  purple saxifrage and climatic
      warming, 183, *see also*,
      climatic warming
  reindeer grazing, 170–1
  reindeer populations, 82, 91,
      107–8, 125, 132, 210–2, 227,
      230
  rock ptarmigan, 143–4
  seed bank, 149, 171
  soil respiration, 82
  voyage of the Fram, 137
  wetland meadows, 77–9
Stellaria longipes, 147 Longstalk
      Starwort
Stercorarius longicaudus, 172 *see*
      Long-tailed Skua
Sterna paradiseae, see Arctic Tern
succession, 49, 92, 93, 104, 110, 118, 257
sunspot activity, 47–49
Supercooling, 129, 153, *see also*
      undercooling
Svalbard, *see* Spitsbergen
  ancient polar bears, 252
  climate, 260
  foxes, 138, 172–3, 252
  geese
      barnacle geese, 173
      pink-footed geese, 252
  muskoxen, 91
  ocean currents, 6
  plant polyploidy, 179, 199, 200
  plant populations, 28
  reindeer, 88–90, 132–3, 210–12,
      230, 250, 252, 255, 257, 261
  rock ptarmigan, 142–6, 257,
      259–62
  species introductions, 259
Svalbard Poppy *see, Papaver dahlianum*
Svalbard Reindeer
      (*R. t. platyrhynchus*), 210
Swan migrations, 194
Swans, 85, 194, 195 *see Cygnus spp.*
Sweet Vernal Grass *see*
      Anthoxanthim odoratum

T
Taiga
  conflict with bog, 100
  245, 246, 248, 249, 250
  fire hazards, 98
  interface with tundra, 94

northward advance, 34
regeneration, 96
succession, 98
Tatra Mountains, 204
Taxus, 12
Taymyr Peninsula, 26, 37, 39, 41, 42,
      56, 64, 137, 202, 224, 225, 258
tectonic activity, 15
Temporal heterothermy, 123
Teutonic Knights, 214
Thamnolia vermicularis, 240
  worm lichen
thermal regulation, 124
thermo-luminescence, 26
thermokarsting, 105, 120, 217
thúfur, 22, 80, 83, 85
Thuja, 12
Timing of goose migrations, 173
torpor, 126, 127, 128, 129, 145, 149
Trampling Scar-frequency
      distribution, 229
trampling scars, 1, 228
treeline, 1, 2, 13, 75, 76, 95, 104, 106,
      157, 159, 160, 255, 257
Trollius europaeus, 206, 253
Trumpeter Swan (*C. buccinator*),
      194, 195
Tundra Swans (*C. columbianis
      columbianis*), 194
Turkish tribes, 60
Tundra
  Azonal types of Tundra, 79
  Tundra-Taiga Interface, 1, 94
  Tussock-forming, 78
  Tussock-Tundra line, 79

U
undercooling, *see* supercooling
uptake of volatile fatty acids, 132
Urocitellus/ Spermophilus, 127 *see*
      Arctic Ground squirrels
Ursa major, (The Great Bear) 1
Ursus arctos, see Brown Bear
Ursus arctos ssp horribilis
Ursus maritimus, see Polar Bear
Usinsk, Komi Republic, Russia
      (66°30′N), 237
Usnic acid, 132

V
Vaccinium uliginosum, 35, 169, 199,
      see Bog Bilberry

vitamin C availability in the Arctic,
      213
vitamin D, 21, 126 , 212
voles, 126, 86, 123, 133, 136
  Bank Vole, 133 *Myodes glareolus*
  Grey-Sided Volc (*Myodes
      rufocanus*, 86
  population cycles, 86, 136
  Taiga Voles (*Microtus
      xanthognathus*, 123
Vulpes lagopus, 172, *see* Arctic Fox
Vulpes vulpes, 125, 137, *see* Red Fox

W
West Siberian Lowlands, 4, 5, 43
wetland meadows, 77
white adipose tissue, 127, 246
White Sea, 58, 70, 71
White spruce, *see Picea glauca*
Whooper Swan, 194
Wild Reindeer (Caribou), 226
  winter starvation for caribou, 227
  Grey Wolf, *Canis lupus*, 119
Wolf
  Metatarsal bone, 54
  White or Polar Wolf, *C. lupus
      arctuos*, 119
  As predator, 110, 227
Wood Bison, *see Bos bison athabaskae*
Wood mouse, 180, *see Apodemus
      sylvaticus*
Woodland Caribou, *see Rangifer
      tarandus caribou*
Woolly Mammoth, 128, 170
Woolly Rhinoceros, 34, 37, 38, 51

X
Xanthoria elegans, 240
      (Elegant Sunburst Lichen)

Y
Yakutia, 34, 36, 37, 69, 140,
      157, 257
Yakuts, 58, 60, 256
Yasak, 69
Younger Dryas, 16, 21, 23, 31, 32, 37,
      56, 59
Yukon 2, 26, 41, 111, 118, 119, 146,
      223, 225, 226, 227, 250

Z
Zeitgeber stimulus, 175